MOLECULAR AND CHEMICAL CHARACTERIZATION OF MEMBRANE RECEPTORS

RECEPTOR BIOCHEMISTRY AND METHODOLOGY

SERIES EDITORS

J. Craig Venter
Department of Molecular Immunology
Roswell Park Memorial Institute
Buffalo, New York

Len C. Harrison
Department of Diabetes and Endocrinology
The Royal Melbourne Hospital
Victoria, Australia

Volume 1
Membranes, Detergents, and Receptor Solubilization
J. Craig Venter and Len C. Harrison, *Editors*

Volume 2
Receptor Purification Procedures
J. Craig Venter and Len C. Harrison, *Editors*

Volume 3
Molecular and Chemical Characterization of Membrane Receptors
J. Craig Venter and Len C. Harrison, *Editors*

Volume 4
Monoclonal and Anti-Idiotypic Antibodies as Probes for Receptor Structure and Function
J. Craig Venter, Claire M. Fraser, and Jon M. Lindstrom, *Editors*

MOLECULAR AND CHEMICAL CHARACTERIZATION OF MEMBRANE RECEPTORS

Editors

J. Craig Venter
Department of Molecular Immunology
Roswell Park Memorial Institute
Buffalo, New York

Len C. Harrison
Department of Diabetes and Endocrinology
The Royal Melbourne Hospital
Victoria, Australia

ALAN R. LISS, INC., NEW YORK

Address all Inquiries to the Publisher
Alan R. Liss, Inc., 150 Fifth Avenue, New York, NY 10011

Printed in the United States of America.

Library of Congress Cataloging in Publication Data
Main entry under title:

Molecular and chemical characterization of membrane receptors.

(Receptor biochemistry and methodology ; v. 3)
Includes index.
1. Cell receptors. 2. Cell membranes. I. Venter, J. Craig. II. Harrison, Len. III. Series.
QH603.C43M64 1983 574.87′5 83-16233
ISBN 0-8451-3702-6

Second Printing, October 1984

Contents

Contributors

Sylvia F. Babelyn [61]
Department of Biochemistry and Molecular Biology, College of Medicine, University of Florida, Gainesville, FL 32610

Heinz Baumann [125]
Department of Cell and Tumor Biology, Roswell Park Memorial Institute, Buffalo, NY 14263

George A. Baumbach [61]
Department of Biochemistry and Molecular Biology, College of Medicine, University of Florida, Gainesville, FL 32610

Robert Blumenthal [209]
Laboratory of Theoretical Biology, National Cancer Institute, NIH, Bethesda, MD 20205

William C. Buhi [61]
Departments of Obstetrics and Gynecology, and Biochemistry and Molecular Biology, College of Medicine, University of Florida, Gainesville, FL 32610

J.P. Coghlan [261]
Howard Florey Institute, University of Melbourne, Parkville, Victoria 3052, Australia

Alan Davis [161]
Boehringer Ingelheim (Canada) Ltd., Burlington, Ontario L7L 5J8, Canada

John B. Denny [61]
Cell Biology Laboratory, Rockefeller University, New York, NY 10021

Darrell Doyle [125]
Department of Biological Sciences, State University of New York at Buffalo, Buffalo, NY 14214

R.T. Fernley [261]
Howard Florey Institute, University of Melbourne, Parkville, Victoria 3052, Australia

Laurence A. Fitzgerald [61]
The Gladstone Foundation Laboratories, San Francisco, CA 94140

Claire M. Fraser [1]
Department of Molecular Immunology, Roswell Park Memorial Institute, Buffalo, NY 14263

James W. Goding [31]
The Walter and Eliza Hall Institute, The Royal Melbourne Hospital, Melbourne, Victoria 3050, Australia

J.J. Gorman [261]
Howard Florey Institute, University of Melbourne, Parkville, Victoria 3052, Australia

The number in brackets is the opening page number of the contributor's article.

Len C. Harrison [xi]
Department of Diabetes and Endocrinology, The Royal Melbourne Hospital, Melbourne, Victoria 3050, Australia

Michael N. Horst [61]
Mercer University, School of Medicine, Macon, GA 31207

Geoffrey J. Howlett [179]
Russell Grimwade School of Biochemistry, University of Melbourne, Parkville, Victoria 3052, Australia

Thomas H. Hudson [115]
Laboratory of Neurochemistry, National Institute of Mental Health, NIH, Bethesda, MD 20205

Gary L. Johnson [115]
Department of Biochemistry, University of Massachusetts Medical Center, Worcester, MA 01605

Chan Y. Jung [193]
Department of Biophysical Sciences, SUNY at Buffalo, School of Medicine, Buffalo, NY 14215

Daniel L. Kaufman [241]
Department of Biology, University of California, Los Angeles, CA 90024

Richard D. Klausner [209]
Laboratory of Biochemistry and Metabolism, National Institute of Arthritis, Diabetes, and Digestive and Kidney Diseases, NIH, Bethesda, MD 20205

Jon Lindstrom [1]
Receptor Biology Laboratory, The Salk Institute, San Diego, CA 92138

H.D. Niall [261]
Howard Florey Institute, University of Melbourne, Parkville, Victoria 3052, Australia

Benjamin Rivnay [209]
Arthritis and Rheumatism Branch, National Institute of Arthritis, Diabetes, and Digestive and Kidney Diseases, NIH, Bethesda, MD 20205

R. Michael Roberts [61]
Department of Biochemistry and Molecular Biology, College of Medicine, University of Florida, Gainesville, FL 32610

Allan J. Tobin [241]
Department of Biology, University of California, Los Angeles, CA 90024

Jos van Renswoude [209]
Laboratory of Biochemistry and Metabolism, National Institute of Arthritis, Diabetes, and Digestive and Kidney Diseases, NIH, Bethesda, MD 20205

J. Craig Venter [xi]
Department of Molecular Immunology, Roswell Park Memorial Institute, Buffalo, NY 14263

Glenn M. Ward [109]
University of Melbourne Department of Medicine, The Royal Melbourne Hospital, Parkville, Victoria 3050, Australia

Contents of Volumes 1 and 2

Preface

Cell hormone and neurotransmitter receptors are involved in the control and/or modulation of virtually every cellular and physiological process known. Changes in hormone receptors are implicated in major diseases including heart disease, hypertension, allergic respiratory diseases, metabolic disorders such as diabetes, and cancer. The recent molecular breakthroughs in the area of hormone and neurotransmitter receptor research have resulted from an interdisciplinary approach with the application of technologies derived from biochemistry, immunology, pharmacology, molecular biology, biophysics, and medicine.

The receptor field is undergoing a period of rapid advancement with a tremendous growth in the number of studies at the molecular level. This transition from receptor analysis using pharmacological techniques and radioligand binding studies toward the molecular resolution of the receptor molecules and their effector proteins requires working with a broad repertoire of new methodologies and techniques.

Receptor Biochemistry and Methodology is a new series devoted to the molecular advances involved in understanding the structure and biochemical basis of receptor action as well as to detailing the necessary methodologies.

The first three volumes are methodological reviews of approaches and technologies required by researchers involved in the isolation, purification, and biochemical characterization of cell surface proteins. Future volumes will deal in depth with single areas including Monoclonal Antibodies (Volume 4) and Radiation Inactivation (Volume 5), or will be devoted to the molecular basis of hormone and neurotransmitter action, with individual volumes on the structure, function, pharmacology, and biochemistry of individual receptors or effector proteins.

In Volume 3, **Molecular and Chemical Characterization of Membrane Receptors,** a broad spectrum of new and sophisticated approaches are detailed, from Chapter 1 on the use of monoclonal antibodies in receptor isolation and characterization to Chapters 10 and 11 on the application of recombinant DNA technology to receptor research. The intervening chapters contain the substantive methods available and essential to the receptor biochemist. These include Chapters 2, 3, and 4 on gel electrophoresis procedures, SDS-polyacrylamide gels, two-dimensional gels and peptide mapping—techniques that are applicable to purified proteins and/or affinity labelled receptors. Chapters 6 and 7 describe procedures for obtaining hydrodynamic parameters of receptor proteins, techniques that require only detergent solubilization of receptors and the binding of reversible ligands. Chapter 5 presents detailed methodology for dealing with an often ignored aspect of receptors, their carbohydrate moieties. All of these techniques require purified or partially purified receptor proteins; however, Chapter 8 introduces the technique of radiation inactivation which can be used to determine the functional molecular size of proteins while still in the native membrane. Chapter 9 is an excellent review of membrane receptor reconstitution, an approach that can be applied to the assay of solubilized membrane proteins, but more importantly, can ultimately provide the means of understanding the relationships between receptor and effector structure and function.

The contents of Volumes 1 to 3 reflect a strong prejudice, developed by us both in the course of our respective purification studies, that one key to success in receptor purification and structural analysis is to utilize many parallel approaches to insure, for example, that the isolated proteins are indeed those desired.

We feel fortunate in obtaining a unique collection of authors who are leaders in their respective fields and thank them for their help in making this series a reality. We also thank Paulette Cohen and the staff of Alan R. Liss, Inc. for their help and advice.

We believe that the reader will find that the articles are not only excellent reference sources but also extremely useful at the laboratory bench. We feel that the information contained in these volumes will be essential to researchers undertaking the molecular characterization of receptors or other integral membrane proteins. In addition, we hope that the reviews and discussions will interest researchers in biochemistry, immunology, pharmacology, physiology, biophysics, and cell and molecular biology as well as researchers and others who deal with the vast array of clinical disorders effected by or controlled through receptors.

J. Craig Venter
Len C. Harrison

Molecular and Chemical Characterization of Membrane Receptors, pages 1–30

1

The Use of Monoclonal Antibodies in Receptor Characterization and Purification

Claire M. Fraser and Jon Lindstrom

Department of Molecular Immunology, Roswell Park Memorial Institute, Buffalo, New York 14263 (C.M.F.) and Receptor Biology Laboratory, The Salk Institute, San Diego, California 92138 (J.L.)

INTRODUCTION

In the few years since Kohler and Milstein first described "continuous lines of cultured cells secreting antibodies of pre-defined specificity" [Kohler and Milstein, 1975], hybridoma technology and monoclonal antibodies have had a tremendous impact on all areas of biomedical research.

A monoclonal antibody is the product of a single clone of B-lymphocytes. Each antibody produced by a single clone has the same amino acid sequence and, therefore, the same physical and chemical properties. It is possible to "immortalize" the production of predefined, specific monoclonal antibodies by fusing the relatively short-lived, normal antibody-producing splenic lymphocytes from immunized mice with cultured myeloma cells to generate clones of hybrid cells (hybridomas) that then secrete normal antibody continuously in culture [Kohler and Milstein, 1975, 1976].

The success of lymphocyte hybridization was due to the development of mutant myeloma cell lines adapted to grow permanently in culture in selective media [Littlefield, 1964]. Cotton and Milstein [1973] adopted the approach of using drug-resistant mutants of myeloma lines, one being a 5′-bromodeoxyuridine-resistant mouse line defective in thymidine kinase, the other an 8-azaguanine resistant rat line defective in hypoxanthine-guanine phosphoribosyl transferase (HGPRT). These lines were, respectively, unable to utilize thymidine and hypoxanthine from the culture medium when the de novo synthesis of these DNA precursors was blocked by aminopterin. Neither of the two cell lines, therefore, survived in selective HAT medium (hypoxanthine, aminopterin, thymidine) but complemented each other when fused, the hybrid cells being the only ones capable of continued growth. The fusion of these two myeloma lines resulted in a hybrid which expressed the product of each parent line (actually, new Ig molecules were produced as a result of mixing of heavy and light chains from the two parent cells, but there was no "scrambling" of V and C regions, indicating that the latter are not determined by cytoplasmic events). Finally, an 8-azaguanine-resistant line was derived from the mouse myeloma line MOPC-21 (P3) which secreted IgG_1 kappa and was designated P3/X63 Ag8. This line was then used by Kohler and Milstein [1975] in their revolutionary experiment to demonstrate that a hybrid formed with normal spleen cells from a BALB/c mouse immunized with sheep red blood cells (SRBC) secreted both IgG_1 myeloma protein and antibody to SRBC.

This apparently simple but elegant experiment has already revolutionized immunology and is having a major impact on many other areas of biology. The basic methodology has been used to generate monoclonal antibodies against a myriad of antigens including haptens, peptide hormones, enzymes, receptors and other proteins, polysaccharides, glycoproteins, lipopolysaccharides, differentiation antigens, histocompatability antigens, and other cell surface and cytoplasmic antigens.

ADVANTAGES OF HYBRIDOMA ANTIBODY PRODUCTION

As outlined in Table I, antibody production by hybridomas has many advantages over the conventional approach of harvesting sera from immunized animals.

1. Antibody production by hybridomas is continuous, theoretically unlimited in quantity, and constant in nature. Therefore, radioimmunoassays and related techniques may be standardized internationally. This is in contrast to the properties of antiserum obtained from immunized animals, which varies in specificity and affinity between animals and also with each harvest from the same animal.

2. Monoclonal antibodies are defined homogeneous reagents directed at only one antigenic determinant. This specificity ensures an unequivocal interpretation when such antibodies are used to purify receptors and map molecular function and structure. Polyclonal, heterogeneous antisera raised in animals contains antibodies to all components of the antigenic mixture and often to distinct antigenic sites on one particular antigen. The properties of polyclonal antisera greatly limit the interpretation and applicability of such preparations in many immunoassays.

3. The antigen used for immunization need not necessarily be pure, since hybridomas are selected for production of desired antibodies. In other words, specific antibodies can be obtained without specific immunization. This "dirty in/clean out" approach has been particularly useful in preparing monoclonal antibodies to some cell membrane receptors, whose low membrane concentration has made

TABLE I. Comparison of Properties of Monoclonal Antibodies and Conventional Antiserum Preparations

Property	Monoclonal antibodies	Antiserum
Composition of heavy and light immunoglobulin chains	Only one heavy chain, only one light chain	Heterogeneous mixture of antibodies of different heavy and light chain composition
Physical properties of antibodies (antigen specificity, affinity for antigen, etc)	Constant and homogeneous due to chemical identity of antibody molecules	Varies with each antiserum harvest due to heterogeneous nature of immunoglobulins in conventional antiserum
Recognizes and binds to:	A single determinant from immunogen	All antigenic determinants on all components of immunogen
Applicability for conventional immunological procedures	May not work	Applicable
Relevant immunoglobulin concentration	10–50 μg/ml (culture media) 1–10 mg/ml (ascites fluid)	0.1–1 mg/ml
Irrelevant proteins	Very small amount from fetal bovine serum (culture media) Some serum proteins and normal mouse immunoglobulins (ascites fluids)	All other serum proteins

complete purification extremely difficult [see chapters in Fellows and Eisenbarth, 1981; and Venter et al, 1983b].

4. Hybridoma technology may allow for the selection of antibodies to molecules that have little or no immunogenicity with respect to serum antibody production in immunized animals. Thus, the growth and expression of sensitized splenic B cells freed from the "immunoregulatory environment" in vivo could be amplified in a hybridoma, thereby increasing the chances for antibody generation.

5. Monoclonal antibodies can be selected and screened for on the basis of a particular desirable property, eg, low affinity, which could be advantageous in recovering receptors after adsorption to an immunoaffinity matrix.

6. C^{14}- or H^3-labeled monoclonal antibodies for use in immunoradiometric assays can be obtained by culture of clonal lines in medium containing labeled amino acids or sugars.

LIMITATIONS OF MONOCLONAL ANTIBODIES

Some of the properties of monoclonal antibodies that distinguish them from serum antibodies limit their applicability for certain experimental protocols. Fortunately, these limitations can usually be overcome provided they are understood.

1. Monoclonal antibodies generally can not precipitate their target antigens as they only cross-link antigens into dimers rather than form an antibody-antigen lattice. For this reason, monoclonal antibodies may not work in the Ouchterlony double-immunodiffusion assay and can not precipitate antigens from solution without addition of a second antibody.

2. Monoclonal antibodies do not readily fix complement, as complement requires at least two bound antibody molecules on neighboring determinants [Howard et al, 1979].

3. Conventional antisera contain a variety of antibody molecules that differ with regard to immunoglobulin class, specificity, and affinities for antigens and physical properties. A monoclonal antibody with a unique structure may differ from the majority of conventional immunoglobulins with regard to a particular property; ie, a monoclonal antibody may be more susceptible to denaturation due to temperature, pH or freezing and thawing than the bulk of serum immunoglobulins [Mason and Williams, 1980]. Furthermore, the rates of association and dissociation of monoclonal antibody-antigen complexes may be considerably slower or faster than preparations of conventional antisera.

4. Conventional antisera contain antibodies to a number of antigenic determinants on a given antigen, but monoclonal antibodies only recognize one determinant. Therefore, in some cases, conventional antisera may provide a more accurate identification of a target antigen in a radioimmunoassay, as any unknown antigen able to compete with all antibodies present in a serum preparation is most likely identical to the known antigen. In contrast, an antigenic determinant recognized by a monoclonal antibody may be present on unrelated antigens, and cross-reactivity elicited between a monoclonal antibody and several antigens may not necessarily be indicative of molecular identity between antigenic species. An appropriate combination of monoclonal antibodies to several determinants on one antigen is sufficient to overcome such a potential problem.

STRATEGIES FOR PREPARING MONOCLONAL ANTIBODIES TO CELL RECEPTORS

The use of monoclonal antibodies as probes of receptor structure and function and as specific reagents in the immunoaffinity purification of receptors offers unique insights heretofore not possible with more conventional biochemical and physiological approaches to receptor characterization. However, the generation of permanent lines of hybridomas secreting monoclonal antibodies with desired properties presents varying degrees of difficulty. Immunization of an animal with an antigen preparation results in the production of a heterogeneous population of immunoglobulin molecules against the antigen. Some of the antibodies produced may possess the desired characteristics, many will not. Immortalization of antibody secreting B-cell clones using hybridoma technology produces a random cross section of the overall population of immune cells. If the immune response to a particular component of an antigenic mixture is weak, this may be reflected in a small number of hybridomas producing antibody within the total population of hybrid cells. Successful production of a monoclonal antibody to a weak immunogen may involve a significant commitment in time and effort to perform several cell fusions and screen thousands of hybridomas for the presence of a desired antibody. Lastly, the properties of a monoclonal antibody identified to a specific antigen or determinant depend upon the experimental design and type of screening assay employed. In other words, if a low-affinity monoclonal antibody is sought, the immunization protocol and screening assays should be designed to maximize the chances of producing and detecting a clone of cells secreting a monoclonal antibody of low affinity. Similarly, if a monoclonal antibody directed to a determinant in the ligand-binding site of a receptor is desired, screening assays should be designed to detect this specificity.

In this chapter we will provide guidelines and suggestions for preparation of receptor-specific monoclonal antibodies with different properties. We will draw heavily on our own experience and present experimental protocols currently in use in our laboratories. This is not meant to imply that we consider our approaches more valid than those of others, but rather that these procedures have been found to work reproducibly in our hands. The following methodology is meant to serve as a basic introduction to hybridoma technology. Modification of all of these procedures have appeared, many of which may be more appropriate for the needs of a particular laboratory. The keys to success with hybridoma technology are simple: Become proficient with tissue culture techniques and use cell lines and fusion protocols that work reproducibly for you.

METHODOLOGY

Specific methodology (solutions, experimental protocols, etc.) will be found in Appendices I and II. The following sections will deal with general theory and approaches for hybridoma production.

Media Preparation

The most commonly used media for growth of myeloma cell lines and hybridomas are Dulbecco's Modified Eagle's Medium (DME) with 4.5 g glucose/liter or RPMI-1640 with glucose added to a final concentration of 4.5 g/liter.

Ready to use media (1× or 10×) is available commercially (GIBCO, Flow Laboratories) or may be prepared from dry powder using tissue culture grade water according to manufacturer's instructions. This requires equipment for filter sterilization.

We prefer to prepare our own media from dry powder, as it is more economical and the date of preparation is known. Media is stored at 4°C in the dark and is discarded if unused after 3 mo. Glutamine (0.3 mg/ml, final concentration) is added to media at biweekly intervals, as it is particularly susceptible to decay during storage of media and essential for cell growth.

Choice of Serum

Fetal bovine serum is presently the preferred supplement for support of hybridoma growth due to its relatively low immunoglobulin content and the observed efficiency of cell fusions in fetal bovine serum as compared to other sources of serum. We routinely use 20% fetal bovine serum for cell fusions and cloning and then reduce the serum concentration to 10% for growth of selected hybridomas.

Lots of fetal bovine serum from different

commercial sources vary tremendously in their ability to support the growth of hybridomas. For this reason, careful testing of lots of serum is recommended. Purchasing a suitable lot of serum in quantity reduces the need for frequent serum screening (see Appendix I).

Tumor Cell Lines

Four major considerations in choosing an appropriate myeloma cell line for fusion should be taken into account. These are drug resistance, the species from which the line was derived, the question of whether or not the line secretes immunoglobulins, and the stability of the hybridomas produced.

As illustrated in Table II, the most commonly used myeloma cell lines are resistant to purine analogs such as 8-azaguanine or 6-thioguanine or to pyrimidine analogs such as 5-bromodeoxyuridine. Drug resistance allows for selection of growth of hybridomas following fusion in a selective medium (see below).

The majority of mouse myeloma cell lines used for production of hybridomas are derived from the BALB/c strain although rat and human lines are now also available. Generally, the wisest choice is to use a myeloma cell line from the same species as that of the immunized animal (which may not be a mouse, in specific cases). In this way, ascites tumors can easily be developed in syngeneic animals when specific hybridomas are selected.

In deciding which myeloma cell lines to use, a major consideration is the expression or lack of expression of myeloma immunoglobulins by the parent tumor cell. Hybridomas derived from immunized spleen cells and myeloma cells secreting heavy and light immunoglobulin chains express four immunoglobulin chains, usually designated G and K (heavy and light chains from parent myeloma) and H and L (heavy and light chains from spleen cell). The codominant expression of these four chains within the hybridoma can lead to the secretion of mixed species of immunoglobulin (such as KHHL, KGGL, KHGL, etc) in addition to the expression of the parental immunoglobulins (KGGK and LHHL) [Cotton and Milstein, 1973; Margulies et al, 1976]. For this reason, it is recommended that one start with a parental myeloma cell that expresses only light immunoglobulin chains (such as NS-1) or is a non-producer (653, SP-2/0). Using myeloma lines which do not produce any myeloma chains results in hybridomas expressing only the immunoglobulin chains of the parental spleen cells.

The stability of hybridomas is in part dependent on the choice of myeloma cell for a fusion partner. As is illustrated in Table II, some of the lines available are myeloma cell

TABLE II. Summary of Myeloma Cells

Cell line	Parental line	Drug resistance	Immunoglobulin chain	Reference
Mouse lines				
P3X63-Ag8 (P3)	MOPC 21	8-azaguanine	$\gamma 1$, kappa	Kohler and Milstein [1975]
P3-NS-1/Ag14 (NS-1)	P3	8-azaguanine	---, kappa	Kohler et al [1976]
P3X63-Ag8.653 (653)	P3	8-azaguanine	---, ---	Kearney et al [1979]
SP2/0-Ag14 (SP2)	P3-spleen cell hybrid	8-azaguanine	---,---	Schulman et al [1978]
FO	SP2-SP2 hybrid	8-azaguanine	---,---	Fazekas de St. Groth and Scheidegger [1980]
S194/5.XXO.BU.5	S194	Bromodeoxyuridine	---,---	Trowbridge [1978]
Rat lines				
Y3-Ag1.2.3	Rat	8-azaguanine	---, kappa	Galfre et al [1977]

lines (P3, NS-1, 653), and others are hybridomas themselves (SP2/0, FO). It has been suggested that the stability of hybridomas (with reference to the continued secretion of monoclonal antibody following fusion) is lower if hybridoma lines are used for fusion as compared to myelomas [Reading, 1982]. Consequently, repeated recloning of spleen cell-hybridoma lines may be necessary to insure continued production of monoclonal antibodies. However, we have used SP2/0 cells for a large number of cell fusions and have lost only 20–30% of antibody-secreting hybrid clones.

Regardless of the choice of parental myeloma cell, the most important factor affecting the success of cell fusions is the correct maintenance of the myeloma cells prior to fusion. The goal is to maintain cultures of myeloma cells in continuous logarithmic growth (3–8 $\times$ 10^5 cells/ml) for at least a week prior to fusion. We prefer to use spinner cultures as opposed to stationary suspension cultures, as this facilitates maintenance of cells at the correct density and allows for more even distribution of essential nutrients in the medium with the growing cells.

Many of the cell lines listed in Table II are supported by the National Cancer Institute and are currently available from The American Type Culture Collection or from the Human Genetic Mutant Cell Repository in Camden, New Jersey, as well as from numerous individual researchers.

Immunization of Animals

While it may be desirable to start with a pure immunogen in order to restrict antibody production and make subsequent screening of hybridomas easier, this strategy is not always possible, especially with receptor proteins. Therefore, it is more feasible to immunize animals with cell membranes or partially purified receptor and invest the effort in screening, even though several fusions may have to be performed before a positive hybridoma is found.

While it is possible that the immune response to a specific receptor present in an impure antigen preparation may be strong and allow for immunization with intact cells or cell membranes, it has generally been our experience that additional purification of receptor preparations used as antigens may be necessary, as, for example, in the case of a receptor present on cell membranes in extremely low concentrations (a few hundred to a few thousand copies per cell). Obviously, the higher the percentage of receptor present in the immunogen, the more likely one is to produce a monoclonal antibody to a specific receptor.

We have found that immunization of animals with intact cells or cell membranes in general results in the derivation of greater numbers of hybridomas (but not necessarily ones that are receptor specific) than does immunization with soluble partially purified receptor preparations. While a large number of hybridomas/fusion may seem desirable, we have found that if screening assays for receptor specific monoclonal antibodies are time consuming, it is more efficient to use partially purified receptor preparations as antigen and screen fewer hybridomas per fusion.

It is suggested that several different immunization protocols be tried, using several animals. If the antigen preparation contains intact cells or cell membranes, we generally inject 1–3 mg protein/mouse without adjuvant. If the antigen consists of solubilized, partially purified receptor material it is emulsified in an equal volume of Freund's complete adjuvant (by repeatedly squirting the suspension through an 18-gauge needle) and 0.1–2 mg protein/mouse are injected. The initial immunization can be given intraperitoneally or subcutaneously at multiple sites across the animal's back. Animals are then rested 3–5 wk and the immunizations repeated. The final immunization is usually given intravenously or intraperitoneally minus adjuvant 3–4 days prior to fusion. This schedule is based on the finding by Kohler and Milstein [1976] that maximum hybridization efficiency occurred 4 days after boosting. We routinely perform all immunizations via the intraperitoneal route and have had considerable success with this method.

Some investigators elect to test antisera from immunized animals for the presence of antibodies prior to performing a cell fusion. In this way animals can receive multiple booster immunizations until a high titer of circulating antibodies is detected. Seven to 10 days following an immunization, a small amount of blood is collected from each animal by nicking the end of its tail and used to screen for the presence of serum antibodies. If a positive antiserum is identified, animals should be rested for 3–4 wk prior to the final immunization and cell fusion. It is not essential that animals be "prescreened" in this manner for the production of circulating antibodies to cell membrane receptors. Furthermore, an apparently negative antiserum response does not necessarily indicate that an animal is not making antibodies to a particular receptor component present in an immunogenic mixture, but may only indicate that the serum antibody titer is very low. A low serum antibody titer may be overcome through cell fusion, screening, and cloning.

Cell Fusions

The factors governing fusion efficiency are not well understood: The frequency of successful fusions is presently no better than one in 10^5 normal cells. However, of the hybridomas formed, 10–50% may produce specific antibodies. This high frequency of antibodies-secreting hybrid cells and the demonstrable absence of T cell-B cell hybrids suggests that recent activated normal spleen cells are especially susceptible to hybridization. They are presumably B-cell blasts, perhaps in a stage of differentiation similar to the tumor line. Elucidation of the properties of both cell types (eg, cell cycle, stage of differentiation) favoring fusion is obviously an important goal.

The first hybridomas were made using Sendai virus as a "fusagen." Currently, however, nearly all workers use polyethyleneglycol (PEG) (MW range 1,000–4,000) [Pontecorvo, 1975]. PEG lots vary considerably in purity and each lot should be tested for its toxicity on the tumor cell line before use. A number of variations in fusion protocol have been described including use of dimethyl sulfoxide with polyethyleneglycol to induce fusion, use of different concentrations and molecular weights of polyethyleneglycol, changing the time that cells are in contact with polyethyleneglycol and altering the ratios of spleen cells and myeloma cells used in cell fusions [Galfre et al, 1977; Gefter et al, 1977; Oi and Herzenberg, 1980; Fazekas de St. Groth and Scheidegger, 1980; Galfre and Milstein, 1981; Kennett et al, 1980]. Many protocols have been used successfully, and, therefore, it is only important to find a procedure that works reproducibly in each laboratory. The protocol found in Appendix I is meant to serve as a starting point to establish a satisfactory routine for cell fusions.

It is common to find that fusion efficiencies increase as laboratory personnel gain experience with the techniques involved. However, the repeated failure to grow hybridomas in HAT medium after cell fusion suggests five possible problems which should be investigated: (1) myeloma cells were not in logarithmic phase of growth prior to fusion, (2) low viability of parental spleen or myeloma cells, (3) cytotoxicity of or overexposure to PEG, (4) nutrient deficiency of medium, and (5) contamination of cultures with mycoplasma [Reading, 1982; Stanbridge and Schneider, 1976].

It is exceptionally difficult to overcome these problems once a cell fusion has been performed and hybridomas have been plated out. The best advice is to begin again with a new series of immunizations and another cell fusion.

Cell Plating

Immediately following a fusion cells are plated onto 96-well microtiter plates in HAT selective medium which only allows for growth of hybrids containing at least one normal spleen cell and one myeloma cell as depicted in Figure 1. Normal spleen cells which have not undergone fusion die within a few days in tissue culture. Continued growth of unfused myeloma cells or myeloma hybrids is prevented by aminopterin-induced inhibition of dihydrofolate reductase, a key enzyme in the de novo synthesis of DNA. In myeloma cells resistant to purine analogs, the de novo pathway is the only route for continued DNA synthesis due to lack of the enzyme hypoxanthine-guanine phosphoribosyl transferase. Fusion between a normal spleen cell and a resistant myeloma cell produces a hybridoma capable of synthesizing DNA by both the de novo pathway and the salvage pathway due to the contribution of HGPRT to the hybrid genome from the normal spleen cell. Therefore, in the presence of aminopterin, hybridomas synthesize DNA via the salvage pathway and are capable of growth in selective medium if hypoxanthine and thymidine are present.

The addition of other cells such as normal spleen cells or peritoneal exudate cells from BALB/c mice to the suspension of hybridomas (as monolayers or mixed with suspension cultures) to provide an extra "feeder effect" is very advantageous. Feeder cells can be added immediately after fusion and/or during the period of stabilization of hybridomas and isolation of specific clones. The feeder cells greatly augment the survival and growth of hybridomas both after fusion and during cloning. The beneficial effect of feeder cells may be due to elaboration of growth stimulating monokines [Reading, 1982]. In addition, the enhanced survival and growth of hybridomas with a feeder layer may be due to the ingestion of cell debris from dying cells and the possible recycling of cellular components by these cells. Repeated medium changes are also unnecessary when hybridomas are grown on a feeder layer, thereby reducing the chances of contamination.

Direct Cloning

Hybridomas can alternatively be plated in agar directly after fusion [Tzartos et al, 1981a; Tzartos and Lindstrom, 1981]. The advantages of "direct cloning" in agar immediately after fusion are that it (1) largely avoids overgrowth by nonproducing clones, (2) saves effort and increases the probability of recovering desirable hybrids because cloning begins from the day of fusion, (3) efficiency of fusion may be greater in agar, and (4) more than one clone can be recovered from a single well. The disadvantage is that some myelomas (like S194/5.XXO.BU.1 [S194]) grow well in agar while others (like 210.R-CY3. Ag1.2.3 [Y3]) do not. The method which we have used for direct cloning is described in Appendix I. Note that there are differences (which are probably not important) in the fusion procedure from that described in Appendix I. The critical feature of the method is plating in soft agar immediately after fusion.

Screening for Antibody Production

The importance of appropriate screening assays in identifying positive hybridomas cannot be overemphasized. The power of hybridoma technology depends on the ability to efficiently screen for desired antibody producing cells. It is imperative that a screening assay be developed in advance of the actual hybridoma screening as time cannot be devoted to screening optimization without risk of losing potentially positive cultures due to overgrowth in 96-well microtiter plates.

Although there are a wide variety of rapid screening assays for detecting the presence of antibodies, not all such assays are directly applicable to monoclonal antibodies to cell membrane receptors. Consequently only the screening assays most suited for use in detecting receptor specific monoclonal antibodies will be discussed here. For more general reviews on screening procedures, the reader is referred to Methods in Enzymology, Vol. 73 [Langone and Van Vunakis, 1981], and Handbook of Experimental Immunology, 3rd edition [Weir, 1978].

While it is not necessary to start with a pure antigen preparation for screening, the higher the purity of the antigen preparation used for

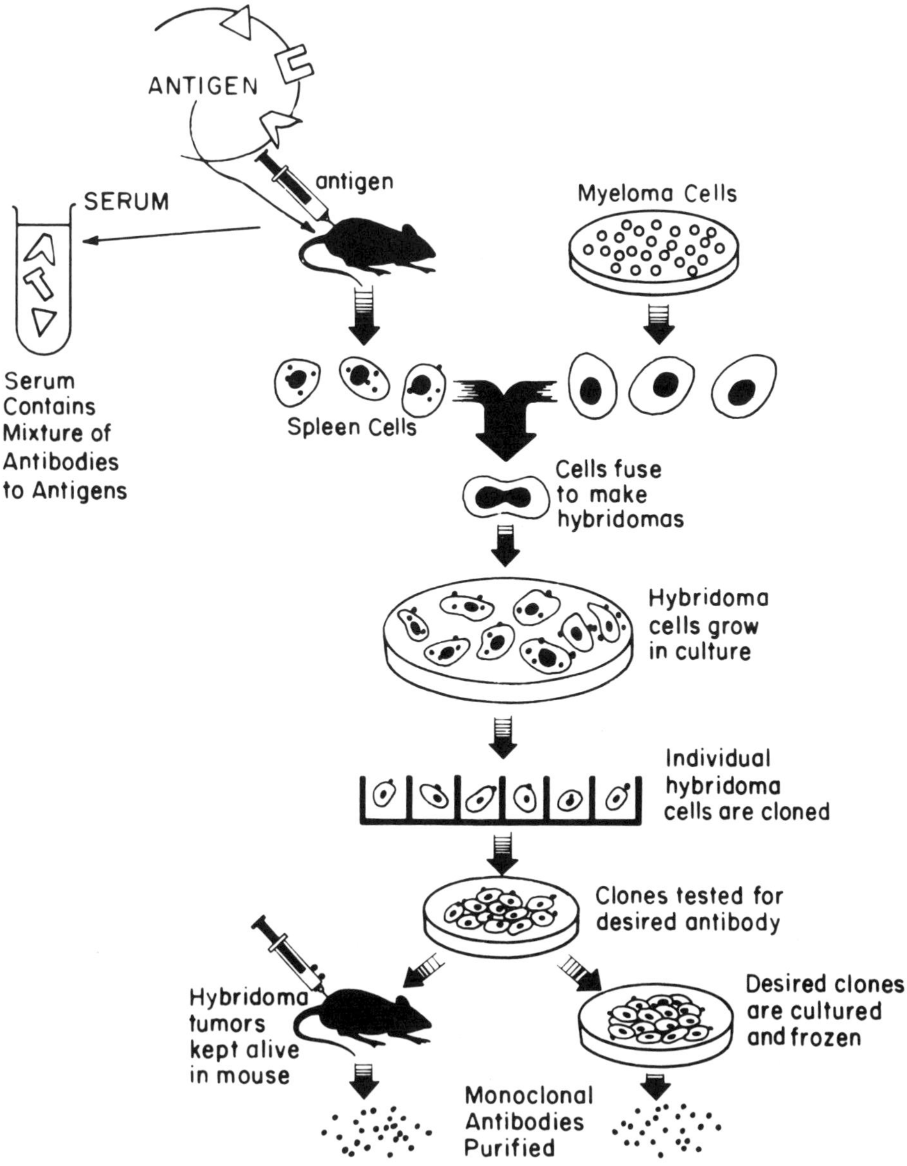

Fig. 1. *Production of monoclonal antibodies. Spleen cells producing antibodies to all determinants present in the antigen preparation are isolated from immunized mice and fused with drug-resistant myeloma cells grown in culture. Hybridomas are placed in media to select for growth of spleen cell-myeloma hybrids and cloned. Clones are screened for the presence of desired antibodies and positive clones are then grown in vivo or in vitro.*

screening, the easier the screening will be. For example, with a pure preparation of receptor material, identification of receptor-specific monoclonal antibodies can be accomplished in one step using a solid-phase radioimmunoassay (RIA) or enzyme-linked immunosorbent assay (ELISA) (see Appendix I). However, if the preparation of receptor material is only 50% pure, a positive reaction in a solid phase RIA or ELISA could indicate either a monoclonal antibody to the receptor or to one of the contaminants. Antibody specificity would, therefore, have to be determined by some other method. Nonetheless, solid-phase RIA or ELISA are of great utility as initial screening assays in order to determine which hybridomas are secreting monoclonal antibodies to any component(s) of the antigen mixture. In this way, one can often reduce the number of hybridomas subsequently screened for production of receptor-specific monoclonal antibodies by first determining how many hybridomas are secreting detectable antibodies [Venter et al, 1983a]. The one major drawback to using this method lies in the fact that this assay works best with monoclonal antibodies of relatively high affinity that remain bound to the antigen during repeated washing procedures. Monoclonal antibodies of low affinity may be missed with this assay.

There are two main types of screening assays which can be used when dealing with impure receptor preparations. The most specific of these assays, the indirect immunoprecipitation assay, requires stable, solubilized receptor preparations or prelabeled solubilized receptor preparations [Fraser and Venter, 1980; Fraser, 1983]. In this procedure, aliquots of soluble receptor are first incubated with samples of culture medium from each well containing hybridomas for an appropriate time (usually 6–24 h) to allow for formation of receptor-monoclonal antibody complexes which are subsequently precipitated by addition of an excess of antimouse immunoglobulin (usually IgG) antiserum. As described in Limitations of Monoclonal Antibodies, monoclonal antibodies usually cannot precipitate target antigens on their own and require addition of a second antibody to obtain a large lattice complex. Receptor quantitation may be performed directly by measuring receptor concentration in the precipitated immune complexes or indirectly by measuring the remaining receptor concentration in the supernate. We routinely measure each fraction and get very good agreement between both compartments. The main advantage of using an indirect immunoprecipitation assay for identifying monoclonal antibodies to receptors is its specificity. One is screening for the ability of a mouse immunoglobulin to immunoprecipitate an isolated soluble, specifically labeled receptor protein from solution.

Depending upon whether soluble receptor preparations are unoccupied or prelabeled with radioligand, it is possible to derive information as to whether or not monoclonal antibodies are specific for epitopes within or outside of the ligand binding domain of the receptor protein. For example, if a receptor is prelabeled either covalently or with a reversible ligand and found to be immunoprecipitated in this type of assay, it is very likely that the monoclonal antibody is directed to an epitope outside of the ligand-binding domain of the receptor as prior occupation of this region with radioligand might be expected to block antibody binding. Perhaps the only drawback to using prelabeled receptor in an indirect immunoprecipitation assay is that monoclonal antibodies specific for the ligand-binding site of receptors may not be detected. However, many soluble receptors are unable to bind ligand following detergent solubilization from membranes and must be prelabeled.

If a particular receptor has not been successfully solubilized from membranes or if the stability of a soluble receptor preparation is too low to survive the time required for an indirect immunoprecipitation assay (approximately 12–24 h), it will be necessary to use an alternative screening assay for detection of monoclonal antibodies. Another type of assay commonly used to identify monoclonal antibodies to cell surface receptors involves screening medium from hybridomas for inhibitory effects on receptor functions such as ligand binding or other receptor-mediated events. For example, spent medium from hybridoma cultures may be assayed for the ability to inhibit specific radioligand binding to a cell membrane receptor. While this type of assay is valid for screening, it may be important to remember that medium may contain components other than mouse immunoglobulins (such as hormones, growth factors, etc) which may also affect ligand binding or other receptor-mediated activity. Therefore, if possible, it is probably best to avoid using this type of assay as a sole screening assay for detection of monoclonal antibodies to receptors. Another limitation of this type of assay is that monoclonal antibodies to regions of the receptor outside of the active site will not be detected.

Cloning of Positive Hybridomas

In general, it is best to clone positive hybridomas as early as possible. If a culture contains more than one clone, each may grow at a different rate and undesirable cells may overgrow desirable cells in a very short time. Furthermore, the nature of hybrid cells is such that for a time after cell fusion, chromosomes from the cell nucleus are randomly lost and this trait conspires against stability of expression of immunoglobulins. However, a stable cell line can be isolated by cell cloning if done early enough.

The first step to be taken when cloning positive hybridomas is to prepare duplicate cultures, which minimizes the risk of loss of a particular clone during the cloning procedure. It is very important to establish frozen stocks of cells as soon as possible because clones of interest may be lost or may stop producing antibody at any point after fusion and cloning. If stock cultures have not been prepared before hybridomas revert, the cell line will be lost.

There are two basic procedures which can be used for cell cloning: cloning in semisolid agar [Williams et al, 1977; Tzartos et al, 1981a] and cloning by limiting dilution [McKearn, 1980]. Mixtures of hybridoma cells can be separated by localized growth of cells in soft agar. After growth of cell colonies to a size visible to the naked eye, they can be removed from the agar with a pasteur pipet and grown in liquid medium. One major disadvantage of the soft agar technique for cell cloning is that cell colonies picked for growth usually do not represent homogeneous cell populations due to contamination with other cells present in the soft agar and must be recloned to insure homogeneity.

Cloning by limiting dilution is usually the method of choice, particularly when the number of antibody secreting hybrids is very low. Cell suspensions containing 500, 50, and 10 cells/ml are prepared and 0.1 ml of each suspension is pipetted into each well of a 96-well microtiter plate containing 5×10^4 mouse peritoneal cells per well as a feeder layer. By using a feeder layer of cells one can overcome the usual cell density-dependent growth exhibited by hybridomas after fusion, and cloning efficiency using this technique is often close to 100%. Hybridomas from dishes showing cell growth in less than 30% of the wells are then rescreened for the presence of the appropriate monoclonal antibodies.

Cryopreservation of Hybridoma Cells

Many protocols for myeloma and hybridoma cell freezing have been described which require sophisticated equipment for programed decrements in temperature during cell freezing. While cell viability upon thawing is usually higher using a programed cell freezer, this is not necessary for successful cryopreservation. Cells can reliably be frozen and recovered using a much simpler procedure that does not require elaborate equipment [Kennett, 1980]. The most important element in successful cryopreservation of stock cultures lies in harvesting healthy cells in mid-logarithmic phase of growth (5×10^5 cells/ml), a criterion identical to that for successful cell fusions.

Large-Scale Production of Monoclonal Antibodies

There are two methods for generating large amounts of monoclonal antibodies from relevant hybridomas: culturing cells in vitro in large volumes of media [Appendix I or Cleveland et al, 1983] and growing hybridomas as ascites tumors in syngeneic or irra diated animals in vivo (Appendix I). Monoclonal antibodies are harvested from the spent culture medium or from the ascites fluid and serum of the tumor bearing animal, respectively. Each approach has its advantages and disadvantages and often these techniques are used concurrently.

The concentration of monoclonal antibody which can be generated by culturing cells in vitro is generally in the range of 10–50 μg/ml. One advantage of this technique for harvesting monoclonal antibodies is that the only antibody of mouse origin present in culture medium is that secreted by the hybridomas. If hybridomas are adapted to grow in low serum concentrations prior to large scale growth in culture, the amount of contaminating bovine immunoglobulins will be very low. Alternatively, synthetic medium that does not include serum may be used [Iscove and Melchers, 1978; Kawamoto et al, 1983; Murakami et al, 1982; Chang et al, 1980]. If sufficient quantities of monoclonal antibodies can be obtained from culture medium, this is the method of choice as a more controlled antibody source.

If larger amounts of monoclonal antibodies are necessary than can practically be obtained from large scale culture, an alternative source is to harvest monoclonal antibodies from the ascites fluid and serum of tumor bearing mice (rats). The concentration of monoclonal antibodies which can be obtained in this manner is generally on the order of 1–10 mg/ml, which is 100–1,000-fold greater than the concentration achieved in culture media. However, serum/ascites fluid contains proteins other than monoclonal antibodies and may contain significant amounts of normal mouse immunoglobulins. Nonetheless, this procedure can be used to generate large quantities of relatively pure monoclonal antibodies in a short time and eliminates the need for large scale tissue culture work.

Enrichment of Monoclonal Antibodies From Culture Media or Ascites Fluid

With small volumes of culture medium, monoclonal antibody concentrations can be increased by using ultrafiltration devices (Millipore, Amicon). For larger volumes of culture medium or for ascites fluid, we generally perform an $(NH_4)_2\ SO_4$ (50%) precipitation to collect immunoglobulins followed by dialysis against an appropriate buffer solution. Additional antibody purification may be achieved by immunoaffinity chromatography of this preparation on antimouse (or antirat) immunoglobulin-Sepharose. While we have used this procedure for routine purification of monoclonal antibodies, it should be remembered that due to the unique chemical nature of each antibody, $(NH_4)_2\ SO_4$ precipitation or elution from immunoaffinity columns may prove to be too harsh for a particular preparation. Unfortunately there is no way to predict whether or not this will be the case and it is advisable to test antibody stability prior to large scale purification.

Monoclonal antibody preparations can generally be stored as conventional antisera. Sterile samples can usually be kept for reasonable periods at 4°C and often contain 0.1% sodium azide as a preservative. For longer storage, it is recommended that monoclonal antibody preparations be kept at −70°C, preferably in small aliquots. Repeated freezing and thawing should be avoided, as this treatment rapidly denatures antibodies, particularly IgM.

Characterization of Monoclonal Antibodies

Sodium dodecyl sulfate polyacrylamide gel electrophoresis (SDS-PAGE) is useful for an initial characterization of the immunoglobulin chains produced by hybridomas. It discriminates between μ- and γ-heavy chains and usually allows discrimination between two light chains, if present [see Roberts et al, this volume; Kennett et al, 1980].

Further characterization of immunoglobulin subclass or light chain class can be done by conventional immunodiffusion (Ouchterlony's test) of concentrated culture medium or purified immunoglobulins against antiserum to specific immunoglobulin chains (Miles, Cappel, Research Products International; see Handbook of Experimental Immunology, 3rd edition [Weir, 1978], for methodologic details).

Mapping Monoclonal Antibodies by Competitive Binding

Which subunit of a protein is bound by a monoclonal antibody can be determined by "Western" blotting or immune precipitation of ^{125}I-labeled subunits dissociated in SDS, if the monoclonal antibody recognizes a determinant formed by a contiguous sequence of amino acids. However, if the monoclonal antibody binds to a determinant formed by noncontiguous amino acids, these approaches will not work. Competitive binding techniques can help in mapping the binding sites for such monoclonal antibodies (Fig. 2). This approach can also be used to distinguish antigenic determinants within a receptor subunit [Tzartos and Lindstrom, 1980; Tzartos et al, 1981a] or to determine the spectrum of antibody specificities in an antiserum by competition for binding of these antibodies to the antigen with a library of mapped monoclonal antibodies [Tzartos et al, 1981b]. We have used several variations of competitive binding techniques in studies of the acetylcholine receptor, of which five variations are described below. Variations on these methods could be used in many systems, and the principles involved are more important than the precise methodologies.

In one approach, the antigen (acetylcholine receptor) was insolubilized by conjugating it to cyanogen bromide-activated agarose beads [Tzartos and Lindstrom, 1980] or by allowing it to bind to *Naja naja* toxin coupled by cyanogen bromide-activated agarose beads [Tzartos et al, 1981a]. Alternatively, we could have insolubilized it by adsorption to plastic immunoassay wells. In any case, the washed insolubilized receptor was exposed to an excess of a monoclonal antibody to be used as a protector. Unbound monoclonal antibody was washed off and the bound monoclonal antibody covalently attached through the addition of 0.7% glutaraldehyde for 15 min. Control receptor preparations were similarly treated and then used to determine the maximum amount of test antibody which could be added and be completely bound. Then this amount of antibody was added to similar aliquots of insolubilized receptor coated with a protecting monoclonal antibody and the amount of antibody which remained in solution was assayed. If the protecting monoclonal antibody inhibited binding of the test monoclonal antibody, it was presumed that they bound at or near the same site. If the protecting monoclonal antibody inhibited binding of a certain fraction of serum antibodies, it was similarly presumed that this fraction bound at or near the same site as the monoclonal antibody. In such experiments, the ability of a single monoclonal antibody to inhibit the binding of a majority of antireceptor monoclonal antibodies and more than half of the antibodies in an antireceptor serum demonstrated the existence of a "main immunogenic region" on receptor [Tzartos and Lindstrom, 1980; Tzartos et al, 1981a].

Another approach to competitive binding studies [Hochschwender and Lindstrom, unpublished] uses an F(Ab) fragment of a monoclonal antibody of known specificity labeled with ^{125}I or glucose oxidase (for an ELISA system) and assays the ability of other monoclonal antibodies or serum antibodies to inhibit its binding. The antigen can be immobilized by binding to plastic immunoassay wells.

A third approach is to separate complexes of antigen and excess protecting monoclonal antibody by sucrose gradient centrifugation [Conti-Tronconi et al, 1981] or high-pressure liquid chromatography (HPLC) [Crawford et al, 1982] and then determine whether the addition of an excess of an unknown monoclonal antibody generates larger complexes. If larger complexes are generated, the two monoclonal antibodies bind to different sites on the antigen. If the same size complexes appear, the two monoclonal antibodies are directed at or near the same site.

A fourth approach is to use protecting monoclonal antibodies which do not bind to fixed Staph A and test antibodies which do, and determine whether saturation of the antigen with protecting monoclonal antibody prevents its precipitation by test antibody plus fixed Staph A [Tzartos et al, 1981a]. A fifth approach, which is a variation on this theme, is to use rat monoclonal antibodies as protectors and determine whether saturation of the antigen with protecting monoclonal antibody prevents its precipitation by human autoantibodies plus antihuman IgG [Tzartos et al, 1981b]. This method was used to show that antireceptor autoantibodies in myasthenia gravis patients were directed at many parts of the receptor molecule, but disproportionately at the main immunogenic region.

Immunoaffinity Purification of Receptors With Monoclonal Antibodies

The considerable specificity and, in some cases, relatively low affinity of monoclonal antibodies make them ideal reagents for simple, rapid immunoaffinity purification of cell surface receptors whose isolation has been hindered to a large extent by extremely low tissue concentrations and by the relatively low stability of detergent solubilized recep-

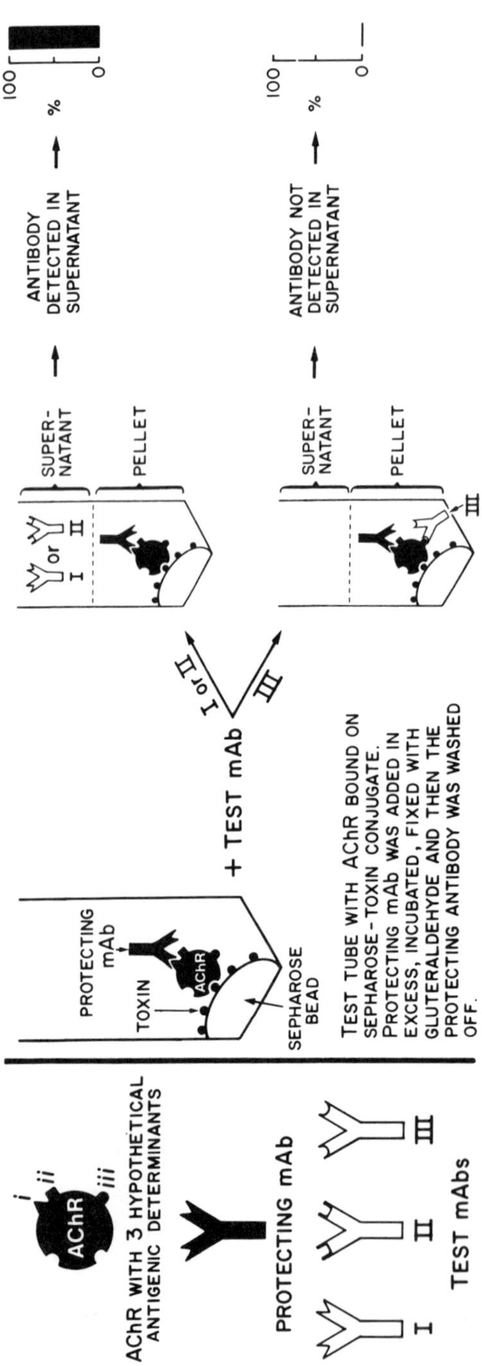

Fig. 2. *Method for mapping mAbs by competitive binding. The method shown is one of several competitive binding techniques used [Tzartos and Lindstrom, 1980; Tzartos et al, 1981a]. Receptor is immobilized on agarose and saturated with a protecting mAb that is subsequently covalently linked with glutaraldehyde. A limiting amount of test mAb is added, all of which would bind in the absence of protecting mAb. The amount remaining unbound is measured. If the test mAb binds at or near the site at which the protecting mAb is bound, binding of the test mAb is inhibited. [Reproduced from Tzartos et al, 1981a].*

tors when subjected to long purification procedures.

The first requirement for immunoaffinity chromatography of integral membrane proteins such as receptors is a crude detergent solubilized or partially purified antigen preparation [see Hjelmeland and Chrambach, this series]. It is important to keep solubilization conditions as mild as possible to avoid protein denaturation, particularly if monoclonal antibodies were produced to an intact functional protein. If the concentration of receptors in the starting membrane preparation is too low (< 10 ng) to be visualized on SDS-polyacrylamide gels with conventional protein stains (Coomassie blue, silver stain), it is important that receptors be radiolabeled with specific ligands or Na ^{125}I for later identification.

The production of immunoaffinity columns requires relatively large amounts of monoclonal antibody (on the order of 5–20 mg), and, therefore, it is suggested that antibody be harvested from ascites fluid of tumor-bearing animals. In general, it is advisable to partially purify monoclonal antibodies from ascites fluid (see Appendix I, section 20) for use as immunoadsorbants. This will increase the specific activity of the antibody preparation by removing albumin and other serum proteins.

There are a number of gels available commercially that are suitable for coupling to monoclonal antibodies. We routinely use cyanogen bromide-activated Sepharose 4B (Pharmacia) which couples the protein via primary amino groups. The coupling procedure is very straightforward and essentially consists of mixing the gel and the protein at the appropriate pH [Dalchau and Fabre, 1982]. We usually couple 5–10 mg monoclonal antibody per ml of swollen gel, the efficiency ranges from 50–80%. Coupling efficiency tends to increase with increasing antibody concentrations up to 25 mg/ml gel.

In general, we add 1–3 ml of antibody-Sepharose to an ISCO disposable polypropylene column. The column is then washed with 5–10 ml of a suitable buffer containing a relatively high salt concentration (such as PBS) followed by 5–10 ml of buffer containing 0.1–0.5% nonionic detergent (such as Triton X-100, CHAPS, digitonin) to equilibrate the column. The steps can easily be accomplished by adding buffer to the top of the column with a pipet.

Soluble antigen is next added to the column (in a usual volume of 1–5 ml although much larger volumes can be applied). The rate of sample application should be as slow as possible to allow for equilibration between antigen and immobilized monoclonal antibody. The eluate is collected in one tube and reserved for quantitation of antigen. After all of the sample has passed through the column, the column should be washed with 10–20 ml of buffer-containing detergent to remove proteins adsorbed nonspecifically to the immunoadsorbent. The first part of the washing should be performed rather slowly to avoid washing out any antigen still equilibrating in the gel. The last half of the washing can be carried out more quickly.

Elution of adsorbed antigen can be effected by using low pH (0.1 M glycine, pH 2.4) or high pH (0.1 M triethylamine) with or without detergent. Eluates should be neutralized immediately when using this type of elution protocol. Columns should be reequilibrated by extensive washing with PBS until the pH of the eluate returns to 7.4. The eluates from each step are assayed for the amount of antigen present using appropriate techniques such as measurement of ligand binding activity, liquid scintillation counting, gamma counting, etc. Under ideal conditions, the antigen should be completely depleted from the applied sample and concentrated following specific elution.

A unique type of elution may be possible in some cases depending on the specificity of the monoclonal antibody coupled to Sepharose. If the antibody is directed to a determinant in the active (ligand binding) site of the receptor molecule and competes with receptor specific ligands for receptor binding, this situation may be exploited in terms of ligand-specific elution of adsorbed receptor protein. Buffer containing detergent plus a physiological concentration of receptor specific ligand may be used in place of pH extremes for elution. This approach offers three distinct advantages over receptor elution with acids or bases: (1) It affords a much milder condition for protein elution, increasing the stability of eluted proteins and the immunoaffinity adsorbent; (2) it offers an additional line of evidence supporting the identity of the purified protein as a receptor based on the specificity of elution with receptor ligands; and (3) it gives another affinity step of purification. We have successfully utilized ligand specific elution of β-adrenergic receptors and Na^+/D-glucose cotransporter from immunoaffinity columns prepared with specific monoclonal antibodies [Fraser and Venter, 1980; Venter et al, 1983a]. Under these conditions, immunoaffinity columns may be reused many times (10–25) and a very high degree of protein purification is possible with only one or two steps.

Potential Problems in Immunoaffinity Purification of Receptors

The antigen is not retained by the column. There are two possible reasons for lack of retention of antigen by immunoaffinity columns. The first is that the affinity of the monoclonal antibody for antigen is too low to be an effective immunoadsorbent. This is unlikely, especially if screening assays have involved indirect immunoprecipitation of solubilized receptors.

It is more likely that this problem may be the result of antibody denaturation during the coupling procedure. There are several possible solutions to this problem including antibody coupling via a different reactive group (such as COO-) or use of noncovalent linkages (such as monoclonal antibody binding to protein A or antiimmunoglobulin-Sepharose).

Antigen cannot be eluted from immunoaffinity columns. It is possible that the monoclonal antibody may be of very high affinity and that changes in tertiary structure of the antigen binding site cannot be effected by conventional elution conditions (pH extremes). In this situation, other elution conditions may be tried such as elution with 2% SDS either alone or by boiling the gel in 2% SDS. If such harsh elution conditions are required it is unlikely that the immunoaffinity adsorbent may be reused due to antibody denaturation.

Monoclonal Antibodies as Probes of Receptor Structure and Function

Numerous reports have appeared in the past few years on the production of monoclonal antibodies to hormone and neurotransmitter receptors, demonstrating the utility of hybridoma technology in the development of antibodies to cell proteins of low concentrations and stability [Venter et al, 1983b]. The use of monoclonal antibodies to receptors as probes of receptor structure and function have afforded insights heretofore not possible with more conventional techniques for receptor characterization. In addition, monoclonal antibodies to receptors have been shown to be powerful tools in the immunoaffinity purification of many cell receptors whose isolation has proven to be a formidable undertaking. The following sections will highlight many of the advances in receptor research made possible with monoclonal antibodies as well as point out potential clinical applications of these antibodies.

Specific examples

β-Adrenergic receptor. In 1980, Fraser and Venter reported on the production of monoclonal antibodies to partially purified turkey erythrocyte β_1-receptors. One antibody competitively inhibited β-adrenergic ligand binding not only to turkey erythrocyte β-receptors but also to β_1- and β_2-receptors from several species and tissues. These results suggested the existence of a common determinant in the ligand binding site of β-receptors which was not present on α-adrenergic receptors.

The ligand binding site specific monoclonal antibody was utilized in the immunoaffinity purification of β_1- and β_2-receptors. The β_1-adrenergic receptor from turkey erythrocytes appeared as a single polypeptide of 70,000 daltons on SDS-polyacrylamide gels [Fraser and Venter, 1980; Venter and Fraser, 1981], in contrast to the β_2-adrenergic receptor from canine lung, which appeared as a protein of 114,000 daltons and was further shown to be composed of sulfhydryl-linked subunits of 59,000 daltons [Fraser and Venter, 1982]. Monoclonal antibodies to β-adrenergic receptors allowed for the first purification of both β_1- and β_2-adrenergic receptors from mammalian and nonmammalian sources.

Estrogen receptor. Greene and colleagues first reported on the generation of rat-mouse hybridomas secreting monoclonal antibodies to estrogen receptors (estrophilin) from calf uterus in 1980 [Greene et al, 1980a]. Antibodies were initially identified by their ability to immunoprecipitate soluble nuclear receptor labeled with ^{3}H-estradiol from solution in a double antibody immunoprecipitation assay. The antibodies reacted with both occupied and unoccupied receptors and did not effect estradiol binding to the receptor, suggesting that the determinant(s) recognized by these antibodies was outside of the ligand binding domain of the receptor molecule. All of these antibodies were specific for calf uterine estrogen receptors, displaying no cross-reactivity with any other estrogen receptor examined.

In a subsequent report, Greene and colleagues [1980b] described monoclonal antibodies to estrogen receptors from MCF-7 human breast cancer cells. In contrast to monoclonal antibodies against calf uterine estrophilin, monoclonal antibodies to human breast cancer estrogen receptors displayed cross-reactivity against other mammalian estrophilin, particularly receptors from primate endometrium. However, these antibodies did not recognize estrophilin from avian tissues, suggesting the existence of a common determinant between primate and possibly mammalian estrogen receptors that is not present on avian receptors.

In addition to their usefulness as probes of molecular homology between estrophilins of different species, an important potential application of monoclonal antibodies to human estrophilin is for the immunoassay of estrogen receptors in breast cancers as a guide to therapeutic management of disease.

Insulin receptor. Kull et al [1982] first reported on the development of a monoclonal antibody to human placental insulin receptors in 1982. The monoclonal antibody was identified by its ability to immunoprecipitate ^{125}I-insulin-receptor complexes but not ^{125}I-insulin from solution. Antibody displayed high specificity for insulin receptors from human placenta and human IM-9 lymphocytes but little or no cross-reactivity with insulin receptors from human erythrocytes, rat liver, rat adipocytes, rat erythrocytes, or horse erythrocytes, suggesting that the human insulin receptor may possess an antigenic determinant not shared by insulin receptors from other species.

In another report, Roth et al [1982] described a monoclonal antibody to human IM-9 lymphocytes. This antibody inhibited insulin binding to its receptor in human IM-9 lymphocytes, human adipocytes, and human placental membranes. The antibody appeared to be species specific as insulin binding to insulin receptors from rat tissues was not affected. Consistent with its ability to block insulin binding to the human insulin receptor, this monoclonal antibody inhibited insulin stimulation of 2-deoxyglucose uptake into human adipocytes and fibroblasts.

It was also shown that this monoclonal antibody could be used to immunoprecipitate ^{35}S-methionine-labeled insulin receptors from IM-9 cells and that prior receptor occupation with insulin prevented receptor immunoprecipitation.

Low-density lipoprotein (LDL) receptor. Monoclonal antibodies raised against the bovine adrenal cortex LDL receptor have demonstrated the utility of monoclonal antibodies as probes of the cellular and genetic factors involved in LDL receptor regulation. The most extensively studied monoclonal antibody was found to cross-react with LDL receptors of bovine and human origin but not of rodent, rabbit, or dog origin. This antibody binds to human LDL receptors in equimolar amounts as does ^{125}I-LDL, suggesting that each receptor contains one antibody and one LDL binding site. These sites appear to be on distinct regions of the LDL receptor as LDL did not reduce the binding of monoclonal antibody to human fibroblasts.

Beisiegel et al [1981] utilized this monoclonal antibody to the LDL receptor to study receptor regulation independent of ^{125}I-LDL binding. In normal human fibroblasts, receptor-bound monoclonal antibody was internalized and degraded at 37°C at a rapid rate similar to that measured for LDL. In a patient with familiar hypercholesterolemia due to a defect in receptor internalization, it was demonstrated that monoclonal antibody bound to the surface of fibroblasts but was not internalized or degraded, consistent with results from previous studies on this form of familial hypercholesterolemia. Fibroblasts from patients with the receptor-negative form of homozygous familial hypercholesterolemia, which bind less than 5% of the level of normal LDL, also displayed less than 5% of monoclonal antibody binding as fibroblasts from normal individuals.

Nicotinic acetylcholine receptors. Monoclonal antibodies have been used for many purposes in studying acetylcholine receptors and the autoimmune response to acetylcholine receptors that causes myasthenia gravis. These studies have been reviewed in detail elsewhere [Lindstrom, 1983a,b; Lindstrom et al, 1983a]. In summary, there are six sorts of applications for monoclonal antibodies in studies of acetylcholine receptors from fish electric organ tissue and mammalian muscle: (1) to identify receptor subunits [Tzartos and Lindstrom, 1980] and identify [Tzartos et al, 1981a] and localize substructures within receptor subunits; (2) to test the role of substructures to which they bind in receptor function [Lindstrom et al, 1983a]; (3) to compare the structures of receptors from different tissues and species [Lindstrom et al, 1983b; Einarson et al, 1982]; (4) to localize subunits within the quaternary structure of the receptor molecule [Kristofferson et al, 1983]; (5) to study the synthesis, conformational maturation, and assembly of receptor subunits [Merlie et al, 1982, 1983a; Sebbane et al, 1983; Merlie and Lindstrom, 1983] and (6) to identify cDNA clones for receptor subunits [Merlie et al, 1983b].

In the study of the autoimmune response to acetylcholine receptors in myasthenia gravis and its animal model, monoclonal antibodies are being used in three ways: (1) to determine the antigenic structure of the receptor [Tzartos and Lindstrom, 1980; Tzartos et al, 1981a,b; Gullick et al, 1981; Gullick and Lindstrom, 1983], (2) to determine the specificities of autoantibodies to receptor [Tzartos et al, 1981b], and (3) to determine the pathological effects of various antibody specificities [Tzartos and Lindstrom, 1980; Conti-Tronconi et al, 1981; Lennon and Lambert, 1981; Gomez et al, 1981; Souroujon et al, 1983]. Finally, monoclonal antibodies raised against acetylcholine receptors from electric organs and muscle are being used in studies of acetylcholine receptors in neurons in four ways: (1) to identify putative acetylcholine receptors in neurons [Swanson et al, 1983]; (2) to localize these neurons [Swanson et al, 1983]; (3) to purify these receptors [Tzartos and Lindstrom, 1980] and (4) to characterize their subunit structure [Swanson et al, 1983].

Thyrotropin receptors. Yavin et al [1981] described monoclonal antibodies raised against thyroid membrane preparations which specifically inhibit ^{125}I-thyrotropin binding to a solubilized preparation of the glycoprotein component of the thyrotropin receptor from several different species but have no effect on ^{125}I-thyrotropin binding to liposomes containing gangliosides. The antibodies interacted with functioning thyroid cells in culture and inhibited thyrotropin stimulated iodide uptake but did not affect adenylate cyclase activity. These findings suggested that the glycoprotein component of the thyroid-stimulating hormone (TSH) receptor contains the physiologically relevant hormone-binding site which mediates at least one TSH response, namely, iodide uptake.

A second report by Valente and co-workers [1982] extended initial TSH receptor characterization with mouse monoclonal antibodies to the study of human monoclonal antibodies to the TSH receptor derived from the lymphocytes of patients with Graves' disease. Mouse myeloma cells were fused with peripheral lymphocytes from patients with active Graves' disease to obtain four TSH receptor-specific monoclonal antibodies. Two antibodies mimic the effects of stimulatory antibodies in Graves disease as they stimulate thyroid function. These antibodies appear to interact with the ganglioside component of the thyrotropin receptor rather than the glycoprotein component. In contrast, the other two monoclonal antibodies obtained by this group mimic the inhibitory activities of certain antibodies in sera from Graves' disease patients. They inhibit thyrotropin mediated responses in tissue preparations and appear to recognize the glycoprotein component of the receptor with little or no cross-reactivity for the ganglioside component. The data confirm the heterogeneous nature of the autoimmune immunoglobulins in Graves' disease and suggest that distinct components of the TSH receptor are important in expression of thyrotropin mediated responses.

Transferrin receptor. Trowbridge and coworkers [1978] identified a cell surface glycoprotein from human hematopoietic cells using a monoclonal antibody obtained from a fusion of BALB/c spleen cells immunized against a human hematopoietic cell line K562. The glycoprotein migrated with an apparent molecular weight of 110,000 daltons on reducing SDS-polyacrylamide gels and of 200,000 daltons on nonreducing gels.

In subsequent studies, Trowbridge and Omary [1981] determined that the biological function of this glycoprotein was that of the transferrin receptor. Transferrin is the major serum iron-transport protein and has been shown to be obligatory for cell growth in vitro. The monoclonal antibody used to purify this receptor protein was found to inhibit the binding of transferrin to its receptor in a line of human T leukemia cells and also shown to inhibit the growth of these cells in vitro [Trowbridge and Lopez, 1982].

These studies demonstrate the utility of monoclonal antibodies in the association of a cell surface molecule with a specific function as well as in structural analysis of cell membrane components.

ACKNOWLEDGMENTS

The authors wish to thank Drs. Len C. Harrison and J. Craig Venter for helpful suggestions regarding the text of this chapter.

Supported by grants AHA-82881 from the American Heart Association (to C.M.F.) and NS-11323 from the National Institutes of Health, by the McKnight Foundation, Muscular Dystrophy Association (Los Angeles and California chapters). Myasthenia Gravis Foundation, and the Alexander Onassis Public Benefit Foundation (to J.L.).

REFERENCES

Beisiegel U, Schneider WJ, Goldstein JL, Andersen RGW, Brown MS (1981): Monoclonal antibodies to the low-density lipoprotein receptor as probes for study of receptor mediated endocytosis and the genetics of familial hypercholesterolemia. J Biol Chem *256:*11923–11932.

Chang TH, Steplewski Z, Koprowski H (1980): Production of monoclonal antibodies in serum free medium. J Immunol Methods *39:*369–375.

Cleveland WL, Wood I, Erlanger BF (1983): Routine large-scale production of monoclonal antibodies in a protein-free culture medium. J Immunol Methods *56:*221–234.

Conti-Tronconi B, Tzartos S, Lindstrom J (1981): Monoclonal antibodies as probes of acetylcholine receptor structure. II. Binding to native receptor. Biochemistry *20:*2181–2191.

Cotton RGH, Milstein C (1973): Fusion of two immunoglobulin-producing myeloma cells. Nature *244:*42–43.

Crawford G, Correa L, Salvaterra P (1982): Interaction of monoclonal antibodies with mammalian choline acetyltransferase. Proc Natl Acad Sci USA *80:*7031–7035.

Dalchau R, Fabre JW (1982): The purification of antigens and other studies with monoclonal antibody affinity columns. The complementary new dimension of monoclonal antibodies. In McMichael AJ, Fabre JW (eds): "Monoclonal Antibodies in Clinical Medicine." London: Academic Press, pp 519–556.

Einarson B, Gullick W, Conti-Tronconi B, Lindstrom J (1982): Subunit composition of fetal calf muscle nicotinic acetylcholine receptor. Biochemistry *21:*5295–5302.

Fazekas de St. Groth SF, Scheidegger D (1980): Production of monoclonal antibodies: Strategy and tactics. J Immunol Methods *35:*1–21.

Fellows RE, Eisenbarth GS (1981): "Monoclonal Antibodies in Endocrine Research." New York: Raven Press.

Fraser CM (1983): Monoclonal antibodies to β-adrenergic receptors and receptor structure. In Cuatrecasas P, Greaves MF (eds): "Receptors and Recognition." London: Chapman and Hall.

Fraser CM, Venter JC (1980): Monoclonal antibodies to β-adrenergic receptors: Use in purification and molecular characterization of β-receptors. Proc Natl Acad Sci USA *77:*7034–7038.

Fraser CM, Venter JC (1982): The size of the mammalian lung β_2-adrenergic receptor as determined by target size analysis and immunoaffinity chromatography. Biochem Biophys Res Comm *109:*21–29.

Galfre G, Howe SC, Milstein C, Butcher GW, Howard JC (1977): Antibodies to major histocompatibility antigens produced by hybrid cell lines. Nature *226:*550–552.

Galfre G, Milstein C (1981): Preparation of monoclonal antibodies: Strategies and procedures. In Langone JJ, Van Vunakis H (eds): "Methods in Enzymology, Vol. 73". New York: Academic Press, Inc., pp. 3–46.

Gefter ML, Margulies DH, Scharff MD (1977): A simple method for polyethylene glycol-promoted hybridization of mouse myeloma cells. Somatic Cell Genet *3:*231–236.

Gomez C, Richman D, Burres S, Arnason B (1981): Monoclonal hybridoma anti-acetylcholine receptor antibodies: Antibody specificity and effect of passive transfer. Ann NY Acad Sci *377:*97–109.

Greene GL, Fitch FW, Jenson EV (1980a): Monoclonal antibodies to estrophilin: Probes for the study of estrogen receptors. Proc Natl Acad Sci USA *77:*157–161.

Greene GL, Nolan C, Engler JP, Jensen EV (1980b): Monoclonal antibodies to human estrogen receptor. Proc Natl Acad Sci USA *77:*5115–5119.

Gullick W, Lindstrom J (1983): Mapping and binding of monoclonal antibodies to the acetylcholine receptor from *Torpedo californica*. Biochemistry 22:3312–3320.

Gullick WJ, Tzartos S, Lindstrom J (1981): Monoclonal antibodies as probes of acetylcholine receptor structure. I. Peptide mapping. Biochemistry *10:*2173–2180.

Howard JC, Butcher GW, Galfre G, Milstein C, Milstein CP (1979): Monoclonal antibodies as tools to analyze the serological and genetic complexities of major transplantation antigens. Immunol Rev *47:*139–174.

Iscove NN, Melchers F (1978): Complete replacement of serum by albumin, transferrin and soybean lipid in cultures of lipopolysaccharide-reactive B lymphocytes. J Exp Med *147:*923–933.

Kawamoto T, Sato JD, Le A, McClure DB, Sato GH (1983): Anal Biochem *130:*445–453.

Kearney JF, Radbruch A, Leisegang B, Rajewsky K (1979): A new mouse myeloma cell line that has lost immunoglobulin expression permits the construction of antibody-secreting hybrid cell lines. J Immunol *123:*1348–1350.

Kennett RH (1980): Freezing of Hybridoma Cells. In Kennett RH, McKearn TJ, Bechtol KB (eds): "Monoclonal Antibodies. Hybridomas: A New Dimension in Biological Analysis." New York: Plenum Press, p 375.

Kennett RH, McKearn TJ, Bechtol KB (1980): "Monoclonal Antibodies. Hybridomas: A New Dimension in Biological Analysis." New York: Plenum Press.

Kohler G, Howe SC, Milstein C (1976): Fusion between immunoglobulin-secreting and nonsecreting myeloma cell lines. Eur J Immunol *6:*292–295.

Kohler G, Milstein C (1975): Continuous cultures of fused cells secreting antibody of predefined specificity. Nature *256:*495–497.

Kohler G, Milstein C (1976): Derivation of specific antibody-producing tissue culture and tumor cell lines by cell fusion. Eur J Immunol *6:*511–519.

Kristofferson D, Desneules P, Hochschwender S, Lindstrom J, Stroud R (1983): Acetylcholine receptor α subunits located using monoclonal antibody F(ab) fragments. (Submitted).

Kull FC Jr, Jacobs S, Su Y-F, Cuatrecasas P (1982): A monoclonal antibody to human insulin receptor. Biochem Biophys Res Commun *106:*1019–1026.

Langone JJ, Van Vunakis H (1981): "Methods in Enzymology, Vol. 73." New York: Academic Press, Inc.

Lennon V, Lambert E (1981): Monoclonal autoantibodies to acetylcholine receptors: Evidence for a dominant idiotype and requirement of complement for pathogenicity. Ann NY Acad Sci *377:*77–96.

Lindstrom J (1983a): Use of monoclonal antibodies in the study of myasthenia gravis. In Eisenbarth GS, Haynes BF (eds): "Monoclonal Antibodies: Probes for the Study of Autoimmunity and Immunodeficiency." (In press).

Lindstrom J (1983b): Use of monoclonal antibodies in the study of acetylcholine receptors and myasthenia gravis. Neurosci Comment (In press).

Lindstrom J, Tzartos S, Gullick W, Hochschwender S, Swanson L, Sargent P, Jacob M, Montal M (1983a): Use of monoclonal antibodies to study acetylcholine receptors from electric organs, muscle and brain and the autoimmune response to receptor in myasthenia gravis. In: 48th Cold Spring Harbor Symposium: Molecular Neurobiology. (In press).

Lindstrom J, Cooper J, Swanson L (1983b): Purification of acetylcholine receptors from the muscles of *Electrophorus electricus*. Biochemistry. (In press).

Littlefield SW (1964): Selections of hybrids from matings of fibroblasts in vitro and their presumed recombinants. Science *145:*709–711.

Margulies DH, Cieplinski W, Dharmgrongartama B, Gefter ML, Morrison SL, Kelly T, Scharff MD

(1976): Regulation of immunoglobulin expression in mouse myeloma cells. Cold Spring Harbor Symp Quant Biol *41:*781–791.

Mason DW, Williams AF (1980): The kinetics of antibody binding to membrane antigens in solution and at the cell surface. Biochem J *187:*1–20.

McKearn TJ (1980): Cloning of hybridoma cells by limiting dilution in fluid phase. In Kennett RH, McKearn TJ, Bechtol KB (eds): "Monoclonal Antibodies. Hybridomas: A New Dimension in Biological Analyses." New York: Plenum Press, p 374.

Merlie JP, Sebbane R, Tzartos S, Lindstrom J (1982): Inhibition of glycosylation with tunicamycin blocks assembly of newly synthesized acetylcholine receptor subunits in muscle cells. J Biol Chem *257:*2694–2701.

Merlie JP, Lindstrom J (1983): Assembly in vivo of mouse muscle acetylcholine receptor: Identification of an α subunit species which may be an assembly intermediate. Cell (In press).

Merlie JP, Sebbane R, Gardner S, Olson E, Lindstrom J (1983a): The regulation of acetylcholine receptor expression in mammalian muscle. 48th Cold Spring Harbor Symposium: Molecular Neurobiology. (In press).

Merlie J, Sebbane R, Gardner S, Lindstrom J (1983b): Regulation of acetylcholine receptor gene expression: Molecular cloning of cDNA specific for α subunit of the receptor from the mouse muscle cell line BC3H-1. Proc Natl Acad Sci USA (in press).

Murakami H, Masui H, Sato GH, Sucoka N, Chow TP, Kanu-Suela T (1982): Growth of hybridoma cells in serum-free medium: Ethanolamine is an essential component. Proc Natl Acad Sci USA *79:*1158–1162.

Oi VT, Herzenberg LA (1981): Immunoglobulin producing hybrid cell lines. In Mishell BB, Shiigi SM (eds): "Selected Methods in Cellular Immunology." San Francisco: W.H. Freeman and Co., p 351.

Paul J (1973): "Cell and Tissue Culture," 4th Edition. London:Churchill Livingstone.

Pontecorvo G (1975): Production of mammalian somatic cell hybrids by means of polyethylene glycol treatment. Somatic Cell Genet *1:*397–400.

Reading CL (1982): Theory and methods for immunization in culture and monoclonal antibody production. J Immunol Methods *53:*261–292.

Roth RA, Cassell DJ, Wong KY, Maddux BA, Goldfine ID (1982): Monoclonal antibodies to the human insulin receptor block insulin binding and inhibit insulin action. Proc Natl Acad Sci USA *79:*7312–7316.

Schulman M, Wilde CD, Kohler G (1978): A better cell line for making hybridomas secreting specific antibodies. Nature *276:*269–270.

Sebbane R, Clokey G, Merlic JP, Tzartos S, Lindstrom J (1983): Characterization of the mRNA for mouse muscle acetylcholine receptor α subunit by quantitative translation in vitro. J Biol Chem *258:*3294–3303.

Souroujon M, Mochly-Rosen D, Gordon A, Fuchs S (1983): Interaction of monoclonal antibodies to Torpedo acetylcholine receptor with the receptor of skeletal muscle. Muscle Nerve *6:*303–311.

Stanbridge E, Schneider E (1976): "The Hazards of Mycoplasma Contamination." New York: Spectrum Publications, Inc.

Swanson L, Lindstrom J, Tzartos S, Schmued L, O'Leary D, Cowan W (1983): Immunohistochemical localization of monoclonal antibodies to the nicotinic acetylcholine receptor in the midbrain of the chick. Proc Natl Acad Sci USA *80:*4532–4536.

Trowbridge IS (1978): Interspecies spleen-myeloma hybrid producing monoclonal antibodies against mouse lymphocyte surface glycoprotein, T200. J Exp Med *148:*313–323.

Trowbridge IS, Lopez F (1982): Monoclonal antibody to transferrin receptor blocks transferrin binding and inhibits human tumor cell growth in vitro. Proc Natl Acad Sci USA *79:*1175–1179.

Trowbridge IS, Omary MB (1981): Human cell surface glycoprotein related to cell proliferation is the receptor for transferrin. Proc Natl Acad Sci USA *78:*3039–3043.

Tzartos S, Lindstrom JM (1980): Monoclonal antibodies used to probe acetylcholine receptor structure: Localization of the main immunogenic region and detection of similarities between subunits. Proc Natl Acad Sci USA *77:*755–759.

Tzartos S, Lindstrom JM (1981): Production and characterization of monoclonal antibodies for use as probes of acetylcholine receptors. In Fellows R, Eisenbarth G (eds): "Monoclonal Antibodies in Endocrine Research." Raven Press, pp 69–86.

Tzartos SJ, Rand DE, Einarson BE, Lindstrom J (1981a): Mapping of surface structures of electrophorus acetylcholine receptor using monoclonal antibodies. J Biol Chem *256:*8635–8645.

Tzartos SJ, Seybold M, Lindstrom J (1981b): Specificity of antibodies to acetylcholine receptors in sera from myasthenia gravis patients measured by monoclonal antibodies. Proc Natl Acad Sci USA *79:*188–192.

Valente WA, Vitti P, Yavin Z, Yavin E, Rotella CM, Grollman EF, Toccafondi RS, Kohn LD (1982): Monoclonal antibodies to the thyrotropin receptor: Stimulating and blocking antibodies derived from the lymphocytes of patients with Graves disease. Proc Natl Acad Sci USA *79:*6680–6684.

Venter JC, Fraser CM (1981): The development of monoclonal antibodies to β-adrenergic receptors and their use in receptor purification and characterization. In Fellows RE, Eisenbarth GS (eds): "Monoclonal Antibodies in Endocrine Research." New York: Raven Press, pp 119–134.

Venter JC, Eddy B, Schmidt U, Fraser CM (1983a): Production of monoclonal antibodies to integral membrane transport and receptor proteins and their use in structural elucidation. In Kohler H, Cazenave PA, Urbain J (eds): "Idiotypic Manipulations in Biological Systems." New York: Academic Press.

Venter JC, Fraser CM, Lindstrom JM (1983b): "Monoclonal and Anti-Idiotypic Antibodies: Probes for Receptor Structure and Function." New York: Alan R. Liss, Inc.

Weir DM (1978): "Handbook of Experimental Immunology, 3rd edition." Oxford: Blackwell Scientific Publications.

Williams A, Galfre E, Milstein D (1977): Analysis of cell surfaces by xenogeneic myeloma-hybrid antibodies: Differentiation antigens of rat lymphocytes. Cell *12*:663–673.

Yavin E, Yavin Z, Schneider MD, Kohn LD (1981): Monoclonal antibodies to the thyrotropin receptor: Implications for receptor structure and the action of autoantibodies in Gravis disease. Proc Natl Acad Sci USA *78*:3180–3184.

APPENDIX I—EXPERIMENTAL PROCEDURES

The reader is referred to an excellent book on tissue culture techniques by Paul [1973].

1. Tissue Culture Equipment

37°C Incubator (with a controlled CO_2 and humidity atmosphere)
Laminar flow hood (or sterile work area)
Phase-contrast microscope
Temperature-controlled water bath
Clinical centrifuge
Liquid N_2 cell storage container
Sterile glassware and tissue culture plastics

2. Serum Screening for Support of Hybridoma Growth

Materials

DME containing 20% fetal bovine serum
96-well micotiter plates
Microliter pipet with sterile tips

Method

a) Prepare DME-containing 20% fetal bovine serum.

b) Add 100 μl of media containing test serum to each well in rows 1–6 of a 96-well microtiter plate and 100 μl of media containing control, proven serum to each well in rows 7–12.

c) Prepare suspensions of hybridoma (or myeloma) cells containing 1000, 500, 250, and 125 cells/ml in DME.

d) Add 50 μl of each cell suspension to 24 consecutive wells on the 96-well microtiter plate (rows A and B, C and D, E and F, G and H) and incubate plate at 37°C in a humidified atmosphere of 90% air and 10% CO_2.

e) Examine wells for cell growth after 5–7 days.

f) Choose serum lot which best supports cell growth.

3. Preparation of Myeloma Cells for Cell Fusion

Materials

50-ml conical centrifuge tubes, sterile
DME containing 2.5 percent fetal bovine serum (37°C)

Method

a) Myeloma cells in logarithmic growth are pelleted in a sterile conical centrifuge tube by centrifugation at room temperature for 10 min at 1,000g. Cells are resuspended in 10 ml of DME containing 2.5% fetal bovine serum, counted and tested for viability using trypan blue exclusion. Cell viability should be at least 75%. Cells are kept at 37°C until fusion. (Note: Cells should be grown for 4–5 days prior to fusion in 20 μg/ml 8-azaguanine to select against revertants.)

4. Preparation of Spleen Cells (Normal and Immune) for Cell Fusion

Materials

DME containing 2.5% fetal bovine serum (37°C)
DME minus serum (4°C)
0.17 M NH_4Cl (4°C)
70% ethanol
Sterile scissors, forceps

Sterile Bellco Cellector, finest wire mesh
50-ml conical centrifuge tubes, sterile
100-mm petri dishes, sterile
Sterile pipets

Method

a) Sacrifice mouse by cervical dislocation.

b) Immerse mouse in 70% ethanol for 1–2 min.

c) Place mouse in a sterile 100-mm petri dish and aseptically remove spleen in a sterile hood. It is good practice to use a separate sterile pair of scissors and forceps to incise skin and another pair to incise peritoneum and remove spleen.

d) Place spleen in another sterile petri dish containing 10 ml DME-2.5% fetal bovine serum.

e) Push the spleen through the wire mesh screen of the Bellco Cellector using a sterile glass pestle. Rinse screen with 5 ml DME-2.5% fetal bovine serum.

f) Transfer the cell suspension to a sterile 50-ml conical centrifuge tube. Centrifuge at 1000g for 10 min.

g) Aspirate supernate, add 5 ml *cold* 0.17 M NH_4Cl to the cell pellet and gently resuspend. Incubate for 10 min at 4°C. This step causes lysis of red blood cells.

h) Add 5 ml cold DME minus serum. Centrifuge at 1000g for 10 min. Aspirate supernate.

i) Resuspend cell pellet in 10 ml DME-2.5% fetal bovine serum. Perform cell count and viability determination. Cell viability should be 80–90%. A spleen usually contains 5–10 × 10^7 cells.

5. Cell Fusion

Materials

Spleen cells from mice collected sterilely, 1 × 10^8 per fusion mix
Myeloma cells grown continuously in log phasc
DME minus serum (37°C)
DME plus 20% fetal bovine serum (37°C)
HT media (37°C)
40% PEG 1000 (w/v) in serum-free DME—prepared fresh (42°C)
Hat media (37°C)
Sterile pipets
Sterile 50-ml conical centrifuge tubes
96-well microtiter plates

Method

a) Combine 10^8 spleen cells with 10^7 myeloma cells in a conical centrifuge tube. Centrifuge at 1,000g for 10 min.

b) Aspirate supernate and resuspend cells in 10 ml. DME minus serum. Centrifuge at 1,000g for 10 min.

c) Aspirate supernate *completely*. Loosen the cell pellet by tapping the bottom of the centrifuge tube.

d) Slowly add 0.2 ml 40% PEG 1000 and gently swirl the tube for 30 s.

e) Centrifuge the cell suspension for 6 min at 500g.

f) Eight minutes after the addition of PEG, *slowly* add 5 ml of DME minus serum to the cell pellet.

g) Add an additional 5 ml of DME-20% fetal bovine serum. Centrifuge cell suspension for 10 min at 1,000g.

h) Aspirate supernate and resuspend cells in 50 ml HT media/10^7 myeloma cells.

i) Add normal spleen cells to a final concentration of 5 × 10^5/ml as a feeder layer.

j) Distribute 0.15 ml of the cell suspension into each well of 96-well microtiter plates.

k) Incubate overnight at 37°C in a humidified atmosphere of 95% air-5% CO_2.

l) The following day, add 0.15 ml of HAT media (2× A) to each well.

m) Replenish the wells with fresh HAT media every 3–4 days until vigorous growth of hybridomas becomes evident. This usually takes 7–10 days.

n) When cultures become macroscopically visible and cover 10–20% of the area of each well, screening should be performed. Culture media usually becomes more yellow by this point.

6. Direct Cloning in Soft Agar

Materials

Spleen or lymph node cells from mice (rats) collected sterilely in serum-free DME, 1×10^8 cells per fusion mix. (A rat spleen gives enough lymphocytes for two such fusions, whereas two mouse spleens may be required for one fusion.)

Myeloma cells grown continuously in log phase (around 5×10^5 cells/ml) in DME plus 10% fetal bovine serum or 10% horse serum and grown for 4 or 5 days prior to the fusion in appropriate medium to select against revertants, 1.5×10^7 cells total.

PEG 1500 at 50% w/v in serum-free DME (37°C)

DME minus serum

DME plus 2% horse serum

HAT medium

Soft agar in HAT medium (43°C): four bottles are used, each made with 12 ml of melted 2.5 percent agar in saline and 108 ml of HAT medium.

Method

a) The day before the fusion, treat 30 plates containing 24 (2 ml) wells with 1 ml per well of DME plus 2% horse serum and leave in the incubator overnight.

b) Aspirate the DME plus 2% serum and precoat the 24-well plates with 0.2 ml/well of soft agar in HAT medium on the morning of the fusion.

c) Mix the lymphocytes and myeloma cells in a 50-ml conical centrifuge tube and centrifuge at 2,000 rpm for 8 min. Remove the supernate and place the tube in a 37°C water bath.

d) Add the PEG dropwise over 1 min while agitating the tube.

e) Add 1 ml of serum free DME dropwise over the next minute with gentle agitation.

f) Add 20 ml of serum free DME over the next 8 min while continuing to agitate the tube in the water bath.

g) Centrifuge as before and aspirate the supernate.

h) Tap the pellet loose and add 2 ml of HAT medium.

i) Add 240 ml of 37°C HAT medium in soft agar and quickly dispense into the 720 precoated wells at 3 drops per well.

j) Put the plates on trays of ice covered with aluminum foil to solidify.

k) Then add 1.5 ml of HAT medium to each well and plate in the incubator.

l) Every 4–5 days over about the next 2 wk, replace half of the HAT medium in each well. We find a Tuesday/Friday schedule convenient.

m) Assay the supernatant media for monoclonal antibodies and mark the positive wells Half replace the media in these wells with HT medium.

n) After 2 or 3 days, reassay the positive wells and mark those which have higher titers than before. Half replace the HT medium. Colonies can be picked from these wells. If necessary, all wells can be reassayed after 4 or 5 wk to screen for slower growing hybridomas.

o) When colonies of at least 0.2 mm are observed, these can be transferred with Pasteur pipettes into 0.2-ml wells containing HT medium and dispersed by pipetting.

p) After 2 or 3 days, the medium in the small wells is assayed to detect positive clones. Positive clones are simultaneously grown to larger volumes for freezing and recloned in agar.

q) Recloning in agar uses 24 (2 ml) well plates. In the first well, plate 1 ml of warm soft agar in DME plus 10% serum and put 0.5 ml in the next 11 wells. Add a drop of cell suspension to the first well, mix, and transfer 0.5 ml to the next well. Similarly, serially dilute the cells through all 12 wells. The next day add 1.5 ml of medium to each well. After about 2 wk, when colonies are visible, assay the media from each well. The most dilute well in which antibody is detected is used as a source of colonies for transfer to 0.2 ml wells as before. The medium of almost all of these wells should contain antibody after

2 or 3 days. Grow one up for mass culture and freezing.

Initial Screening Assays for Detection of Monoclonal Antibodies to Antigen Components

These assays are designed to detect the presence of monoclonal antibodies to any component of the antigenic mixture used for immunization. Consequently, with an impure antigen preparation the specificity of each monoclonal antibody cannot be determined directly using these assays. It is suggested that these types of assays be used initially if large numbers of hybridomas have been produced and cannot be conveniently screened using receptor specific assays. Hybridomas positive in these assays may then subsequently be screened for receptor-specific monoclonal antibodies.

Antigen(s) is immobilized by attachment to plastic such as PVC microtiter plates and incubated with culture medium from actively growing hybridomas. Medium and unbound antibody is washed off the plates and antibody bound to immobilized antigen is detected by addition of a second mouse immunoglobulin specific antiserum which is radiolabeled or conjugated with an enzyme.

7. Solid-Phase Radioimmunoassay and/or Enzyme-Linked Immunosorbent Assay. Immobilization of Soluble Antigens and Membranes

Materials

Flexible polyvinyl chloride microtiter plates, 96-well, round bottom

PBS

Antigen solution: 1–100 μg/ml in PBS

BSA solution: 1% in PBS

Method

a) Add 100 μl of antigen solution to each well of a microtiter plate.

b) Incubate at 4°C overnight.

c) Empty the wells by flicking the plates quickly.

d) Add 100 μl of 1% BSA solution and incubate for 1–2 h at room temperature. BSA blocks the remainder of the protein binding sites on the plastic which may nonspecifically bind immunoglobulins from culture media.

e) Empty the wells by flicking the plates.

f) Wash wells once with PBS.

8. Solid-Phase RIA

Materials

Antigen-coated microtiter plates

Supernatants from hybridoma cultures

PBS

[^{125}I]-rabbit or sheep antimouse (or anti-rat) IgG (New England Nuclear, Amersham)

Method

a) Add 50 μl of spent culture medium to each well of an antigen coated microtiter plate. Include a negative control consisting of tissue culture medium.

b) Cover plate and incubate at 4°C for 1–2 h.

c) Empty plates by flicking quickly.

d) Wash twice with 100 μl PBS.

e) Add 100 μl of radiolabeled second antibody (20,000–50,000 cpm).

f) Cover plate and incubate at 4°C for 1–2 h.

g) Empty plates.

h) Wash once with 100 μl PBS.

i) Dry the plates and cut each well with scissors or a hot wire.

j) Transfer each well to a clean tube and count in a gamma counter.

9. ELISA

Materials

Antigen-coated microtiter plates

Supernatants from hybridoma cultures

PBS

Rabbit anti-mouse IgG (Miles, Cappel)

Alkaline phosphatase conjugated goat anti-rabbit IgG (Cappel)

p-nitrophenylphosphate

Diethanolamine buffer

Method

a) Add controls and test medium to wells in a total volume of 100 μl. Incubate for 4 h at 37°C *or* overnight at 4°C.

b) Shake solution from tray and wash 4–5× with PBS. Add 100 μl of rabbit antimouse IgG (1:100) and incubate 2 h at 37°C.

c) Shake solution from tray and wash 4–5× with PBS. Add 100 ml of alkaline phosphatase-conjugated probe in PBS (goat antirabbit IgG; 1:1,000). Incubate overnight at 4°C. If background is a problem do last wash with 0.05% Tween 20 in saline.

d) Prepare substrate by dissolving 60 mg p-nitrophenylphosphate in 100 ml of diethanolamine buffer, pH 9.8 (substrate will not last more than 3 days).

e) Wash trays 4–5× in PBS. Add 100 μl of substrate and incubate until color develops, read optical density (O.D.). Time varies in each preparation. The reaction can be stopped by adding 25–50 μl of 3N NaOH.

Note: Unbound intact cells ($\sim 10^6$/well) may also be used in these assays but plates should be centrifuged in centrifuge plate carriers prior to careful aspiration of supernate in steps 3, 4, 7, and 8.

10. Indirect Immunoprecipitation Assay

Materials

Solubilized receptor preparation [See Hjelmeland and Chrambach, this series]
Supernatants from hybridoma cultures
Normal mouse (or rat) serum
Antiserum to mouse (or rat) immunoglobulin (IgG) (Miles or Cappel Laboratories)

Method

a) An aliquot of solubilized receptor material containing sufficient receptor activity is incubated with 50 μl of spent culture medium plus 50 μl of 100-fold diluted normal mouse serum for 4–6 h at 4°C.

b) An excess of antiserum to mouse IgG (10–25 μg) is added to each tube and samples are incubated for 16–24 h at 4°C.

c) Immune complexes are pelleted at 5,000g for 15 min.

d) Supernates are removed and counted for residual radioactivity or ligand binding activity and pellets are counted for precipitated radioactivity or ligand-binding activity.

11. Assays to Detect Biological Effects of Monoclonal Antibody

It is impossible to describe one particular assay of this type, as details will vary based on each receptor-effector system in question. Nonetheless, these assays are designed for identification of monoclonal antibodies by their ability to affect the biological activity (such as ligand binding, enzyme activation, receptor internalization, cell growth) associated with a particular receptor. Spent medium from hybridomas is added to a biologically active preparation of receptor (such as intact cells or membranes) and after an appropriate incubation in the presence of culture medium, the sample is assayed for a loss in activity. Such findings may be taken as preliminary evidence for the identification of specific monoclonal antibodies. This method does not require an active solubilized receptor preparation in contrast to the indirect immunoprecipitation assay. However, these types of assays are more subject to interference from substances in the culture medium, particularly those which may be present in fetal bovine serum.

12. Cloning by Limiting Dilution

Materials

HAT medium
96-well microtiter plates
Mouse peritoneal cells

Method

a) Prepare suspensions of positive hybridomas in HAT medium containing 500, 50, and 10 cells/ml. Prepare 10 ml of each cell suspension.

b) Add 0.1 ml/well to each well of a 96-well microtiter plate. Prepare one plate for each cell dilution.

c) Add 0.1 ml/well of a suspension of peritoneal cells containing 5×10^5 cells/ml.

d) After 10–14 days inspect plates for macroscopic clusters of hybridomas. Assay supernates for monoclonal antibody from plates which show cell growth in less than 30% of the wells.

e) Expand positive hybridomas.

13. Harvesting Mouse Peritoneal Cells (for Use as a Feeder Layer in Cloning)

Materials

50-ml conical centrifuge tubes, sterile
10-ml syringe, sterile
21-g hypodermic needle, sterile
DME containing 2.5% fetal bovine serum (37°C)
BALB/c mouse, 4–8 wk old
70% ethanol

Method

a) Sacrifice mouse by cervical dislocation.

b) Wet fur thoroughly with 70% ethanol.

c) Lift skin up from abdominal wall with sterile forceps and make a longitudinal incision.

d) Strip back fur completely with thumb and forefinger to expose peritoneum.

e) Inject 5 ml DME into the peritoneal cavity along the midline of the abdomen. If the intestines are accidentally punctured, the mouse, syringe, and needle must be discarded to avoid infection.

f) The injected medium is circulated within the abdominal cavity by gently shaking the mouse.

g) Insert a needle into left side of the peritoneal cavity and apply lateral pressure so as to form a pocket of fluid behind the spleen.

h) Aspirate as much fluid as possible and transfer to a conical centrifuge tube.

i) Centrifuge cells at 1,000g for 10 min, aspirate supernate, and resuspend in HAT medium. Count.

Note: Average yield of peritoneal cells/mouse is on the order of $1–5 \times 10^6$.

14. Cloning in Soft Agar

(See section 6)

15. Expansion of Positive Hybridoma Cultures

Materials

HT medium
24-well cluster dishes (Costar)
Mouse peritoneal cells in HT medium (5×10^6/ml)
Sterile Pasteur pipets

Method

a) Add 1 ml of a suspension of mouse peritoneal cells to each well of a 24-well cluster dish.

b) Gently resuspend hybridomas from positive wells in a 96-well microtiter plate with a sterile pasteur pipet and transfer contents of each well to one well in a 24-well cluster dish.

c) Mix contents of each well in cluster dishes and add an equal volume of cell suspension back to the original well of a microtiter plate to prepare master plate.

d) Incubate at 37°C in a humidified atmosphere of 95% air - 5% CO_2.

e) When confluent, split cells onto larger plates. At this point, HT stock may be omitted from the medium.

16. Cryopreservation of Hybridomas

Materials

Hybridomas in log phase growth
Fetal bovine serum (4°C)
Dimethyl sulfoxide (which is itself sterile due to its toxicity)
Sterile screw cap freezing vials, 2 ml
Sterile conical centrifuge tubes, 50 ml
Liquid nitrogen storage container

Method

a) Centrifuge healthy, log phase cells at 1,000 g for 10 min.

b) Aspirate supernate from cell pellet and add 1 ml of 90% fetal bovine serum: 10% DMSO per 5×10^6 cells.

c) Transfer 1 ml of freezing mixture to sterile freezing vial and close vials air-tight.

d) Place vials at −70°C for one day.

e) Transfer vials to liquid nitrogen storage on the following day.

Materials

DME plus 20% fetal bovine serum (4°C)
Tissue culture dishes or spinner flasks
Sterile conical centrifuge tube, 50 ml

17. Thawing Hybridoma Cultures

Method

a) Remove vial from liquid nitrogen storage and thaw contents quickly in a 37°C water bath.

b) Transfer contents of vial to a conical centrifuge tube containing 50 ml cold DME medium.

c) Centrifuge at 1,000g for 10 min, aspirate supernate, and resuspend cells at a density of $\sim 10^5$/ml.

d) Plate out cells and place in CO_2 incubator at 37°C.

18. Large-Scale Culture of Hybridomas

Materials

Spinner flasks (at least 1-liter capacity)
DME containing 5% fetal bovine serum (37°C)
DME containing 2.5% fetal bovine serum (37°C)

Method

a) Before attempting large-scale culture of hybridomas as a source of monoclonal antibodies it is best to adapt cells to growth in medium containing 5% serum. This can be done by feeding a healthy log phase culture with DME-5 percent serum to gradually reduce the serum concentration.

b) For large-scale culture, add 10–20% of the desired end volume of a culture of vigorously growing log phase cells to a spinner flask and dilute in half with DME-2.5% serum.

c) Continue diluting the culture with DME-2.5% to keep cells in log phase growth.

d) When the spinner flask is full, allow cells to reach the stationary phase of growth, leave it for an additional 24–48 h and harvest media by centrifugation of cells. (Note: Do not use these cells for freezing stock cultures.)

19. Preparation of Ascites Tumors in Syngeneic Animals

Materials

Young mice (or rats)
Pristane mineral oil (2,6,10,14-tetramethylpentadecane, Aldrich)
DME minus serum (37°C)
Hypodermic needle, 20-gauge

Method

a) Prior to injection of cells, mice (rats) receive 0.5 ml (1.0 ml) of Pristane intraperitoneally. Animals are then rested 2–8 wk.

b) Hybridomas (at least 10^7 cells/animal) in 0.5 ml DME are injected intraperitoneally.

c) When ascites fluid accumulates as evidenced by abdominal swelling (usually 2–4 wk after inoculation) it can be removed by "tapping" the animal. This involves inserting a hypodermic needle (20-gauge) into the abdominal cavity of an anesthetized animal and collecting the fluid that drips out. Tapping can be performed every 2–3 days.

d) Alternatively, animals may be sacrificed, the abdominal cavity exposed, and the ascites fluid collected with a syringe and hypodermic needle.

e) Ascites fluid should be clarified by centrifugation at 40,000g for 20 min.

f) Hybridoma cells in the pellet may be reinjected into mice (rats) or may be frozen as from culture.

20. Purification of Monoclonal Antibodies From Culture Medium or Ascites Fluid

Materials

Saturated $(NH_4)_2SO_4$ solution

0.5 M Tris buffer, 1 mM EDTA, pH 7.4
Culture medium or ascites fluid

Method

a) Slowly add an equal volume of saturated $(NH_4)_2SO_4$ solution to medium or ascites fluid with stirring at 4°C. Let sit 1 h.

b) Centrifuge at 40,000g for 30 min. Discard supernate.

c) Dissolve the pellet in 15 ml 0.5 M Tris buffer and dialyze this solution against 0.5 M Tris buffer for 36 h at 4°C.

d) Centrifuge solution at 40,000g for 30 min and discard any pellet.

e) Aliquot the solution into small vials and store at −70°C. One can stop here and use this antibody preparation which usually contains approximately 50% pure monoclonal antibody or further purify the antibodies from solution on an immunoaffinity adsorbent of goat (rabbit) antimouse IgG coupled to Sepharose.

21. Biosynthetic Labeling of Monoclonal Antibodies With Radioactive Amino Acids

Materials

Hybridomas in log phase growth
DME minus leucine (GIBCO) (37°C)
^{14}C-leucine (New England Nuclear)

Method

a) Suspend 2 × 10^6 cells/ml in leucine-deficient DME containing 5% fetal bovine serum. Add 5 μCi ^{14}C-leucine per ml of culture medium and incubate for 16–24 h at 37°C in 95% air-5% CO_2.

b) Harvest the radioactive supernate by centrifugation of cells for 10 min at 1,000g. Incorporation of radiolabel is generally 50–75%.

c) Purification of monoclonal antibodies considerably reduces the radioactive background due to free labeled amino acids.

22. Iodination of Monoclonal Antibodies

[See Ward, this series; Kennett et al, 1980]

APPENDIX II

Solutions

1. Phosphate buffered saline (PBS)—dissolve the following in 1 liter of double-distilled H_2O:

NaCl	8.0 g
KCl	0.2 g
$Na_2HPO_4 \cdot 2H_2O$	1.15 g
KH_2PO_4	0.2 g

Filter sterilize (0.2-μm filter).

2. Glutamine stock (100×; Gibco)—dissolve 3 g L-glutamine (GIBCO, Flow) in 100 ml PBS. Filter sterilize and store frozen in 10-ml aliquots. Glutamine decomposes rapidly in solution when stored at 4°C and is routinely replenished in culture medium at 2-wk intervals.

3. HT stock (100×; Sigma)—136.1 mg of hypoxanthine and 38.7 mg of thymidine are added to 50 ml of double-distilled water, and 0.1 M NaOH is added dropwise until dissolved. Volume is adjusted to 100 ml and the solution is filter sterilized and stored frozen in 10-ml aliquots.

4. A stock (100×; Sigma)—17.6 mg aminopterin are added to 50 ml of double-distilled water, and 0.1 M NaOH is added dropwise until dissolved. Volume is adjusted to 100 ml, the solution is filter sterilized, and stored frozen in 10-ml aliquots. Aminopterin stock solutions should be protected from light.

5. 8-Azaguanine stock (100×; Sigma)—200 mg of 8-azaguanine are added to 50 ml of double-distilled water, and 0.1 M NaOH is added dropwise until dissolved. Volume is adjusted to 100 ml, the solution is filter sterilized and stored frozen in 10 ml aliquots.

6. 1 M Hepes, pH 7.4 (Calbiochem)—dissolve 47.7 g Hepes in 100 ml double-distilled water. Adjust pH to 7.4 with 10 N NaOH and bring to final volume of 200 ml with double-distilled H_2O. Filter sterilize.

7. IPO Stock (100×; Sigma)—dissolve 2,000 U bovine insulin, 500 mg sodium pyruvate, and 1.5 g oxalacetate in 100 ml double-distilled water. Filter sterilize and store frozen in 10-ml aliquots.

8. 0.17 M NH_4Cl—dissolve 4.55 g NH_4Cl

in 500 ml double-distilled water. Adjust pH to 7. Filter sterilize.

9. 40% PEG 1000 (Fisher)—weigh out 4 g of PEG 1000 and sterilize by autoclaving. Keep above 40°C to prevent solidification. Add 6 ml DME minus serum (42°C) and adjust pH to 7 by checking the color of the phenol red. Keep at 42°C until use. Prepare fresh before each fusion.

10. Diethanolamine buffer (For ELISA)—dissolve 48 ml diethanolamine and 24.5 mg $MgCl_2$ in 400 ml double-distilled water. Adjust pH to 9.8 with 1N HCl and bring to final volume of 500 ml.

Media Preparation

1. Myeloma Cell Medium

380 ml DME - high glucose (GIBCO)

100 ml Fetal bovine serum (GIBCO, Flow, Sterile Systems)

5 ml 8-azaguanine stock

5 ml Glutamine stock (every 2 wk)

10 ml 1 M Hepes, pH 7.4

2. Hybridoma Selective Medium (HAT—for fusion and cloning)

380 ml DME-high glucose

100 ml Fetal bovine serum

5 ml HT stock

5 ml A stock

5 ml IPO stock

5 ml Glutamine stock (every 2 wk)

3. Hybridoma Medium (HT)—for growing positive, cloned hybridomas

435 ml DME—high glucose

50 ml Fetal bovine serum

5 ml HT stock

5 ml IPO stock

5 ml Glutamine stock (every 2 wk)

Once hybridomas have been passaged several times in HT media, the HT stock can be omitted from the medium.

Molecular and Chemical Characterization of Membrane Receptors, pages 31–60

2

Immunoprecipitation and Electrophoretic Analysis of Membrane Proteins

James W. Goding

The Walter and Eliza Hall Institute of Medical Research, P.O. Royal Melbourne Hospital, Victoria 3050, Australia

INTRODUCTION

The elucidation of the signaling mechanisms of hormone receptors represents a major challenge. As a preliminary step towards that goal, basic information is needed about receptor polypeptide composition and membrane insertion, leading to amino acid sequences and three-dimensional structure. We are still a long way from realizing even these limited objectives. Compared to the study of water-soluble proteins, progress in obtaining detailed structural information about membrane proteins has been slow.

Most of the current understanding of membrane proteins has arisen from the study of simple model systems, such as the erythrocyte membrane [Tomita et al, 1978]. Human erythrocytes are readily obtained in large numbers. Their membranes are easily purified and have a relatively simple polypeptide composition. Similarly, viral envelope proteins have proven an ideal model system because their synthesis is rapid, and synthesis of host cell proteins is often suppressed [Rothman et al, 1977].

The study of membrane proteins in complex nucleated cells with slow turnover has proven much more difficult and has placed much greater technical demands on analytical systems. Perhaps the outstanding example of an area where progress has been achieved is that of the histocompatibility antigens [Strominger et al, 1981], whose fundamental importance in immune responses and graft rejection has provided the necessary impetus and funding for a major research commitment. Substantial progress has also been made in understanding membrane immunoglobulins (antigen receptors) of lymphocytes [Rogers et al, 1980]. In this case, structural analysis leading to molecular cloning and DNA sequencing was greatly facilitated by the fact that the receptor forms share most of their sequences with secreted antibodies.

In marked contrast, detailed information about the structure of hormone receptors is still scarce. It is probably safe to generalize that most hormone receptors are proteins or glycoproteins, but in most cases that is all that can be said at the present time. The best-characterized peptide hormone receptor is that for insulin [Czech et al, 1981], perhaps because of its medical importance. Even in this case, structural characterization is at an early stage.

There are two major reasons for this slow progress: The first is the fact that membrane proteins are, by their very nature, insoluble in aqueous solution. This problem is now reasonably well solved as a result of a better understanding of the forces which hold membranes together [Tanford, 1980] and the parallel development of membrane solubilization by detergents [Helenius and Simons, 1975; Helenius et al, 1979]. A second reason for slow progress is the very low abundance of membrane proteins. Although accurate figures are not available, membrane proteins are thought to comprise only 1–2% of total cellular protein. An individual hormone receptor might represent 0.1–1.0% of this figure (typically 10^4–10^5 molecules per cell).

Structural studies of hormone receptors thus require methods capable of very high resolution and sensitivity. These requirements are met by a combination of strategies. The problem of sensitivity may be overcome by isotopic labeling. The problem of purification of an individual protein is best dealt with by methods that exploit biological affinities rather than classical chemistry. The problem of resolution is met by a variety of electrophoretic procedures, each designed for a precise purpose.

USE OF ANTIBODIES AS PROBES FOR CELL SURFACE PROTEINS

The problem of low abundance of membrane proteins and the need for detergents to maintain their solubility have already been mentioned. In addition, membrane glycoproteins frequently exhibit substantial charge heterogeneity due to variations in sialic acid content. These factors make membrane glycoprotein purification by standard physical procedures such as gel filtration, ion exchange chromatography, and isoelectric focusing a very difficult task.

The introduction of methods which depend on biological affinities [Cuatrecasas et al, 1968; Schwartz and Nathenson, 1971] have provided a general solution to this problem. In the case of hormone receptors, purification by affinity chromatography on a solid matrix to which the relevant ligand has been coupled has sometimes resulted in spectacular purifications [eg, Jacobs et al, 1977]. The use of affinity chromatography is discussed in detail elsewhere in this series [Strosberg]. For a variety of reasons, however, affinity chromatography on the physiological ligand may be disappointing. In addition (and this is especially true for immunologists), the physiological ligand may not be known.

If specific antibodies to the receptor can be generated, they may provide the selectivity needed for its isolation and characterization. This approach has been used widely in recent years [reviewed by Kessler, 1981]. The main limitation of the approach is that the production of specific antisera often depends on having already purified the receptor! There are, however, a few exceptions. Naturally occurring antireceptor antibodies occur in some disease states [see Harrison, this series]. For example, antibodies to the insulin receptor occur in acanthosis nigricans [Harrison et al, 1979], and antibodies to the acetylcholine receptor occur in myasthenia gravis [Lindstrom, 1979].

A second way in which specific antisera may be generated without purifying the receptor involves the exploitation of genetic polymorphisms in receptor structure [Goding, 1978]. If two different inbred strains of mice differ in their structural genes for a given receptor, cross-immunization with receptor-bearing cells from the other strain may result in highly specific antibodies [eg, Goding et al, 1976]. Finally, highly specific monoclonal antibodies against any protein may be made without purification by the use of hybridoma technology [see Fraser and Lindstrom, this series].

In addition to the use of antireceptor antibodies, receptors may sometimes be isolated by the use of antibodies to their ligands. This point will be discussed in more detail in a subsequent section of this chapter.

Production of Conventional Antisera

The general principles of antiserum production have been reviewed recently [Hurn and Chantler, 1980]. In general, the antigen should be administered to a different species.

Rabbits are popular because they are convenient to maintain, but the response to a given antigen varies enormously between individuals, and for each immunization a minimum of three rabbits should be used. Goats and sheep produce excellent antibodies, and their variability is much less.

In the 1960s and early 1970s it was common to immunize with 5–20 mg of antigen. It is now clear that very good antisera may result from injection of as little as 50 μg of protein emulsified in Complete Freund's Adjuvant (see below), and possibly even considerably less [Vaitukaitis, 1981]. Low doses of antigen have the advantages of economy, production of higher-affinity antibody, and less risk of an antibody response to impurities.

If it desired to make antisera against cell surface proteins, some thought should be given to the likely quantities of cells needed.

$$M = \frac{n \times m \times p}{6.023 \times 10^{23}}$$

where M = mass of receptor protein, n = number of receptors per cell, m = number of cells, p = molecular weight of receptor, and 6.023×10^{23} = Avogadro's number.

Thus, for 100 μg of transferrin receptors, assuming 10^5 per cell and a molecular weight of 200,000, 3×10^9 cells will be required.

The antigen should be emulsified in Complete Freund's Adjuvant [see Hurn and Chantler, 1980], using a 1:1 mix of the aqueous and oily phases. Emulsification proceeds easily if two *glass* syringes are connected together via a disposable plastic three-way stopcock, and the contents squirted back and forth for about 5 min. (The plungers of plastic syringes swell in the oil and become virtually immovable.) The emulsion (2–3 ml) may be given subcutaneously or intramuscularly. Intradermal injection has sometimes been recommended, but it generates large ulcers and is unnecessarily cruel.

Proteins treated with sodium dodecyl sulfate (SDS) may still be highly antigenic, and the resulting antibodies react well with the native protein [Stumph et al, 1974]. Acrylamide is said to be a good adjuvant [Weintraub and Raymond, 1963]. Recently, many authors have used SDS gel electrophoresis as a final purification step, and injected the macerated gel slices (emulsified in Complete Freund's Adjuvant) directly into the animal [Spielman et al, 1974; Tjian et al, 1975; Springer et al, 1977b; Silver and Elgin, 1977; Lane and Robbins, 1978; Carroll et al, 1978; Granger and Lazarides, 1979].

Monoclonal Antibodies

One of the fundamental requirements for the production of specific conventional antisera is that the immunizing antigen be pure. Even with "pure" antigen, the resulting antiserum will contain a great mixture of antibodies of various affinities. Some of these antibodies may cross-react with antigenic determinants which are shared with other proteins (eg, carbohydrate on glycoproteins). There is thus a fundamental limit to the specificity that is attainable by conventional serology.

In 1976, Köhler and Milstein demonstrated that individual clones of normal antibody-forming cells could be immortalized by fusion with myeloma cells, and the resulting "hybridoma" cells continued to secrete specific antibody [Köhler and Milstein, 1976]. This discovery revolutionized serology, because it eliminated the requirement for purity of antigen. Unlimited quantities of pure, homogenous, absolutely specific antibodies could be obtained, even if the desired antigen was only a minor component of the immunizing antigen. All that was required was a way of selecting the desired clone.

Monoclonal antibodies have also had a revolutionary impact on membrane research, because they provided a way of generating a specific "handle" on any desired membrane protein, regardless of its abundance [Williams et al, 1977].

It is beyond the scope of this chapter to describe the production of monoclonal antibodies (see Fraser and Lindstrom [this series] for a detailed account). The production of monoclonal antibodies has been the subject of many reviews [Goding, 1980b; Galfrè et al, 1981; Köhler, 1981; Hämmerling et al, 1981]. There are, however, a few points concerning the use of monoclonal antibodies that are directly relevant to this chapter. An appreciation of the differences between conventional and monoclonal serology is vital if the full power of the latter is to be realized.

The very fact of monoclonality means that each antibody will behave differently. Conventional antisera contain antibodies of a wide range of affinities, and Scatchard plots are usually curvilinear. In many assays, the final result tends to be dominated by the higher-affinity subpopulation. Accordingly, inactivation of a subset of antibodies (eg, by freeze/thaw cycles, lyophilization, chemical derivitization, or changes in pH or salt concentration) tends to be masked by those antibodies remaining.

In contrast, monoclonal antibodies will (in theory at least) show linear Scatchard plots, but the affinity may range anywhere between 10^{-6} and 10^{-11}, depending on the individual antibody. Similarly, each monoclonal antibody may be expected to have its own characteristics of stability to denaturation and changes in affinity with temperature, pH, or salt concentration [Mason and Williams, 1980; Herrman and Mescher, 1979]. Thus, binding of a monoclonal antibody may be effectively abolished by changes which would not be noticed when conventional sera are used. In addition, certain monoclonal antibodies may precipitate spontaneously at low temperature (cryoglobulins) or low salt concentration (euglobulins).

A second important difference between monoclonal and conventional antibodies concerns their specificity. An "antigenic determinant" may be defined as a region of an antigen which is recognized by a particular monoclonal antibody. Conventional antisera usually recognize many (perhaps 10–100) different determinants on a typical protein antigen. If the antigen is altered in some way (eg, genetic polymorphism, radioiodination, lack of glycosylation or partial denaturation), it will often still be detectable by conventional antisera because a subpopulation of antibodies will "see" the surviving determinants. In contrast, a given monoclonal antibody may or may not bind in such cases.

RADIOACTIVE LABELING OF RECEPTORS

Immunoprecipitation analysis of receptors virtually always involves labeling them with radioactive isotopes as a first step. Radioactive labeling provides a degree of sensitivity that cannot be matched by any other technique, although it can be approached by silver staining of gels [Marshall and Latner, 1981]. In addition to sensitivity, radioactive labeling has the very important advantage that it allows the introduction of extraneous non-radioactive proteins, such as those in antisera, without their being detected in the readout.

There are a large variety of labeling procedures [see also Ruoho et al; Harrison, this series; Hubbard and Cohn, 1976; Kaplan et al, 1981]. However, the two which are by far the most widely used are lactoper-oxidase-catalyzed cell surface radioiodination and biosynthetic labeling.

Cell Surface Radioiodination

In the presence of small amounts of hydrogen peroxide and iodide ions, lactoperoxidase catalyzes the iodination of tyrosine [Morrison, 1980]. When intact cells are used as substrate, iodination is confined to the plasma membrane [Phillips and Morrison, 1971; Marchalonis et al, 1971; Hubbard and Cohn, 1976; Morrison, 1980]. There are many variations of the basic method.

Whichever method is chosen, it is absolutely crucial to have cells of high viability (more than 97%) to minimize cytoplasmic labeling. Although claims have been made that actin is present on the surface of lymphocytes [Owen et al, 1978], in my experience this finding is diagnostic of cytoplasmic labeling.

The older methods of viability assessment, such as trypan blue or eosin exclusion, are not very accurate, and modern methods are preferable. I use the acridine orange/ethidium bromide method [Lee et al, 1975]. One drop of cells is placed on a glass slide. Then, one drop of a mixture of one part per million acridine orange and ethidium bromide in phosphate-buffered saline is added. A coverslip is placed over the drop, and the slide is immediately examined under the fluorescence microscope using filter combinations for fluorescein. Living cells stain bright green, and dead cells bright orange. Erythrocytes do not stain at all. Ethidium bromide and acridine orange are light sensitive and should be stored in a foil-wrapped bottle. They are both potentially carcinogenic.

It must be stressed that the finding of labeled proteins after cell surface iodination does not of itself constitute proof that they are membrane proteins. This seemingly obvious point has been repeatedly ignored [see Sidman, 1981]. A particularly common problem is the iodination of serum components adherent to cells or to the surface of the tube.

The Lactoperoxidase Method

The version given below has given consistently high incorporation with a variety of cell types.

Wash the cells thoroughly (three times) in phosphate-buffered saline (PBS: 7.7 g NaCl, 2.27 g Na_2HPO_2, 0.54 g KH_2PO_4 to 1 liter H_2O) to remove extraneous proteins. Resuspend from 10^7 to 5×10^7 cells in 200 μl PBS at room temperature (The number of cells is not critical; I use 10^7 for large cells and 5×10^7 for small cells; the volume should not be changed, however.) Just before beginning the iodination, dilute 10 μl of 30% H_2O_2 to 10 ml PBS (ie 1:1,000). The optical density of the 1:1,000 stock should be about 0.7 at 230 nm [Phillips and Morrison, 1971]. Then, prepare serial threefold dilutions in PBS (add 100 μl of 1:1,000 H_2O_2 to 200 μl PBS, mix, and transfer to next tube), giving final dilutions of 1:3,000; 1:9,000, and 1:27,000. Add 50 μl of 0.2 mg/ml lactoperoxidase to the cells, then 0.5 mCi carrier-free sodium iodide (0.05 mM; 100 μCi/μl). Then, at 1-min intervals, add successively 10 μl of 1:27,000, 1:9,000; 1:3,000, 1:1,000 H_2O_2. One minute after the last addition of H_2O_2, wash the cells twice in PBS containing 1 mM KI to terminate the reaction. (The KI should be freshly prepared to prevent oxidation to I_2, which would kill the cells.)

Comments. The addition of "carrier iodide" during the iodination may increase the incorporation of label, but its use is not recommended, because it increases cytoplasmic labeling. The presence of azide ions or Merthiolate will prevent the reaction from working. The enzyme concentration is not critical. The most important variable is the peroxide concentration. At low peroxide concentrations, the reaction is inefficient. High peroxide concentrations are toxic for the enzyme and result in lowered incorporation. However, incorporation of iodine into tyrosine is irreversible. Gradually increasing the concentration of H_2O_2 ensures that the optimal concentration will be reached at a point towards the end, and ensures that the reaction will work well under a variety of conditions. The PBS recipe has somewhat more phosphate than usual, to buffer the NaOH in the NaI.

A good iodination will result in 8–30% of input counts becoming associated with the cells, and 1–2% of input counts becoming bound to protein as assessed by precipitation with trichloracetic acid (TCA). While these results may suggest a low efficiency, they represent the best that can be achieved by the method and usually allow adequate sensitivity for subsequent steps.

The lactoperoxidase/glucose oxidase method. The lactoperoxidase method as described above may sometimes cause difficulties because of the high peroxide concentration [Hubbard and Cohn, 1976]. These authors have proposed an alternative procedure in which micromolar concentrations of hydrogen peroxide are continuously generated by the action of glucose oxidase.

Mouse L cells (5×10^6 per ml) are washed in PBS and resuspended in PBS plus 20 mM glucose. Then 4–8 milliunits (0.01 mg) of lactoperoxidase and 0.1 μg of glucose oxidase (Sigma type V, from *Aspergillus niger*) and 100 μCi ^{125}I are added. Cells are kept on ice for 10 min, and the reaction is terminated by washing with PBS.

The Iodogen™ method. Recently an alternative to lactoperoxidase for cell surface iodination has been published [Fraker and Speck, 1978]. This method uses a water-insoluble compound, 1, 3, 4, 6-tetrachloro-3α,-6α-diphenylglycoluril (IODOGEN™, Pierce Chemical Co) coated onto the walls of glass tubes or coverslips [Markwell and Fox, 1978; Watt et al, 1979]. The advantages of the method are that no extraneous protein is added during iodination, and that the reaction does not involve H_2O_2, which could cause cytoplasmic labeling in cells containing peroxidases [Watt et al, 1979].

The concentration of the glycoluril is important. Excessive amounts (eg, more than 5–10 μg per tube) are likely to kill the cells. Coated tubes should be stored protected from light.

Selectivity of surface iodination. Lactoperoxidase-catalyzed surface iodination predominantly labels exposed tyrosine residues. In a few rare cases, proteins which were thought to be present at the exterior of the membrane have failed to label. In these cases, an alternative labeling procedure is necessary. However, this problem has been somewhat exaggerated in the literature, and the method remains the least selective of all current procedures [Hubbard and Cohn, 1976; Kaplan et al, 1981].

Assessment of isotope incorporation by TCA precipitation. When the fraction of counts in trichloracetic acid precipitable material is very low (as is usually the case after labeling of receptors), conventional TCA precipitations in tubes will not give meaningful results. It is therefore preferable to collect the precipitate on a membrane filter, as follows:

1. Take 1-μl samples of cell lysate (three replicates); pipette into 0.5 ml PBS containing 0.5% fetal calf serum, 10 mM NaN_3, 1 mM KI, and 1 mM methionine.
2. Add 0.5 ml 10% TCA. Mix.
3. Let stand at least 30 min at room temperature or overnight at 4°C. Solution will go very slightly cloudy.
4. Filter though Millipore type HA 0.45-μm filter (25-mm diameter, catalogue No. HAWP 02500) attached to a filtration/precipitation apparatus (ICN Model E8B, catalogue No. 803012, 15/16 paper size), into a side-arm flask attached to a suction pump.
5. Wash filter with 10 ml 5% TCA.
6. For ^{125}I, place in tube for counting.
7. For ^{35}S, ^{14}C, or ^{3}H, dry filter under heat lamp and place in Triton-based scintillation fluid for counting.

Biosynthetic Labeling of Receptors

As a general rule, it is possible to label receptors to the highest specific activity by radioiodination. This method has the added advantage that labeling is confined to the membrane. Biosynthetic labeling with radioactive amino acids may be needed if a given receptor fails to label with ^{125}I, or for a variety of other reasons. However, biosynthetic labeling places a much greater burden on the subsequent purification procedures, because the cytoplasmic proteins are much more numerous and in general more rapidly synthesized.

In general, the aim is to achieve the highest possible incorporation of counts at the least expense. There are several factors that will influence the choice of amino acid [see Coligan and Kindt, 1981]. One of the most important factors is the specific activity of the amino acid used. The "hottest" amino acid currently available is ^{35}S-methionine (often available at 1,200 Ci/mmol). The half-life of ^{35}S is 87 days, and energy of emission 167 KeV. The extremely high specific activity of ^{35}S-methionine often more than compensates for the relatively low abundance of methionine in most proteins, and ^{35}S-methionine will usually be the amino acid of choice [see Vitetta et al, 1976].

The next-highest specific activity amino acids are those labeled with tritium. These have the advantages of very long half-life (12 y) and relatively low cost. The very low energy of tritium (18 KeV) was once a serious problem, but the advent of simple fluorography techniques (Bonner and Laskey [1974]; or EN^{3}HANCE^{TM} made by New England Nuclear) have made autoradiography with ^{3}H a trivial matter.

Labeling with ^{14}C-amino acids (t½ = 5,760 y; energy = 155 KeV) is seldom justified. They are of lower specific activity than ^{3}H amino acids, and are much more expensive. The gamma-emitting analogue of methionine, ^{75}Se-methionine, has some advantages over ^{35}S-methionine [Gutman et al, 1978].

Metabolic considerations. Labeling with essential amino acids (Table I) is generally better than for nonessential. Useful indications of which amino acids are likely to result in the highest specific activity have been published by Vitetta et al [1976] and Coligan et al [1981]. Cysteine receives its sulfur from methionine, and if ^{35}S-cysteine is used, it is best to use ^{35}S-methionine as well. Additions of an inhibitor of transamination greatly increases incorporation of aspartic acid [Coligan et al, 1981]. The uptake and incorporation of amino acids does not exactly parallel their relative abundance in cellular proteins. Good results are obtained with methionine, arginine, leucine, lysine, and tyrosine. Even though cysteine can be obtained at as high specific activity as methionine and is usually more abundant, it generally gives poorer labeling.

In some circumstances, use of cysteine and methionine can result in nonbiosynthetic incorporation into protein [Gutman et al, 1978; Suissa, 1981]. The problem seems to be due to an unidentified contaminant in the isotope preparations, and is not observed when the proteins are reduced.

TABLE I. Essential and Nonessential Amino Acids Recommended for High Incorporation by Biosynthetic Labeling

Essential	Nonessential
Arginine	Alanine
Leucine	Cysteine
Lysine	Proline
Methionine	
Tyrosine	

Practical aspects. Metabolic labeling requires a medium omitting the radioactive amino acid. A very convenient source is the Selectamine Kit (Gibco), which allows any combination of amino acids to be used.

Methionine and cysteine are supplied in aqueous solution containing 0.1% (1.2 × 10^{-2}M) 2-mercaptoethanol. The maximum concentration of 2-mercaptoethanol that most cells can tolerate is around 8 × 10^{-4}M (this corresponds to 500 μCi/ml). Similarly, tritiated or ^{14}C amino acids often come in water plus 2% ethanol. The ethanol also limits the amount of isotope that can be added to cultures. In general, a concentration of 0.1% ethanol should not be exceeded. If higher concentrations of isotope are required, the 2-mercaptoethanol or ethanol can be removed by lyophilization. Since these agents are added to stabilize the amino acid, lyophilization should be done just prior to labeling, and only in the quantities needed for the experiment.

A typical labeling protocol involves culture of cells at 10^7/ml in RPMI-1640 or Eagle's medium lacking the desired amino acid, and including 50–500 μCi/ml of the isotope and 5–10% dialyzed fetal calf serum. Since membrane proteins turn over relatively slowly, labeling should be for 4–12 h or longer, depending on how healthy the cells are in culture. At the end of the culture, it is advisible to include a large excess of the labeling amino acid in nonradioactive form. Failure to do so may result in increased "background" during the subsequent analysis.

It is advisable to check the level of incorporation of label by TCA precipitation at the end of each labeling (see above).

Storage of radioactive amino acids. ^{35}S-methionine and cysteine may be stored at −70°C. When the vial is first opened, the isotope should be aliquoted into small tubes (say 0.5 mCi/tube). Do not repeatedly freeze and thaw. ^{3}H and ^{14}C amino acids are usually supplied in H_2O plus 2% ethanol to scavenge free radicals. Store at 4°C, protected from light. In general, aqueous solutions of ^{3}H compounds should not be stored frozen because molecular clustering by freezing may accelerate the rate of self radiolysis.

The rate of decomposition of radiochemicals is usually faster than their radioactive half-life (eg, ^{14}C, 1–3% per year; ^{3}H, 1–3% per month) even under optimal storage.

Carbohydrate labeling of membrane glycoproteins. Since many, if not most, externally disposed membrane proteins are glycoproteins, labeling may also be performed with radioactive sugars, such as glucosamine, mannose, or fucose. Sialic acid is not taken up by cells. Labeling with radioactive sugars must be performed in glucose-free medium. It has the advantage that labeling is mainly confined to glycoproteins, which are in general membrane or secretory proteins. Hence, the signal:noise ratio will be higher. The specific activities that can be achieved are generally low.

The carbohydrate moieties of surface glycoproteins may also be labeled with ^{3}H by oxidation with periodate or galactose oxidase followed by reduction with tritiated borohydride [Gahmberg and Andersson, 1977]. However, the specific activities obtained are usually low.

SOLUBILIZATION OF RECEPTORS

Solubilization by Proteolysis

Cell surface proteins can sometimes be detached from the membrane in water-soluble form by controlled proteolytic cleavage [Turner et al, 1975]. The resulting fragment may be quite large and may retain ligand binding [Omary and Trowbridge, 1981] or other biological activity such as certain antigenic determinants [Turner et al, 1975; Kessler et al, 1979]. However, solubilization by proteolysis is often difficult to control, and the yield of homogenous high molecular weight material may be poor. Although the method may be useful in certain specific instances, it cannot be recommended as a generally useful procedure.

Solubilization by Detergents

Membrane solubilization by detergents is discussed in detail elsewhere in this series [see Hjelmeland and Chrambach; Strauss, this series; Helenius and Simons, 1975; Helenius et al, 1979].

The essential points are as follows: Detergents are amphiphilic molecules which exist in aqueous solution as monomers and micelles. The micelles consist of aggregates of 2–100 monomers, with their hydrophobic portions buried in the center and their hydrophilic portions at the surface. At low detergent concentrations most molecules exist as monomers. With increasing total detergent concentration, the concentration of monomers rises until at a certain concentration called the *critical micelle concentration* (CMC), the monomer concentration ceases to rise, and further increase in detergent concentration results from an increase in micelle concentration. The micelle size is largely independent of detergent concentration, but is influenced to a variable extent by the type of detergent, the salt concentration, and the pH.

Solubilization of integral membrane proteins occurs by replacement of the planar lipid bilayer with a micelle of detergent [Helenius and Simons, 1975]. A great deal of knowledge concerning use of detergents has been obtained empirically, and there is a need for individualization of choice of detergent for solubilization of a particular membrane protein.

It is the monomer concentration that determines the solubilizing power. Thus, the concentration of detergent should at all times exceed the CMC so that the monomer concentration is as high as possible [Helenius et al, 1979]. Membrane lipid and detergent may be regarded as competing with each other for the hydrophobic regions of the membrane proteins, so the total mass of detergent used should be at least ten times the mass of cell lipid. These two requirements are easily met in analytical experiments.

IMMUNOPRECIPITATION OF CELL SURFACE PROTEINS

Success in immunoprecipitation depends very largely on having specific antisera or monoclonal antibodies, together with highly radioactive, soluble membrane proteins. If these requirements are met, it is unlikely that the remaining steps will present difficulties. Immunoprecipitation analysis of membrane proteins can be considered as occurring in four distinct phases: binding of antibody, precipitation, washing, and elution.

Binding of Antibody

As a general rule, antibodies will bind almost equally well at 37°C, room temperature, and 4°C. It is advisable to perform all steps at 4°C, however, to minimize the chance of proteolytic degradation. For the same reason, work should be performed as rapidly as possible.

It is preferable to perform the precipitation step on the same day as the labeling step, because freezing and thawing detergent extracts may cause protein denaturation or disruption of protein-protein interactions. If it is impossible to do this, the detergent extract may be frozen, preferably by snap-freezing. The extracts should always be centrifuged after thawing to remove any insoluble denatured protein.

How much antibody should be used? The answer to this question will depend on the titer of the antiserum, its affinity, and the amount of receptor to be precipitated. As a rough guide, 1–10 μl of a high-affinity rabbit antiserum to mouse immunoglobulin is more than sufficient to precipitate all the membrane immunoglobulin from 10^7 B lymphocytes (approximately 100,000 receptors per cell). It will seldom be necessary to use more than 50 μl of any reasonable antiserum to precipitate most types of receptors from 10^7 cells.

The length of time of incubation at 4°C with the antiserum is usually not important. For high-affinity sera, 30–60 min are adequate for all receptors to be bound. Occasionally, weak or low-affinity antibodies may require longer, and overnight incubations may be used, although I have never found this to be necessary.

The incubations should be carried out at physiological salt concentration and pH. As mentioned previously, some monoclonal antibodies may precipitate spontaneously in the cold or at low ionic strength.

Isolation of Receptors by Antibodies to Their Ligands

Often, antibodies to a receptor may not be available. In these cases, receptors may sometimes be isolated by use of antibodies to their ligand. For example, the receptor for diphtheria toxin was isolated by binding to the toxin followed by antitoxin antibodies [Proia et al, 1979]. Similarly, the receptor for transferrin has been isolated by antitransferrin antibodies [Wada et al, 1979].

Immunoprecipitation

The quantities of receptor material that are present in typical immunoprecipitation experiments (picograms-nanograms) are usually insufficient to allow good lattice formation and precipitation. It is therefore virtually universal practice to use an additional step to facilitate precipitation.

If the first step utilizes rabbit IgG, the second step may be sheep or goat antirabbit IgG antibodies. Typically, 5–15 times the volume of the first antiserum should be used, depending on the titer of the second antibody. Incubation with the second antibody should proceed for at least an hour, and preferably overnight at 4°C to allow a large lattice to form.

More recently, it has become common practice to use a solid-phase second step, such as heat-killed and fixed staphylococci [Kessler, 1975, 1976; Goding, 1978; Kessler, 1981], or protein A-Sepharose. The use of solid-phase precipitation has many advantages. The overall amount of IgG that must be loaded onto the gels is much less, resulting in less protein overloading and less nonspecific precipitation. Much shorter incubation times are possible, as the binding of immune complexes to staphylococci or protein A-Sepharose is virtually instantaneous. After adding the staphylococci or beads and mixing for a few seconds, washing may be commenced.

Ten microliters of a 10% (v/v) suspension of protein A-Sepharose (Pharmacia) or the same volume of 10% (v/v) staphylococci (Pansorbin, Calbiochem; IgGsorb, New England Enzyme Center, or numerous other suppliers) are sufficient to bind all the IgG in 1 μl of serum.

In some cases, it may be necessary to "preclear" the cell lysate by adding staphylococci and then removing them prior to adding antibodies, to deplete the lysate of molecules which bind to the staphylococci alone.

Not all forms of IgG bind to staphylococcal protein A [Goding, 1978; Kessler, 1981]. Table II summarizes the protein A specificities. In situations where the major IgG subclass does not bind (eg, sheep, goat), surprisingly good recoveries of immune complexes may be found. It is thought that the formation of multivalent complexes may stabilize weak binding [Kessler, 1981]. Alternatively, a "second antibody" (eg, rabbit antisheep IgG) may be added (see above) prior to addition of staphylococci. An incubation time of 1 h at 4°C is adequate for the second antibody.

Alternative solid-phase matrices. There are a number of other forms of beads which might be used for immunoprecipitation of cell surface proteins (eg, Affigel 10, Bio-Rad; Immunobeads, Bio-Rad; Covaspheres, Covalent Technology). These beads have not been widely used for this purpose, and may exhibit differing levels of nonspecific binding. As a general rule, beads with significant numbers of charged groups or hydrophobic regions such as "spacer

TABLE II. Binding of Immunoglobulin Classes and Subclasses to Protein A in Various Species

Species	Classes which bind	Reference	Comments
Man	G1,2,4	Hjelm [1975]	Some IgG3 may bind weakly [Skvaril, 1976]. A minority of human IgM myelomas bind [Lind et al, 1975; Saltveldt and Harboe, 1976].
	A2	Saltveldt and Harboe [1976]	
	some IgM	Saltveldt and Harboe [1976]	
Rabbit	IgG	Goding [1976]	Yield from serum is quantitative. Uncomplexed IgG elutes at pH 4.
Mouse	G2a, G2b, G3	Ey et al [1978]	Highly pH dependent; subclasses can be separated by pH gradient. Binding of G1 maximal at pH 8.5, but not quantitative.
	some G1	Ey et al [1978]	
Sheep	G2	Kessler [1981]	G2 is a minor subclass.
Goat	G2	Kessler [1981] Delacroix and Vaerman [1979]	G2 is a minor subclass. Some binding of G1 at alkaline pH.
Rat	G2a(pH > 8.6), G2c(pH > 7)	Ledbetter and Herzenberg [1979]	Some variability in literature [see Medgyesi et al, 1978]. Classes which bind are minority of rat IgG.
Guinea pig	G1, G2	Forsgren [1968]	
Bovine	G2	Goudswaard et al [1978]	IgG1 binds weakly.

arms" will result in high levels of nonspecific binding.

Washing Immune Complexes on Solid-Phase Protein A

After immune complexes have been bound to staphylococci, unbound material must be washed away. The two main points that require consideration are the amount of washing and the composition of the wash buffer. A typical protocol is as follows: Centrifugation of the staphylococci (3,000g for 10 min at 4°C) is performed and the radioactive supernate discarded. A convenient volume of wash buffer (5 ml) is added, and the bacteria are resuspended with a Pasteur pipette. The bacteria are again centrifuged, and the wash process is repeated.

It is realistic to expect that less than 50 μl liquid will remain after each wash. Each wash will therefore provide a theoretical dilution of 100-fold. Two washes will provide a dilution of 10^4, and three of 10^6. Three washes are usually adequate.

Composition of the wash buffer. Early studies used isotonic buffers at neutral pH, containing 0.5% Triton X-100 to maintain solubility of membrane proteins. In recent years, it has become apparent that modified wash buffers may lower the nonspecific binding to the staphylococci.

Nonspecific binding is highly dependent on pH and salt concentration, suggesting a predominantly electrostatic mechanism (Goding, unpublished data). This "background" binding is worst at acidic pH and low salt, suggesting that the bacteria may act as a cation exchanger. This would be consistent with the loss of positive charges during formaldehyde fixation.

Fortunately, alkaline pH tends to stabilize the binding of immunoglobulins to protein A [Ey et al, 1978], and the strength of interaction seems to be independent of salt concentration over a wide range (Goding, unpublished data). Significantly lower backgrounds may be obtained with a wash buffer composed of 0.05 M Tris-HCl, pH 8.3, with 0.6M NaCl and 0.5% Triton X-100 or Nonidet P-40 [Goding and Herzenberg, 1980]. Other variations of the wash buffer composition are listed by Kessler [1981].

It must be emphasized, however, that the combination of high salt and alkaline pH could result in loss of antigen in some cases. This problem is unlikely to be encountered with conventional antisera, but certain monoclonal antibodies may fail to bind antigen under these conditions [Herrman and Mescher, 1979].

Elution of Receptors From Antibodies

Elution of receptors from immune complexes requires denaturing conditions. Many different agents are suitable [MacSween and Eastwood, 1981]. The major constraint will usually be the subsequent analytical step.

If the analysis is to consist of sodium dodecyl sulfate (SDS) polyacrylamide gel electrophoresis, the dry staphylococcal pellet should be resuspended in SDS-sample buffer (typically 100–200 μl), placed in a boiling-water bath for 5 min, and the staphylococci removed by centrifugation. It is usual to load 30–50 μl onto the gel for analysis. The remainder may be kept frozen for subsequent analysis.

If the next step is isoelectric focusing (IEF) or nonequilibrium pH gradient electrophoresis (NEPHGE), elution is performed by addition of 200 μl of a buffer containing 9.5M urea, 2% pH 3–10 Ampholines, 0.5% Triton X-100, and 50 mM dithiothreitol (DTT). Heating is both unnecessary and undesirable, as it may result in carbamylation of proteins and artifactual charge heterogeneity [Bobb and Hofstee, 1971]. The bacteria are removed by centrifugation as described previously.

In some cases, it may be desired to retain as much of the native state of the receptor as possible. If the receptor has a single proteolytic cleavage site, it is sometimes possible to liberate a fragment by gentle proteolysis of the precipitate [Pischel and Little, 1979; Goding, 1980]. This method is particularly attractive when monoclonal antibodies are used, because different antibody attachment sites may allow release of individual portions of the receptor [Goding, 1980a].

SDS-POLYACRYLAMIDE GEL ELECTROPHORESIS

For detailed practical information the reader is referred to Roberts et al [this volume].

Laemmli System

Electrophoresis of proteins in polyacrylamide gels in the presence of SDS results in separation which is almost totally based on molecular weight [Weber and Osborne, 1969]. Numerous buffer systems are available, but the most widely used is that devised by Laemmli [1970]. The Laemmli system consists of a main or resolving gel and a shorter "stacking" gel, the function of which is to concentrate the sample to a narrow band before it enters the main gel. As a result the bands are extremely sharp, and their width is independent of the volume of the sample. The length of the stacking gel must be greater than the length of the sample, or else stacking will be incomplete.

The sharpness of bands on SDS gels of high acrylamide concentration is often inadequate, particularly for low molecular weight proteins. Resolution can be improved significantly by using gradient gels, in which the concentration of acrylamide increases as the protein migrates down the gel. This has the effect of slowing down the leading edge of a band and allowing the trailing edge to "catch up". Excellent results have been reported using a 10–15% linear gradient [Dobberstein et al, 1979].

SDS-Urea Gels

In general, SDS gels do not give good resolution of proteins in the molecular weight (MW) range 15,000–30,000. The bands tend to be broad, and the graph of log MW versus mobility nonlinear. Recently, a procedure was published claiming to give better results [Anon, 1981]. The system is basically a continuous phosphate-buffered SDS gel with incorporation of urea. An "upper gel" may perform some sort of "stacking" function. Extra "stacking" is obtained by use of a sample buffer of lower ionic strength (which increases the voltage gradient across the sample). The conductivity is higher than that of Laemmli gels, and does not change during the run.

Two-Dimensional (Nonreduced, Reduced) SDS Gels

The subunit structure of proteins which consist of disulfide-bonded polypeptide chains may be assessed by a form of two-dimensional SDS-polyacrylamide gel electrophoresis in which proteins are intact in the first dimension, but reduced in the second [Wang and Richards, 1974; Phillips and Agin, 1977; Hynes and Destree, 1977; Goding and Harris, 1981]. Those proteins which are not made up of disulfide-bonded subunits will lie on the diagonal, while the subunits of disulfide-bonded proteins will drop below the diagonal (Fig. 1.).

The interpretation of nonreduced-reduced SDS gels requires care. In particular, the possibilities of oxidation [Phillips and Agin, 1977] artifactual disulfide bonds and disulfide rearrangement must be considered if meaningful results are to be obtained.

An artificially produced inter-chain disulfide bond has been found in HLA antigens [Springer et al, 1977a] and Ia antigens [Cullen et al, 1979]. The formation of new disulfide bonds can usually be prevented by treating the cells with iodoacetamide before or during lysis [Springer et al, 1977a; Cullen et al, 1979]. N-ethyl maleimide is also commonly used to block sulfhydryl groups, and the reaction is usually thought to be irreversible [Means and Feeney, 1971]. There is, however, one isolated report suggesting that this compound can actually promote disulfide interchange [Pinter et al, 1978].

The other major problem is reoxidation during electrophoresis. This is easily dealt with by the inclusion of a charged reducing agent, sodium thioglycolate (5 mM) in the cathode buffer of the reducing gels.

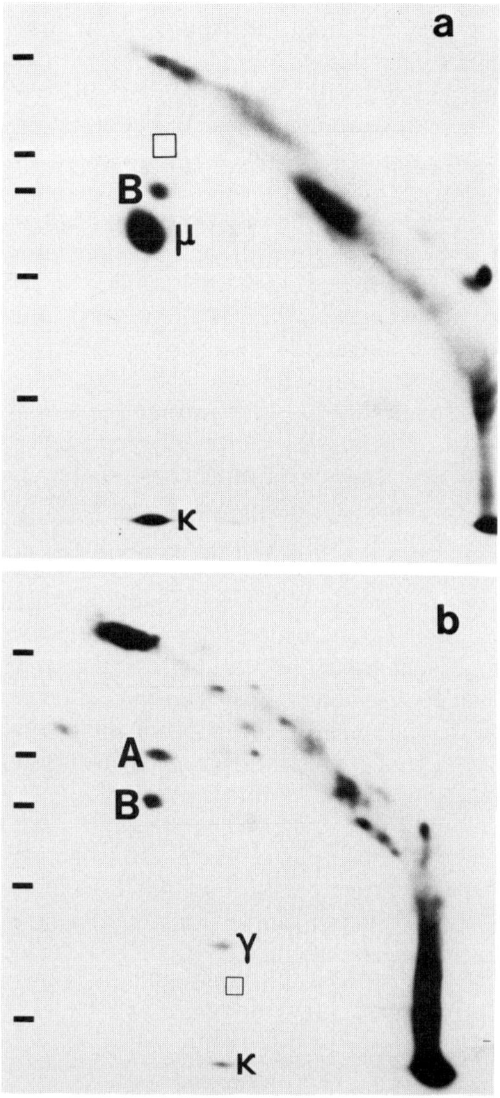

Fig. 1. *Two-dimensional (nonreduced, reduced) electrophoresis of radioiodinated membrane proteins from IgM-bearing lymphoma cells (a) and IgG-bearing lymphoma cells (b). Electrophoresis proceeds from left to right (unreduced; separation according to size) and then from top to bottom (after reduction; separation according to size). Proteins which are not made up of disulfide-bonded proteins (such as immunoglobulins) fall below the diagonal. Polypeptide "A" represents the PC-1 antigen found on secreting cells, while polypeptide "B" is the receptor for the iron transport protein transferrin. The open square in a indicates the expected postion of polypeptide A, and the open square in b indicates the position of secreted γ-chains (not radioactive). The horizontal bars indicate the mobilities of molecular weight standards myosin (200,000), β-galactosidase (116,000), phosphorylase b (95,000), bovine serum albumin (68,000), and ovalbumin (43,000) [Goding, 1982].*

Native/SDS Gel Electrophoresis

Another two-dimensional method for assessment of subunit structure consists of polyacrylamide electrophoresis under native conditions (which separate on the basis of both charge and size), followed by SDS gel electrophoresis. The first dimension is performed in tube gels, the second in slabs. Typical conditions for the first dimension are 4–8% acrylamide in 40 mM Tris/20 mM sodium acetate/2 mM ethylenediamine tetraacetic acid (EDTA), pH 7.4 [Morrow et al, 1980]. Most (but not all) membrane proteins are acidic and will be expected to move towards the anode at neutral pH.

If intrinsic membrane proteins are to be analyzed, the buffers must include Triton X-100 or Nonidet P-40 to maintain solubility. These detergents have rather large micelles (approximately 90,000 daltons), which may result in excessive slowing due to sieving. An alternative detergent would be octyl glucoside, which has a micelle size of 20,000 daltons (Helenius, personal communication).

It is important to recognize that multiple gels of varying acrylamide concentrations are necessary to demonstrate the associations [Morrow et al, 1980].

Isoelectric Focusing/SDS Gel Electrophoresis

The combination of isoelectric focusing (IEF) and SDS gel electrophoresis [O'Farrell, 1975; Scheele, 1975] results in an extremely high resolution system which is capable of separating almost all cellular proteins [O'Farrell, 1975]. As originally described, the method was only suitable for acidic proteins, but these comprise the great majority of cellular proteins. A later modification of the technique which allows analysis of basic proteins involves nonequilibrium pH gradient electrophoresis (NEPHGE), in which the separation process is fundamentally very similar to IEF, but the duration is shorter and the proteins do not stop moving [O'Farrell et al, 1977]. Garrels [1979] has described a number of modifications and refinements to the original system.

Most externally disposed membrane proteins and receptors are glycoproteins. When these forms of analyses are applied to glycoproteins, they do not run as discrete spots, but often as fuzzy smears or "families." This heterogeneity is at least partly a reflection of their variable carbohydrate content (Figs. 2, 3).

Peptide Mapping in Acrylamide Gels

Peptide mapping in polyacrylamide gels is a convenient way of analyzing relatedness and minor differences between proteins. The strategy involves enzymic or chemical fragmentation of the protein, followed by analysis of the breakdown products by size, charge or both. It is not essential for the fragmentation to proceed to completion. In fact, larger fragments are more suited to analysis by polyacrylamide gels, and may be just as informative as smaller ones.

Peptide mapping by isoelectric focusing. Silver et al [1979] used isoelectric focusing to analyze tryptic fragments of murine Ia and human DR (Ia-like) antigens. Digestion was taken to completion. Cells were labeled biosynthetically, solubilized in Triton X-100, and the glycoproteins isolated on lentil lectin-agarose. The Ia or DR antigens were immunoprecipitated by antibody and analyzed by SDS-polyacrylamide gel electrophoresis. Component chains were eluted from 1-mm gel slices by incubation in 0.5 ml H_2O overnight, lyophilized, and reconstituted to a final volume of 1.0 ml 0.9 M Tris, pH 8.8, and 0.24% SDS. Proteins were reduced with 0.1 M dithiothreitol and alkylated with 0.25 M iodoacetamide. One milligram of human IgG was added as a carrier, and 150 μl 100% trichloracetic acid (TCA) was added. The resulting precipitate was washed twice in 5% TCA and twice in acetone, air dried, dissolved in 1.5 ml 70% formic acid, and lyophilized. The sample was then resuspended in 0.1 M ammonium bicarbonate, pH 8.5, and 0.1 mg trypsin-TPCK added. After incubation for 1 h at 37°C, an additional 0.1 mg trypsin-TPCK was added, and the incubation continued for a further 2 h. Then, 100 μl of a buffer containing 10 M urea, 4% Triton X-100, 2% pH 3–10 ampholines, and 5% 2-mercaptoethanol was added, and the sample lyophilized and redissolved in 100 μl water. The sample was loaded onto the cathode end of a 3.6% acrylamide tube gel (7 × 100 mm) containing the same buffer, and focusing was at 0.1 W for 6.5 h. Gels were then sliced and counted for radioactivity.

Peptide mapping by limited proteolysis. Cleveland et al [1977] have devised an ingenious peptide-mapping procedure based on SDS acrylamide gel analysis of the products of incomplete proteolysis by a variety of enzymes which retained activity in SDS. The authors showed that the breakdown pattern of an individual protein was remarkably reproducible and was characteristic of that protein. Because of its simplicity and power, the method has become very popular.

Digestion may be carried out in SDS sample buffer in tubes. If the polypeptide has been isolated by immunoprecipitation, the whole precipitate should be heated in a boiling-water bath for 2–5 min and allowed to cool before adding the protease. Typical conditions involve 1–50 μg/ml of protease (papain, chymotrypsin, or staphylococcal V8 protease) at 20°C for 30–90 min, but should be adjusted to give a good range of breakdown products. Digestion may be terminated by heating to 100°C as above (optional), and the sample is analyzed on a 15% Laemmli gel. Alternatively, the digest may be analyzed on a two-dimensional (charge/size) gel [Whalen et al, 1979].

Other enzymes that may be of use include thermolysin, collagenase, and elastase. Trypsin has been used only rarely and at very high concentrations, presumably due to inactivation by SDS and reducing agents.

Alternatively, the protease may be mixed and immediately added to the sample well of a standard 15% Laemmli gel, and both electrophoresed into the stacking gel. When the bromphenol blue dye marker nears the main gel, the current is turned off for 30–90 min to allow digestion to proceed, and current is then resumed until the dye reaches the bottom of the gel.

A third variant involves cutting out protein bands from polyacrylamide gels. These may

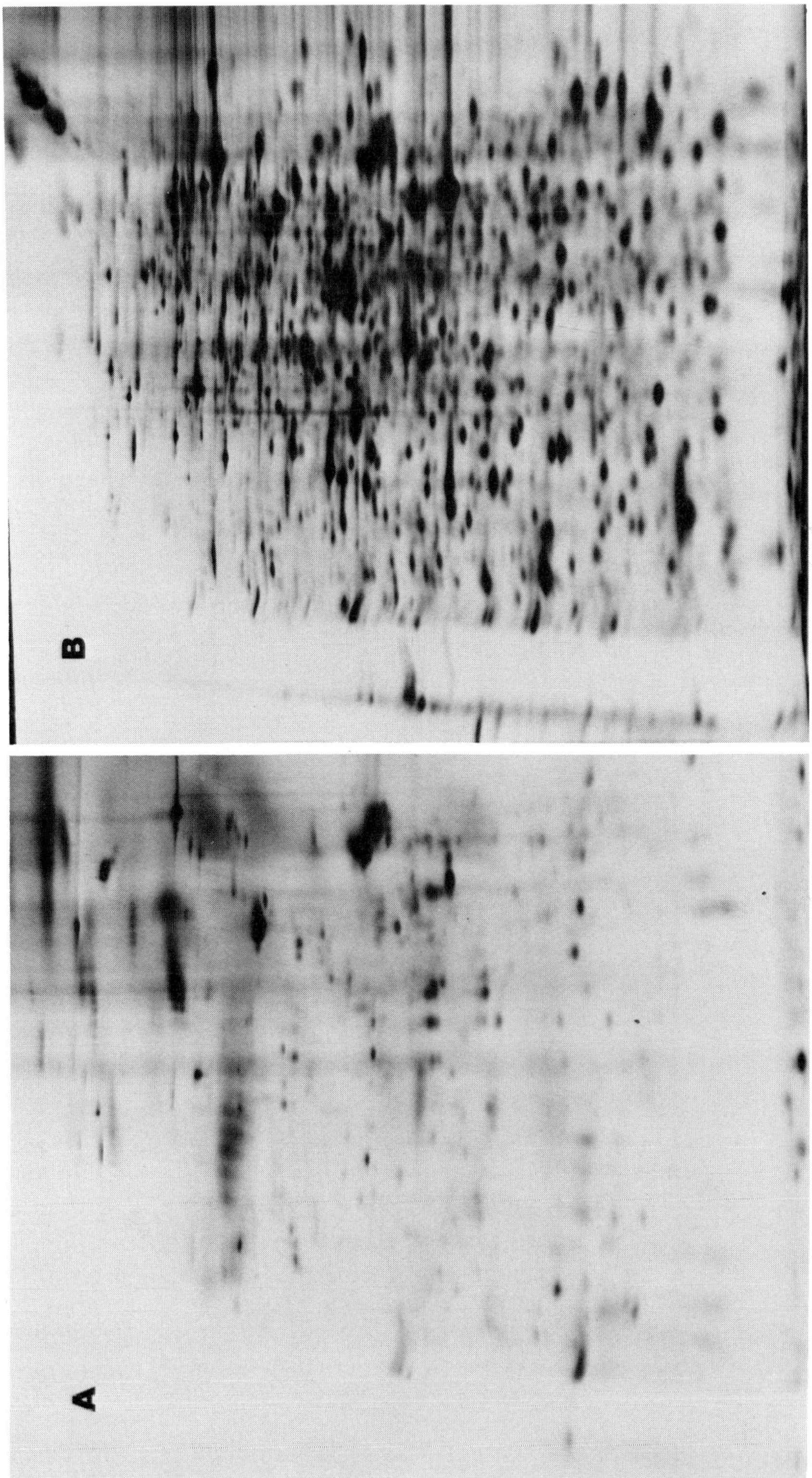

Fig. 2. *Two-dimensional (charge-size) electrophoresis of glycoproteins (A) or nonglycoproteins (B) synthesized by a T-cell hybridoma. Acidic proteins lie to the right. Note the extensive vertical and horizontal fuzziness of spots in A.*

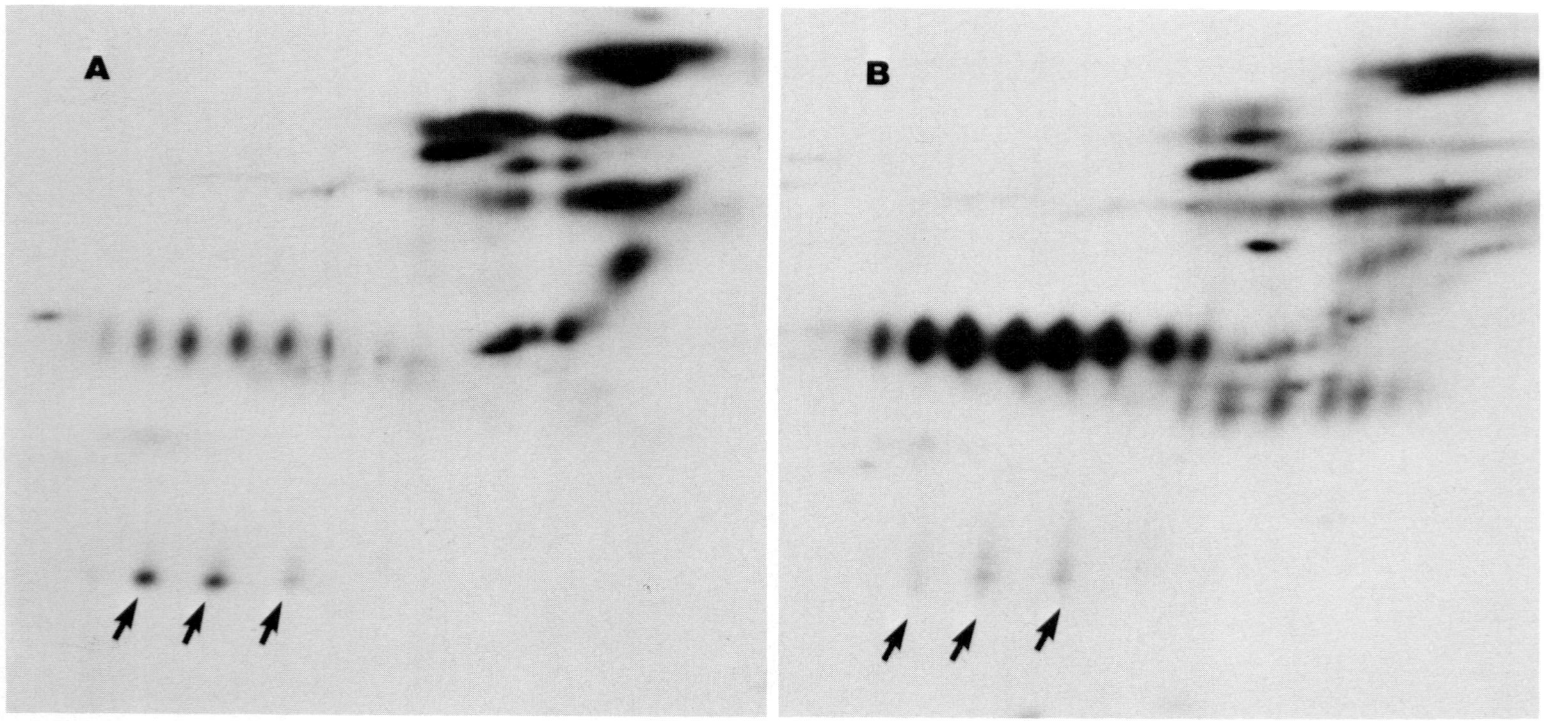

Fig. 3. Two-dimensional (charge-size) electrophoresis of ^{125}*I-labeled membrane proteins of a T-cell lymphoma line. Acidic proteins lie to the right. Note the extensive charge heterogeneity of many proteins, which is mostly due to sialic acid. A shows those proteins which bind to lentil lectin-agarose. B shows the proteins which do not bind. The arrows indicate the T-cell protein Thy-1. The more heavily glycosylated forms seen in B do not bind to lentil lectin (data of J.W. Goding; see Ledbetter and Herzenberg [1979]).*

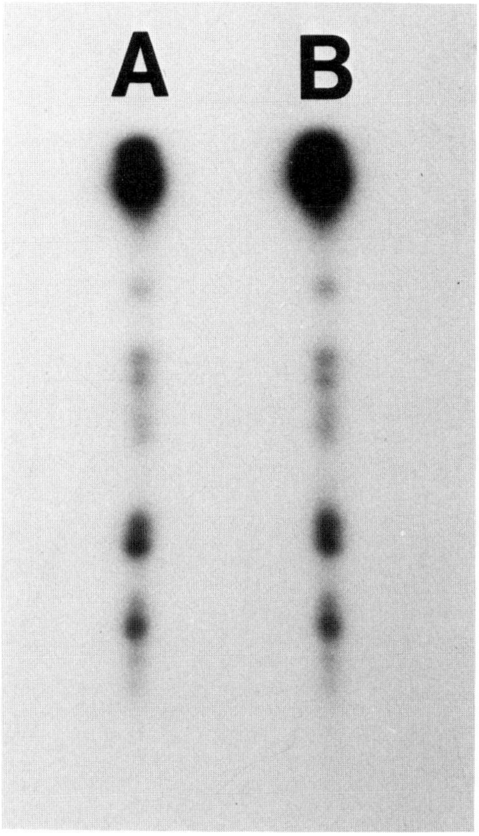

Fig. 4. *Peptide mapping by limited proteolysis. The polypeptides precipitated by transferrin-agarose (lane A) and monoclonal antibody OKT-9 (lane B) were digested with staphylococcal V8 protease, and the fragments analyzed in a 15% polyacrylamide gel in the presence of socium dodecyl sulfate. The patterns of breakdown products in lanes A and B are identical, indicating that the two procedures have isolated the same protein (the receptor for transferrin) (from Goding and Burns 1981]).*

be stained or radioactive, and the gel may be wet (in which case it should be equilibrated with sample buffer) or dried. It is thus possible to analyze a complex protein mixture on a two-dimensional gel, dry the gel, autoradiograph to identify a spot, and then cut out the spot and perform peptide mapping [eg, Handman et al, 1981; Wiegers and Dernick, 1981] (Fig. 4). The dried gel spot is placed directly into the sample well of a 15% Laemmli gel and rehydrated for 10–30 min with sample buffer containing protease (1–50 μg/ml). Electrophoresis proceeds normally, except that the current is temporarily turned off as above.

Occasionally, proteins will be found that are extremely resistant to proteolysis. These may often be mapped by inclusion of 4–5 M urea in the stacking gel [Handman et al, 1981].

Yet another variant of the basic Cleveland procedure has been described. [Bordier and Crettol-Järvinen, 1979; Lam and Kasper, 1980]. A complex mixture of proteins was separated in a tube SDS gel which was then placed onto a slab SDS gel and overlaid with protease. The proteolysis was executed in the stacking gel by temporarily turning off the current. This procedure is capable of performing peptide maps on a large number of proteins simultaneously, but its usefulness is limited by the ability of the first dimension to separate all components in the mixture. The procedure would be useful where the starting complexity was not very great and might be useful in demonstrating multiple related forms of a protein, such as precursor/product relationships.

Peptide mapping by chemical cleavage. Chemical cleavage of polypeptides is also a very useful way of generating large fragments for assessment of homology (Table III). Intact cylindrical gels or gel slices or spots may be treated prior to loading onto second-dimension slab gels [Lam and Kasper, 1980; Lonsdale-Eccles et al, 1981]. Whichever way is chosen, it is very important that the ionic environment (especially the pH) be returned to something approaching the SDS sample buffer prior to loading the analytical gel. This may be achieved by an intermediate equilibration in a strong buffer (eg, 1 M Tris) prior to equilibration in SDS sample buffer.

Of course, chemical cleavage may also be carried out in liquid solution. Hoh et al [1979] digested myosin isoenzymes with CNBr and analyzed the fragments on two-dimensional (isoelectric focusing/SDS) gels.

IMMUNOCHEMICAL ANALYSIS OF PROTEINS AFTER ELECTROPHORESIS—"WESTERN BLOTS"

Immunoprecipitation analysis as described in earlier portions of this chapter has some fundamental limitations. It is seldom possible to know with certainty whether a radioactive band on a gel represents the polypeptide recognized by the antibody, or, alternatively, whether it is precipitated because it is physically attached to another (possibly nonradioactive) molecule which bears the relevant antigenic determinant. A second possible problem with immunoprecipitation analysis is that it may be difficult to radiolabel the antigen to sufficiently high specific activity.

These problems can sometimes be overcome by probing the proteins with antibodies *after* electrophoretic separation [Burnette, 1981]. The heterogenous mixture of nonradioactive proteins is separated by electrophoresis, and proteins electrophoretically eluted by a transverse electric field onto a membrane which binds protein tightly. The membrane is then probed with a radioactive antibody, washed, and autoradiographed.

Choice of Transfer Membrane

A large number of transfer matrices have been described. The matrix may bind protein directly, or may be coupled with antibody. In the latter case it must be assumed that the antigen has more than one antigenic site per molecule, to allow subsequent probing with a second antibody.

Some versions of the technique used diazobenzyloxymethyl (DBM)-cellulose paper coupled with antibody [Erlich et al, 1979], and transfer was by passive blotting. A similar approach was used by Renart et al [1979], who blotted proteins directly onto DBM paper. The use of cyanogen bromide activated paper [Clarke et al, 1979] is simpler and probably just as effective. Regardless of whether DBM- or CNBr-activated paper is used, it is important to avoid extraneous nucleophils (eg, Tris, other amine-containing

TABLE III. Chemical Cleavage of Polypeptides*

Reagent	Cleavage site	Conditions	Availability	Reference
Cyanogen bromide	Met	20 mg/ml in 70% formic acid	Commercial	Gross [1967]
Acid	Asp-Pro	10% acetic acid, adjusted to pH 2.5 with pyridine 40°C, 24 h	Commercial	Landon [1977]
Hydroxylamine	Asn-Gly	2 M hydroxylamine in 0.2 M Na_2CO_3 (pH 9.0) 20 h	Commercial	Bornstein and Balian [1977]
N-Bromosuccinimide	Tryp	See reference	Commercial	Ramachandran and Witkop [1967]
2-nitro-5-thiocyano-benzoic acid (NTCB)	Cys	Reduce with 10 mM dithiothreitol 37°C 30 min, treat with 70 mM NTCB at 37° 20 hours	Commercial	Jacobson et al [1973]
N-chlorosuccinimide	Tryp	See reference	Commercial	Lischwe and Ochs [1982]

*For further details see Allen [1981].

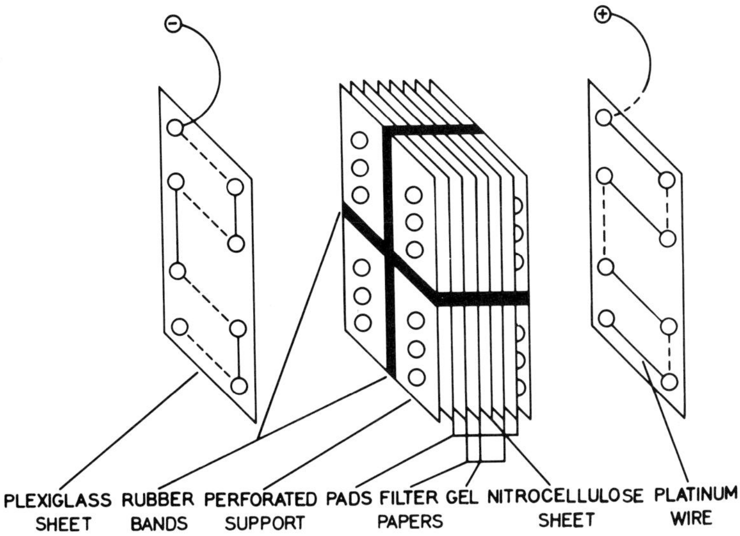

Fig. 5. *The "Western" technique. Following conventional polyacrylamide gel electrophoresis, a nitrocellulose membrane is laid over the gel and sandwiched between filter paper, porous pads, and Plexiglass supports. A transverse electric field drives proteins out of the gel and onto the nitrocellulose membrane.*

compounds, or azide) in the transfer buffer. A suitable buffer is 50 mM phosphate, pH 7.5. After transfer, all unreacted sites must be quenched (10% ethanolamine, pH 9.0, for 1–2 h).

Towbin et al [1979] introduced two major improvements. The first involved electrophoretic elution of protein from the gel, which was much more rapid and complete. The second was to replace DBM or CNBr-activated paper with nitrocellulose, which eliminated the chemistry of activation of the paper.

Unlike DBM or CNBr-activated paper, the binding of protein to nitrocellulose is not covalent. Nonetheless, it is extremely strong, and is resistant to 5% SDS, 3 M NaSCN or pH 2.2 [Burnette, 1981; Legochi and Verma, 1981]. It seems likely that electrophoretic transfer onto nitrocellulose membranes will become the preferred technique.

Practical Procedure

After electrophoresis, the gel is removed from the apparatus and placed in a dish containing 20% methanol in 20 mM Tris base, 150 mM glycine, pH 8.0 (buffer A). (The methanol prevents the gel from shrinking or expanding—see Burnette [1981].)A sandwich is then built up, consisting of a perforated Plexiglass support, a porous polyethylene sheet (Bel-Art) or Scotch-Brite scouring pads, three sheets of heavy filter paper (eg, Whatman 3 MM), the gel, the membrane, more filter paper, porous pad, and perforated support sheet (Fig. 5). All components should be prewetted in buffer A, and special care should be taken to avoid trapping air bubbles between the gel and the nitrocellulose membrane. The whole assembly is held together with rubber bands, and placed in a tank containing buffer A and platinum electrodes. Suitable tanks are the Canalco gel destainer or the Bio-Rad Trans-Blot apparatus. Alternatively, a simple plastic box or bucket would be perfectly adequate and much cheaper. The nitrocellulose sheet must be on the anode side of the gel.

The field strength must be determined empirically, but typical values are 6–8 V/cm for

16–22 h [Burnette, 1981]. Excessive current may result in overheating and distorted bands.

The membrane is then removed from the gel, washed in 150 mM NaCl, 10 mM Tris, pH 8.0, containing 5% bovine serum albumin (buffer B) the saturate remaining protein-binding sites, and placed in a dish containing the antiserum (typically 1:100–1:10,000 in buffer B). After 30–60 min, the membrane is removed, washed in buffer B, and placed in buffer B containing 10^5 cpm/ml of ^{125}I-labeled staphylococcal protein A or affinity-purified antiimmunoglobulin. After a further 30–60 min, the membrane is removed, washed extensively in buffer B, dried and processed for autoradiography using intensifying screens [see Roberts et al, this volume].

The antibody and protein A solutions may be used many times without being used up. In addition, it is possible to make multiple replicas of the same gel by partial transfer [Legochi and Verma, 1981].

The potential sensitivity of the method is extremely high. Burnette [1981] has shown that it is easily capable of detecting murine leukemia virus proteins from as few as 1,000 cells.

Enzymic Color Reaction Readout

The method may also be adapted to provide a nonradioactive color readout. Antigens are detected by peroxidase-conjugated antibodies, which were detected by incubation for 30 min in 0.0025% O-dianisidine, 0.01% H_2O_2 in 10 mM Tris-HCl, pH 7.4 [Towbin et al, 1979]. The product is virtually insoluble in water [Avrameas et al, 1978].

Problems With Western Blots

The antibody response to most protein antigens involves thousands of clonal products directed against hundreds of antigenic determinants. While it is clear that the reactivity of immune serum made against native proteins may be much weaker when tested on denatured proteins [Arnon, 1973], it has been demonstrated that some determinants are still recognizable [Stumph et al, 1974]. Antisera against the native protein will usually contain at least some antibodies which recognize the denatured protein, and vice versa. It is the fact that allows the Western technique to work.

In marked contrast, monoclonal antibodies against the native protein may or may not recognize the denatured product. It may be expected that many clones (perhaps the majority) will totally fail to recognize the denatured antigen, although evidence on this point is anecdotal. A systematic study would be worthwhile, although it may be anticipated that each protein will behave individually. The usefulness of monoclonal antibodies for Western blots thus remains unclear, and conventional polyclonal antisera will usually be preferred.

Some variability in the efficiency of transfer has been noted. In general large proteins (Mr > 50,000) may require higher field strengths and longer times for complete transfer. An additional source of variability is the brand and batch of nitrocellulose. Apparently, some brands have been treated with nonionic detergent, which inhibits protein binding. Boiling the membranes in distilled water or washing in ethanol or methanol prior to use may help.

DETERGENT BINDING BY MEMBRANE PROTEINS—ANALYTICAL AND PREPARATIVE EXPLOITATION

Principles

Membrane proteins can be classified into three main groups: integral, peripheral, and proteolipid [see Schlesinger, 1981].

Integral membrane proteins contain hydrophobic sequences which allow them to penetrate the interior of the lipid bilayer. They may be solubilized only by detergents (which replace the planar membrane lipid sheet by a curved detergent micelle), or by proteolytic removal of the hydrophobic segment. Integral membrane proteins are usually glycosylated, and if present, the carbohydrate is always on the extracellular face of the membrane.

In contrast, peripheral membrane proteins do not penetrate the lipid bilayer and are usually unglycosylated. They lie mostly on the cytoplasmic face of the membrane, from which they may be detached in a water-soluble form by changes in ionic strength or pH.

Finally, the proteolipids represent an interesting new group. The term "proteolipid" was originally defined as a class of membrane proteins which remained in the organic phase after chloroform/methanol extraction. Recently, however, Schlesinger [1981] has proposed that the term be redefined to describe a new class of membrane proteins which possess covalently attached lipid (usually palmitic acid). Examples include Sindbis virus envelope proteins [Schlesinger, 1981], receptors for transferrin [Omary and Trowbridge, 1981], and possibly the Thy-1 antigen [Campbell et al, 1981]. The requirements for solubilization of proteolipids are not yet well defined, although the use of detergents appears mandatory. It seems likely that a detergent micelle would form around the lipid molecule.

Assessment of Detergent Binding

The binding of large amount of nonionic detergent is characteristic of integral membrane proteins [Helenius and Simons, 1977]. In contrast, isolated peripheral membrane proteins and other water-soluble proteins show little, if any, interaction with nonionic detergent. Thus, the ability to bind detergent may be exploited as a test to determine whether a protein is an integral membrane protein [eg, Parkhouse et al, 1980; Goding, 1980], and may be used to separate integral membrane proteins from water-soluble proteins [Bordier, 1981].

Membrane receptors are usually present in extremely small numbers, and as a result it is not usually possible to measure detergent binding by direct methods. In recent years, several new methods have been devised to assess detergent binding on picogram-nanogram amounts of protein.

Charge Shift Electrophoresis

The solubilization of integral membrane proteins in Triton X-100 does not alter their net charge, because the detergent is nonionic. However, in the presence of small amounts of added deoxycholate (DOC), the binding of a mixed micelle of DOC and Triton X-100 results in acquisition of increased negative charge. Similarly, the binding of mixed micelles of cetyl trimethyl ammonium bromide (CTAB) and Triton X-100 results in acquisition of an increased positive charge. These differences in charge are easily detected as changes in electrophoretic mobility [Helenius and Simons, 1977]. Integral membrane proteins show large changes in mobility in the three detergent systems (Fig. 6), while water-soluble proteins possess very similar mobilities in all three.

It is important to note that some highly acidic or basic proteins may bind CTAB or DOC via electrostatic interactions, and binding by hydrophobic interactions is only indicated when large *bidirectional* shifts in mobility occur when tested in all three systems [Helenius and Simons, 1977].

Two-Phase Fractionation in Triton X-114

Nonionic detergents of the Triton series exist as monomers and micelles in aqueous solution. When the temperature is raised, the micelle size increases until at the "cloud point' the solution suddenly turns turbid. If the solution is then centrifuged, it separates into a detergent-enriched lower layer and a detergent-depleted upper layer (Fig. 7). For Triton X-100, the cloud point is 64°C. For Triton X-114, it is about 20°C.

Triton X-114 is effective in solubilizing membranes, and is a "mild" detergent which does not usually denature proteins. When cells

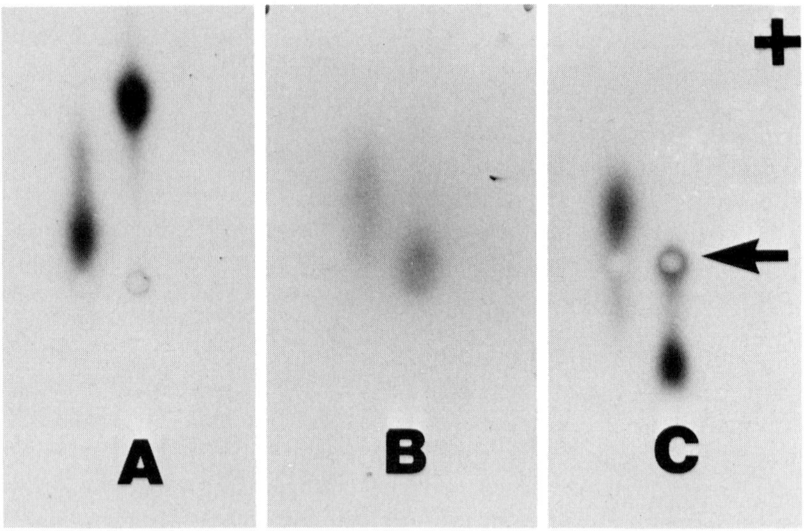

Fig. 6. *Charge shift electrophoresis. Fragments of lymphocyte surface IgD receptors were electrophoresed in agarose in the presence of Triton X-100 plus sodium deoxycholate (A), Triton X-100 alone (B), or Triton X-100 plus cetyl trimethyl ammonium bromide (C). The left lane in each panel contains the Fab portion of IgD, which is water soluble and does not bind detergent. The right lane in each panel contains the Fc portion, which penetrates the lipid bilayer and therefore binds detergent. The differing mobilities reflect the effect of the charge of the bound detergent. The arrow indicates the origin (from Goding [1980]).*

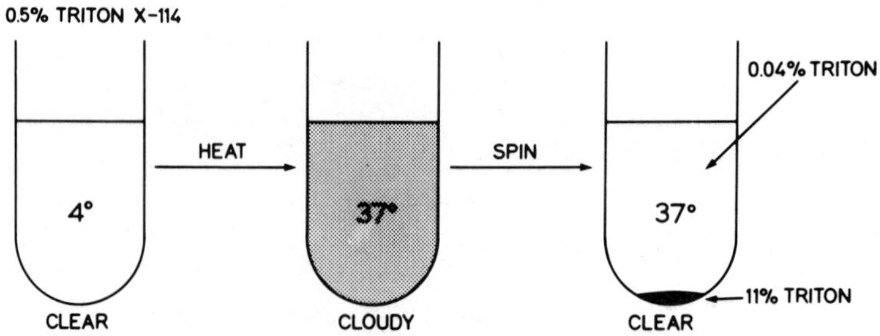

Fig. 7. *Separation of Triton X-114 into detergent-enriched and detergent-depleted phases.*

are solubilized in Triton X-114 at 4°C and the solution heated to 37°C, integral membrane proteins are found almost exclusively in the detergent phase, while water-soluble proteins are in the aqueous phase [Bordier, 1981] (Figs. 8, 9).

The rationale for the separation is simple. Water-soluble proteins do not interact with "mild" nonionic detergents, while integral membrane proteins typically bind one micelle of detergent around their hydrophobic transmembrane segment [Helenius and Simons, 1975]. The detergent replaces the planar lipid bilayer with a curved micelle. Thus, integral membrane proteins will follow the detergent in two-phase systems. The separation is analogous to taking the membranes apart and putting them back together.

The method has been validated for several integral membrane proteins and several water-

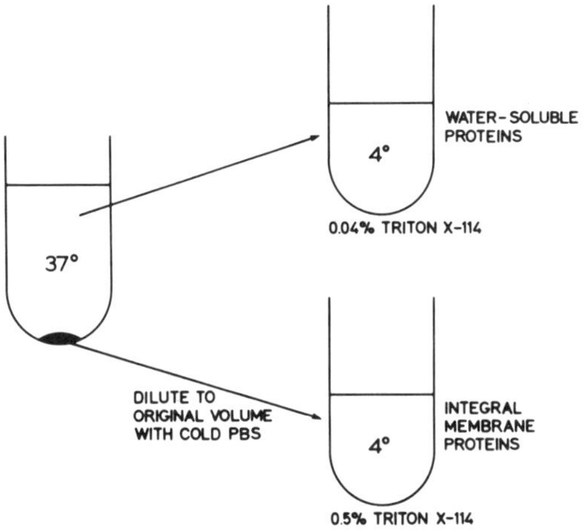

Fig. 8. *Separation of integral membrane proteins from water-soluble proteins by fractionation in Triton X-114 (after Bordier [1981]).*

soluble proteins [Bordier, 1981]. However, when applied to whole cell extracts, there appears to be a certain degree of aggregation and nonspecific loss of some proteins. Purified water-soluble proteins behave as expected, but there may be partial contamination of the detergent phase with cytoplasmic proteins when whole cells are fractionated (Goding, unpublished data). Nonetheless, the methods looks very promising and time will tell whether these problems can be overcome.

Detergent Exchange and Removal: Formation of Protein Micelles

Detergent molecules are in equilibrium with solubilized proteins, and the interchange of detergent monomers between free micelles and micelles attached to protein is quite rapid. As a result, it is relatively easy to exchange one detergent for another. This may be achieved by layering the solution over a su-

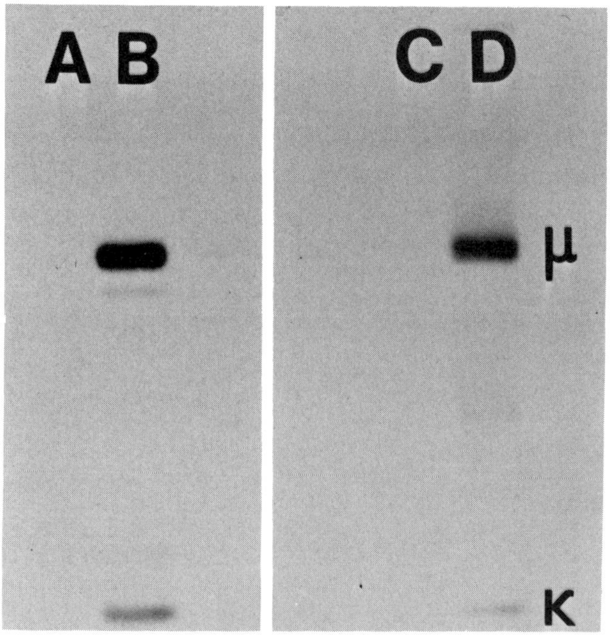

Fig. 9. *Two-phase partition of secretory and membrane IgM by the method of Bordier [1981]. Radioiodinated IgM was separated in Triton X-114 and then analyzed by SDS-polyacrylamide gel electrophoresis. Lane A) Secretory IgM; detergent phase. Lane B) Secretory IgM; aqueous phase. Lane C) Membrane IgM: aqueous phase. Lane D) Membrane IgM: detergent phase.*

crose cushion containing the new detergent, followed by high-speed centrifugation into the new detergent [Helenius, personal communication]. Other simple methods include binding the protein to an insoluble surface (eg, beads couples with lectin, antibody or ion exchanger groups) followed by washing in a buffer containing the new detergent.

Detergent removal will, as a general rule, pose many problems. The use of hydrophobic beads such as Bio-Beads SM-2 (Bio-Rad; Holloway [1973] generally results in unacceptable losses of membrane protein and cannot be generally recommended. Dialysis of Triton X-100 is extremely slow and inefficient, due to its low critical micelle concentration. Removal of octyl glucoside or deoxycholate by dialysis is much more effective, and exchange into these detergents followed by dialysis may be used. However, removal of detergent by dialysis, ion exchange or gel filtration generally results in macroscopic aggregation [Helenius and Simons, 1975].

In some instances, however, detergent removal by sucrose gradient centrifugation has resulted in the formation of water-soluble protein oligomers ("protein micelles") in which the hydrophobic portions are aggregated together away from the aqueous phase, and the hydrophilic portions face outwards [Simons et al, 1978; Terhorst et al, 1981].

REFERENCES

Allen G (1981): "Sequencing of Proteins and Peptides." Amsterdam: Elsevier, pp 43–71.

Anon (1981): A procedure for gel electrophoresis of low molecular weight range proteins. Bethesda Research Laboratories Catalog '81/'82, pp 15–16.

Arnon R (1973): Immunochemistry of enzymes. In Sela M (ed): "The Antigens." New York: Academic Press, Vol 1, pp 88–159.

Avrameas S, Ternynck T, Guesdon J-L (1978): Coupling of enzymes to antibodies and antigens. In Engvall E, Pesce AJ (eds): "Quantitative Enzyme Immunoassay." Oxford: Blackwell, pp 7–23.

Bobb D, Hofstee BHJ (1971): Gel isoelectric focusing for following the successive carbamylations of amino groups in chymotrypsinogen A. Anal Biochem 40:209–217.

Bonner WM, Laskey RA (1974): A film detection method for tritium-labeled proteins and nucleic acids in polyacrylamide gels. Eur J Biochem 46:83–88.

Bordier C, Crettol-Järvinen A (1979): Peptide mapping of heterogeneous protein samples. J Biol Chem 254:2565–2567.

Bordier C (1981): Phase separation of integral membrane proteins in Triton X-114 solution. J Biol Chem 256:1604–1607.

Bornstein P, Balian G (1977): Cleavage at Asn-Gly bonds with hydroxylamine. Methods Enzymol 47:132–145.

Burnette WN (1981): "Western blotting:" Electrophoretic transfer of proteins from sodium dodecyl sulfate-polyacrylamide gels to unmodified nitrocellulose and radiographic detection with antibody and radioiodinated protein A. Anal Biochem 112:195–203.

Campbell DG, Gagnon J, Reid KBM, Williams AF (1981): Rat brain Thy-1 glycoprotein. The amino acid sequence, disulphide bonds and an unusual hydrophobic region. Biochem J 195:15–30.

Carroll RB, Goldfine SM, Mehero JA (1978): Antiserum to polyacrylamide gel-purified simian virus 40 T antigen. Virology 87:194–198.

Chamberlain JP (1979): Fluorographic detection of radioactivity in polyacrylamide gels with the water-soluble fluor, sodium salicylate. Anal Biochem 98:132–135.

Clarke L, Hitzeman R, Carbon J (1979): Selection of specific clones from colony banks by screening with radioactive antibody. Methods Enzymol 68:436–442.

Cleveland DW, Fischer SG, Kirschner MW, Laemmli UK (1977): Peptide mapping by limited proteolysis in sodium dodecyl sulfate and analysis by gel electrophoresis. J Biol Chem 252:1102–1106.

Coligan JE, Kindt TJ (1981): Determination of protein primary structure by radiochemical techniques. J Immunol Methods 47:1–12.

Cuatrecasas P, Wilchek M, Anfinsen CB (1968): Selective enzyme purification by affinity chromatography. Proc Natl Acad Sci USA 61:636–643.

Cullen SE, Kindle CS, Littman DR (1979): Structural comparison of murine Ia antigens determined by the I-A and I-E subregions. J Immunol 122:855–859.

Czech MP, Massague J, Pilch PF (1981): The insulin receptor: Structural features. Trends Biochem Sci 6:222–225.

Delacroix D, Vaerman JP (1979): Simple purification of goat IgG1 and IgG2 subclasses by chromatography on protein A-Sepharose at various pH. Mol Immunol 16:837–840.

Dobberstein B, Garoff H, Warren G (1979): Cell-free synthesis and membrane insertion of mouse H-2D^d histocompatibility antigen and β2-microglobulin. Cell 17:759–769.

Erlich HA, Levinson JR, Cohen SN, McDevitt HO (1979): Filter affinity transfer. A new technique for the in situ identification of proteins in gels. J Biol Chem 254:12240–12247.

Ey PL, Prowse SJ, Jenkin CR (1978): Isolation of pure IgG1, IgG2a and IgG2b immunoglobulins from mouse serum using protein A-Sepharose. Immunochemistry 15:429–436.

Forsgren A (1968): Protein A from Staphylococcus aureus. VI. Reaction with subunits from guinea pig γ1- and γ2-globulin. J Immunol 100:927–930.

Fraker PJ, Speck JC (1978): Protein and cell membrane iodinations with a sparingly soluble chloroamide, 1,3,4,6-tetrachloro-3α,6α-diphenylglycoluril. Biochem Biophys Res Commun 80:849–857.

Gahmberg CG, Andersson LC (1977): Selective radioactive labeling of cell surface sialoglycoproteins by periodate-titiated borohydride. J Biol Chem 252: 5888–5894.

Galfrè G, Milstein C (1981): Preparation of monoclonal antibodies: Strategies and procedures. Methods Enzymol 73:3–46.

Garrels JI (1979): Two-dimensional gel electrophoresis and computer analysis of proteins synthesized by clonal lines. J Biol Chem 254:7961–7977.

Goding JW (1976): Conjugation of antibodies with fluorochromes: Modifications to the standard procedures. J Immunol Methods 13:215–226.

Goding JW (1978): Allotypes of IgM and IgD receptors in the mouse: A probe for lymphocyte differentiation. Contemp Topics Immunobiol 8:203–243.

Goding JW (1978): Use of staphylococcal protein A as an immunological reagent. J Immunol Methods 20:241–253.

Goding JW (1980a): Structural studies of murine lymphocyte surface IgD. J Immunol 124:2082–2088.

Goding JW (1980b): Antibody production by hybridomas. J Immunol Methods 39:285–308.

Goding JW (1982): Violation of symmetry in immunoglobulins: Hybrid molecules on the surface of plasma cells. Ann NY Acad Sci 399:82–96.

Goding JW, Burns GF (1981): Monoclonal antibody OKT-9 recognizes the receptor for transferrin on human acute lymphocytic leukemia cells. J Immunol 127:1256–1258.

Goding JW, Herzenberg LA (1980): Biosynthesis of lymphocyte surface IgD in the mouse. J Immunol 124:2540–2547.

Goding JW, Harris AW (1981): Subunit structure of cell surface proteins: Disulfide bonding in antigen receptors, Ly-2/3 antigens and transferrin receptors of murine T and B lymphocytes. Proc Natl Acad Sci USA 78:4530–4534.

Goding JW, Warr GW, Warner NL (1976): Genetic polymorphism of IgD-like cell surface immunoglobulin in the mouse. Proc Natl Acad Sci USA 73:1305–1309.

Goudswaard J, van der Donk JA, Noordzij A, van Dam RH, Vaerman J-P (1978): Protein A reactivity of various mammalian immunoglobulins. Scand J Immunol 8:21–28.

Granger BL, Lazarides E (1979): Desmin and vimentin coexist at the periphery of the myofibril z disc. Cell 18:1053–1063.

Gross E (1967): The cyanogen bromide reaction. Methods Enzymol 11:238–255.

Gutman GA, Warner NL, Harris AW, Bowles A (1978): Use of [^{75}Se] selenomethionine in immunoglobulin biosynthetic studies. J Immunol Methods 21:101–109.

Hammerling GJ, Hammerling U, Kearney JF (1981): "Monoclonal Antibodies and T-Cell Hybridomas." Amsterdam: Elsevier.

Handman E, Mitchell GF, Goding JW (1981): Identification and characterization of protein antigens of Leishmania tropica isolates. J Immunol 126:508–512.

Harrison LC, Flier JS, Roth J, Karlsson FA, Kahn CR (1979): Immunoprecipitation of the insulin receptor: A sensitive assay for receptor antibodies and a specific technique for receptor purification. J Clin Endocrinol Metab 48:59–65.

Helenius A, Simons K (1975): Solubilization of membranes by detergents. Biochim Biophys Acta 415:29–79.

Helenius A, Simons K (1977): Charge shift electrophoresis: Simple method for distinguishing between amphiphilic and hydrophilic proteins in detergent solution. Proc Natl Acad Sci USA 74:529–532.

Helenius A, McCaslin DR, Fries E, Tanford C (1979): Properties of detergents. Methods Enzymol LVI:734–749.

Herrman SH, Mescher MF (1979): Purification of the H-2K^{k} molecule of the murine major histocompatibility complex. J Biol Chem 254:8713–8716.

Hjelm H (1975): Isolation of IgG3 from normal human sera and from a patient with multiple myeloma by using protein A-Sepharose 4B. Scand J Immunol 4:633–640.

Hoh JFY, Yeoh GPS, Thomas MAW, Higginbotham L (1979): Structural differences in the heavy chains of rat ventricular myosin isoenzymes. FEBS Lett 97:330–334.

Holloway PW (1973): A simple procedure for removal of Triton X-100 from protein samples. Anal Biochem 53:304–308.

Hubbard AL, Cohn ZA (1976): Specific labels for cell surfaces. In Maddy AH (ed): "Biochemical Analysis of Membranes." London: John Wiley, pp 427–501.

Hurn BAL, Chantler M (1980): Production of reagent antibodies. Methods Enzymol 70:104–142.

Hynes RO, Destree A (1977): Extensive disulfide bonding at the mammalian cell surface. Proc Natl Acad Sci USA 74:2855–2859.

Jacobs S, Schechter Y, Bissell K, Cuatrecasas P (1977): Purification and properties of insulin receptors from rat liver membranes. Biochem Biophys Res Commun 77:981–988.

Jacobson GR, Schaffer MH, Stark GR, Vanaman TC (1973): Specific chemical cleavage in high yield at the amino peptide bonds of cysteine and cystine residues. J Biol Chem 248:6583–6591.

Kaplan G, Plutner H, Mellman I, Unkeless J (1981): Studies on externally disposed plasma membrane proteins. Trinitrobenzene sulfonic acid derivatization and immune precipitation. Exp Cell Res 133:103–114.

Kessler SW (1975): Rapid isolation of antigens from cells with a staphylococcal protein A-antibody adsorbent: Parameters of the interaction of antibody-antigen complexes with protein A. J Immunol 115:1617–1624.

Kessler SW (1976): Cell membrane antigen isolation with the staphylococcal protein A-antibody adsorbent. J Immunol 117:1482–1490.

Kessler SW (1981): Use of protein A-bearing staphylococci for immunoprecipitation and isolation of antigens from cells. Methods Enzymol 73:442–459.

Kessler SW, Woods VL, Finkelman FD, Scher I (1979): Membrane orientation and location of multiple and distinct allotypic determinants of mouse lymphocyte IgD. J Immunol 123:2772–2778.

Köhler G, Milstein C (1976): Derivation of specific antibody-producing tissue culture and tumor lines by cell fusion. Eur J Immunol 6:511–519.

Köhler G (1981): The technique of hybridoma production. In Lefkovits I, Pernis B (eds): "Immunological Methods." New York: Academic Press, Vol II, pp 285–298.

Laemmli UK (1970): Cleavage of structural proteins during the assembly of the head of the bacteriophage T4. Nature 227:680–685.

Lam KS, Kasper CB (1980): Sequence homology analysis of a heterogenous protein population by chemical and enzymic digestion using a two-dimensional sodium dodecyl sulfate-polyacrylamide gel system. Anal Biochem 108:220–226.

Landon M (1977): Cleavage at aspartyl-prolyl bonds. Methods Enzymol 47:145–149.

Lane DP, Robbins AK (1978): An immunological investigation of SV40 T antigens. I. Production, properties and specificity of a rabbit antibody to purified simian virus 40 large-T antigen. Virology 87:182–193.

Ledbetter JA, Herzenberg LA (1979): Xenogeneic monoclonal antibodies to mouse lymphoid differentiation antigens. Immunol Rev 47:63–90.

Lee S-K, Singh J, Taylor RB (1975): Subclasses of T cells with different sensitivities to cytotoxic antibody in the presence of anesthetics. Eur J Immunol 5:259–262.

Legochi RP, Verma DPS (1981): Multiple immunoreplica technique: Screening for specific proteins with a series of different antibodies using one polyacrylamide gel. Anal Biochem 111:385–392.

Lind I, Harboe M, Fölling I (1975): Protein A reactivity of two distinct groups of human monoclonal IgM. Scand J Immunol 4:843–848.

Lindstrom J (1979): Autoimmune response to acetylcholine receptors in myasthenia gravis and its animal model. Adv Immunol 27:1–50.

Lischwe MA, Ochs D (1982): A new method for partial peptide mapping using N-chlorosuccinimide/urea and peptide silver staining in sodium dodecyl sulfate-polyacrylamide gels. Anal Biochem 127:453–457.

Lonsdale-Eccles JD, Lynley AM, Dale BA (1981): Cyanogen bromide cleavage of proteins in sodium dodecyl sulfate/polyacrylamide gels. Diagonal mapping of proteins from epidermis. Biochem J 197:591–597.

MacSween JM, Eastwood SL (1981): Recovery of antigen from staphylococcal protein A-antibody adsorbents. Methods Enzymol 73:459–471.

Marchalonis JJ, Cone RE, Santer V (1971): Enzymic iodination: A probe for cell surface proteins of normal and neoplastic lymphocytes. Biochem J 124:921–927.

Markwell MAK, Fox CF (1978): Surface-specific iodination of membrane proteins of viruses and eukaryotic cells using 1,3,4,6-tetrachloro-3α,6α-diphenylglycoluril. Biochemistry 17:4807–4817.

Marshall T, Latner AL (1981): Incorporation of methylamine in ultrasensitive silver stain for detecting protein in thick polyacrylamide gels. Electrophoresis 2:228–235.

Mason DW, Williams AF (1980): The kinetics of antibody binding to membrane antigens in solution and at the cell surface. Biochem J 187:1–20.

Means GE, Feeney RE (1971): "Chemical Modification of Proteins." San Francisco: Holden-Day, pp 110–114.

Medgyesi GA, Füst G, Gergely J, Bazin H (1978): Classes and subclasses of rat immunoglobulins: Interaction with the complement system and staphylococcal protein A. Immunochemistry 15:125–129.

Morrison M (1980): Lactoperoxidase-catalyzed iodination as a tool for investigation of protein. Methods Enzymol 70:214–220.

Morrow JS, Speicher DW, Knowles WJ, Hsu CJ, Marchesi VT (1980): Identification of functional domains of human erythrocyte spectrin. Proc Natl Acad Sci USA 77:6592–6596.

O'Farrell PH (1975): High resolution two-dimensional electrophoresis of proteins. J Biol Chem 250:4007–4021.

O'Farrell PZ, Goodman HM, O'Farrell PH (1977): High resolution two-dimensional electrophoresis of basic as well as acidic proteins. Cell 12:1133–1142.

Omary MB, Trowbridge IS (1981): Covalent binding of fatty acid to the transferrin receptor in cultured human cells. J Biol Chem 256:4715–4718.

Owen MJ, Auger J, Barber BH, Edwards AJ, Walsh FS, Crumpton MJ (1978): Actin may be present on the lymphocyte surface. Proc Natl Acad Sci USA 75:4484–4488.

Parkhouse RME, Lifter J, Choi YS (1980): Chemical characterization of the Fab and Fc fragments from surface immunoglobulin. Nature 284:280–281.

Phillips DR, Morrison M (1971): Exposed protein on the intact erythrocyte. Biochemistry 10:1766–1771.

Phillips DR, Agin PP (1977): Platelet plasma membrane glycoproteins. Evidence for the presence of non-equivalent disulfide bonds using nonreduced-reduced two-dimensional gel electrophoresis. J Biol Chem 252:2121–2126.

Pinter A, Lieman-Hurwitz J, Fleissner E (1978): The nature of the association between the murine leukemia virus envelope proteins. Virology 91:345–351.

Pischel KD, Little JR (1979): Limited trypsin cleavage distinguishes MHC and thymus leukemia antigens. J Immunol 123:551–556.

Proia RL, Hart DA, Holmes RK, Holmes KV, Eidels L (1979): Immunoprecipitation and partial characterization of diphtheria toxin-binding glycoproteins from surface of guinea pig cells. Proc Natl Acad Sci USA 76:685–689.

Ramachandran LK, Witkop B (1967): N-bromosuccinimide cleavage of peptides. Methods Enzymol 11: 283–299.

Renart J, Reiser J, Stark GR (1979): Transfer of proteins from gels to diazobenzyloxymethyl-paper and detection with antisera: A new method for studying antibody specificity and antigen structure. Proc Natl Acad Sci USA 76:3116–3120.

Rogers J, Early P, Carter C, Calame K, Bond M, Hood L, Wall R (1980): Two mRNAs with different 3′ ends encode membrane-bound and secreted forms of immunoglobulin μ chain. Cell 20:303–312.

Rothman JE, Knipe DM, Lodish HF (1977): Membrane assembly: Synthesis and intracellular processing of the vesicular stomatitis viral glycoprotein. J Supramol Struct 7:353–370.

Saltvedt E, Harboe M (1976): Binding of IgA to protein A-containing staphylococci: Relationship to subclasses. Scand J Immunol 5:1103–1108.

Scheele G (1975): Two-dimensional gel analysis of soluble proteins. Characterization of guinea pig exocrine pancreatic proteins. J Biol Chem 250:5375–5385.

Schlesinger MJ (1981): Proteolipids. Annu Rev Biochem 50:193–206.

Schwartz BD, Nathenson SG (1971): Isolation of H-2 alloantigens solubilized by the detergent NP-40. J Immunol 107:1363–1367.

Sidman CL (1981): Lymphocyte surface receptors and albumin. J Immunol 127:1454–1458.

Silver J, Walker LE, Reisfeld RA, Pellegrino MA, Ferrone S (1979): Structural studies of murine I-E and human DR antigens. Mol Immunol 16:37–42.

Silver LM, Elgin SCR (1977): Distribution patterns of three subfractions of drosophila nonhistone chromosomal proteins: Possible correlations with gene activity. Cell 11:971–983.

Simons K, Helenius A, Leonard K, Servas M, Gething MJ (1978): Formation of protein micelles from amphiphilic membrane proteins. Proc Natl Acad Sci USA 75:5306–5310.

Skvaril F (1976): The question of specificity in binding human IgG subclasses to protein A-Sepharose. Immunochemistry 13:871–872.

Spielman H, Erickson RP, Epstein CJ (1974): The production of antibodies against mammalian LDH-1. Anal Biochem 59:462–467.

Springer TA, Robb RJ, Terhorst C, Strominger JL (1977a): Subunit and disulfide structure of monomeric and dimeric forms of detergent-soluble HLA antigens. J Biol Chem 252:4694–4700.

Springer TA, Kaufman JF, Siddoway LA, Mann D, Strominger JL (1977b): Purification of HLA-linked B lymphocyte alloantigens in immunologically active form by preparative sodium dodecyl sulfate-gel electrophoresis and studies on their subunit association. J Biol Chem 252:6201–6207.

Strominger JL, Engelhard VH, Korman AJ, Kostyk TG, Krangel MS, Lancet D, Lopez de Castro JA, Mann DL, Orr HT, Parham PR, Parker HL, Pober JS, Robb RJ, Shackelford DA (1981): Biochemical analysis of products of the MHC. In Dorf ME (ed): "The Role of the Major Histocompatibility Complex in Immunobiology." New York: Garland Press, pp 115–172.

Stumph WE, Elgin SCR, Hood LE (1974): Antibodies to proteins dissolved in sodium dodecyl sulfate. J Immunol 113:1752–1756.

Suissa M (1981): Nonenzymatic labeling of proteins by preparations of [^{35}S] methionine. Anal Biochem 115:67–71.

Swanstrom R, Shank PR (1978): X-ray intensifying screens greatly enhance the detection by autoradiography of the radioactive isotopes ^{32}P and ^{125}I. Anal Biochem 86:184–192.

Tanford C (1980): "The Hydrophobic Effect: Formation of Micelles and Biological Membranes." New York: John Wiley.

Terhorst C, LeClair K, Ma A, Slayter H (1980): Protein micelles of human histocompatibility antigens (HLA-A and HLA-B). Mol Immunol 18:103–112.

Tjian R, Stinchcomb D, Losick R (1974): Antibody directed against bacillus subtilis σ factor purified by sodium dodecyl sulfate slab gel electrophoresis. J Biol Chem 250:8824–8828.

Tomita M, Furthmayr H, Marchesi VT (1978): Primary structure of human erythrocyte glycophorin A. Isolation and characterization of peptides and complete amino acid sequence. Biochemistry 17:4756–4770.

Towbin H, Staehelin T, Gordon J (1979): Electrophoretic transfer of proteins from polyacrylamide gels to nitrocellulose sheets: Procedure and some applications. Proc Natl Acad Sci USA 76:4350–4354.

Turner MJ, Cresswell P, Parham P, Mann DL, Sanderson AR, Strominger JL (1975): Purification of papain-solubilized histocompatibility antigens from a cultured human lymphoblastoid line, RPMI-4265. J Biol Chem 250:4512–4519.

Vaitukaitis JL (1981): Production of antisera with small doses of immunogen: Multiple intradermal injections. Methods Enzymol 73:46–52.

Vitetta ES, Capra JD, Klapper DG, Klein J, Uhr JW (1976): The partial amino acid sequence of an H-2K molecule. Proc Natl Acad Sci USA 73:905–909.

Wada HG, Hass PE, Sussman HH (1979): Transferrin receptor in human placental brush border membranes. J Biol Chem 254:12629–12635.

Wang K, Richards FM (1974): An approach to nearest neighbor analysis of membrane proteins. Applications to the human erythrocyte membrane of a method employing cleavable cross-linkages. J Biol Chem 249:8005–8018.

Watt SM, Burgess AW, Metcalf D (1979): Isolation and surface labeling of murine polymorphonuclear neutrophils. J Cell Physiol 100:1–21.

Weber K, Osborn M (1969): The reliability of molecular weight determinations by dodecyl sulfate-polyacrylamide gel electrophoresis. J Biol Chem 244:4406–4412.

Weintraub M, Raymond S (1963): Antiserums prepared with acrylamide gel used as an adjuvant. Science 142:1677–1678.

Whalen RG, Schwartz K, Bouveret P, Sell SM, Gros F (1979): Contractile protein isozymes in muscle development: Identification of an embryonic form of myosin heavy chain. Proc Natl Acad USA 76:5197–5201.

Wiegers KJ, Dernick R (1981): Peptide maps of labeled poliovirus proteins after two-dimensional analysis by limited proteolysis in sodium dodecyl sulfate. Electrophoresis 2:98–104.

Williams AF, Galfrè G, Milstein C (1977): Analysis of cell surfaces by xenogeneic myeloma-hybrid antibodies: Differentiation antigens of rat lymphocytes. Cell 12:663–673.

APPENDIX

CHARGE SHIFT ELECTROPHORESIS

Stock Solutions

2X buffer (0.1 M glycine; 0.2 M NaCl; pH 9.0. 7.5 g glycine, 11.7 g NaCl, 1l H_2O. Adjust to pH 9.0 with 5 M NaOH (about 3 ml needed). Add 10 ml Triton X-100.

2X agarose. 1.0 g agarose to 50 ml H_2O. Heat until boiling.

Individual buffers. (1) 300 ml 2X buffer plus 0.3 g CTAB. (2) 300 ml 2X buffer plus 1.5 g DOC. (3) 300 ml 2X buffer (no additions).

Agarose slides. Add 15 ml hot 2X agarose to 15 ml hot 2X buffer. Pipette 6 ml onto 50 × 75-mm glass slides and allow to cool.

Tank buffers. Dilute individual buffers with equal volume H_2O.

Sample preparation. Sample should be in buffer containing Triton X-100, but preequilibration with DOC or CTAB is not necessary.

Electrophoresis. Punch a 2-mm diameter hole in centre of agarose. Load 2-μl sample. Run at constant current, 40 mA per slide (4 V/cm) for 90 min.

Fixing. Immerse slide in 1% amido black in methanol:acetic acid:water 40:16:40 for 10 min. Wrap slide in wet filter paper to prevent cracking and dry at 37°C. Destain in 7% acetic acid. Dry. Autoradiograph using intensifying screen. Failure to perform the staining and fixation procedure will cause all the protein in the agarose to be lost into the filter paper.

FRACTIONATION OF INTEGRAL MEMBRANE PROTEINS IN TRITON X-114

The procedure below is adapted from Bordier [1981].

Precondensation of Triton X-114

Triton X-114 (TX-114) is from Fluka AG, CH-9470 Buchs, Switzerland (catalog No. 93421; 250 ml).

1. Take 2 ml TX-114, add to 100 ml cold phosphate-buffered saline (PBS; pH 7.4). Add 1.6 mg butylated hydroxytoluene (antioxidant).

2. Mix well until homogenous, in a cylinder. Seal with parafilm.

3. Leave cylinder in 37°C incubator overnight.

4. Next day, you will find a lower layer (10–20 ml) which is enriched in detergent,

and an upper layer (remainder) depleted in detergent.

5. Remove upper layer by suction using water pump.

6. Make lower layer up to 100 ml with PBS.

7. Mix well, and repeat steps 3–6.

8. Repeat 7.

9. Save lower phase (11.4% TX-114 stock).

Stock Solutions

Solution A: 6% sucrose, 0.06% TX-114 in PBS. 10 ml PBS, 0.6 g sucrose, 50 μl 11.4% TX-114.

Solution B: 0.5% TX-114 in PBS. 10 ml PBS, 500 μl 11.4% TX-114.

Procedure

1. Solubilize cells in 0.5 ml 0.5% TX-114 in PBS on ice for 30–60 min.
2. Spin out nuclei at 4°C.
3. In an ice bucket, have a 10-ml tube ready containing 500 μl solution A (sucrose cushion).
4. Take supernatant from step 2, and carefully load directly onto sucrose cushion (step 3).
5. Place tube in 37°C water bath for 3 min. (Contents will go cloudy, especially upper half.)
6. Spin at 400g in 37°C room (time and speed not critical).
7. Remove supernatant into 10-ml tube on ice. (The detergent-rich oily drop in the bottom is very small, estimated 50–100 μl).
8. Resuspend oily drop in 500 μl cold PBS, on ice.
9. Repeat steps 4–7.
10. Pool second detergent-depleted supernatant with first (kept on ice).
11. Resuspend oily drop in 2.0 ml cold PBS. Leave on ice. This tube now contains the integral membrane proteins.
12. To the pooled depleted supernates, add 50 μl 11.4% TX-114. Mix. Place in 37°C water bath 3 min. Spin at 37°C as above. Discard oily pellet.

The supernate will contain the water-soluble proteins.

AUTORADIOGRAPHY

Following electrophoresis, gels are stained in 0.1% Coomassie Blue in 50% TCA for 30–60 min and destained in ethanol:acetic acid:water (12:7:81). After drying on a Bio-Rad drier, gels may be labeled using adhesive stickers and a fountain pen containing ink with a trace of ^{14}C or ^{35}S. This will allow the gel identification and molecular weight standards to be visible on the film, and also allows precise alignment of the gel with the film after development.

For ^{32}P and ^{125}I, the use of intensifying screens will greatly increase the sensitivity [Swanstrom and Shank, 1978] (Fig. 10). It is important to choose the type of screen and film carefully, as by no means are all suitable. A good combination consists of Kodak X-Omat R or S film and Dupont Cronex Lightning Plus screens. Exposure must be at −70°C. "Preflashing" the film will ensure that the film darkening is linear for faint bands [Swanstrom and Shank, 1978], but it does not

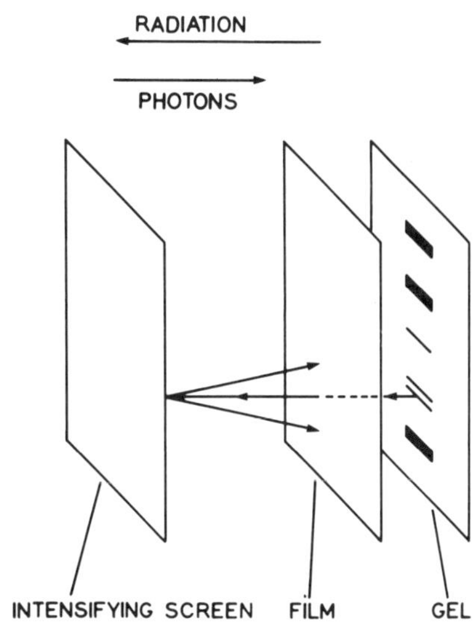

Fig. 10. *Use of intensifying screens to increase the sensitivity of autoradiography. The radiation from the gel passes through the film and strikes the screen, causing the emission of multiple photons, which expose the film.*

greatly alter the intensity of stronger bands and can often be omitted.

For ^3H, it is essential to impregnate the gel with a scintillant such as 2,5-diphenyloxazole (PPO) [Bonner and Laskey, 1974], EN^{3}HANCE^{TM} (New England Nuclear) or sodium salicylate [Chamberlain, 1979]. In my experience, EN^{3}HANCE^{TM} is easiest and most satisfactory.

The use of scintillants will increase the sensitivity of detection of ^{35}S and ^{14}C, especially when thick gels (>1 mm) are used. An improvement of 3–5-fold is typical. Note that intensifying screens cannot be used for ^3H, ^{35}S, or ^{14}C because their energy is too low to penetrate through the film and reach the screen.

Molecular and Chemical Characterization of Membrane Receptors, pages 61–113

3

Analysis of Membrane Polypeptides by Two-Dimensional Polyacrylamide Gel Electrophoresis

R. Michael Roberts, George A. Baumbach, William C. Buhi, John B. Denny, Laurence A. Fitzgerald, Sylvia F. Babelyn, and Michael N. Horst

Departments of Biochemistry and Molecular Biology (R.M.R., G.A.B., W.C.B., J.B.D., L.A.F., S.F.B.) and Obstetrics and Gynecology (W.C.B.), College of Medicine, University of Florida, Gainesville, Florida 32610, and Department of Basic Medical Science, School of Medicine, Mercer University, Macon, Georgia 31207 (M.N.H.)

INTRODUCTION

Two-Dimensional Electrophoresis

The analysis of complex mixtures of proteins by two-dimensional polyacrylamide gel electrophoresis is now part of the standard biochemical repertoire of a great many laboratories worldwide and can no longer be regarded as an esoteric or even technically difficult procedure. In general, methods based on that of O'Farrell [1975], employing a separation based on polypeptide charge in the first dimension and molecular size in the second, have proved most useful and allow the precise and reproducible resolution of many hundreds of components on a single polyacrylamide gel.

The technique of two-dimensional electrophoresis has been increasingly applied to studies on the composition of cellular membranes because these structures are usually complex and contain a wide variety of polypeptide subunits which cannot adequately be resolved in only a single dimension [Ames and Nikaido, 1976; Harell and Robinson, 1979; Jones, 1979; Horst and Roberts, 1979a; Imada et al, 1978; Horst et al, 1980a,b; Sato et al, 1977; Rosenblum et al, 1982]. However, membrane polypeptides can present special difficulties, since many are relatively hydrophobic and some are insoluble even in the presence of nonionic detergents. The main aim of this chapter is to describe in detail methods for solubilization and two-dimensional electrophoretic analysis of membrane protein and glycoprotein components by procedures that have been used successfully in our laboratories for the last 7 yr. In addition, we describe a variety of procedures for detecting such components following their analysis on polyacrylamide gels. Finally, we discuss some general criteria that may be used to determine which of the many polypeptides present in a plasma membrane preparation are normally exposed at the outer surface of a cultured cell.

Plasma Membrane Preparation

In the studies described here we have employed two cell lines which can be grown easily in the laboratory either in spinner flasks, roller bottles, or culture flasks using standard tissue culture media. They are Chinese hamster ovary cells (CHO cells [Kao and Puck, 1967]) and Sp2/0-Ag14 cells (Sp 2/0 cells [Schulman et al, 1978]). The former is a transformed fibroblast, now widely used because it is a stable hypodiploid from which a large number of mutant lines have been derived. It is easily grown, with a doubling time of between 12 and 18 h depending upon the culture conditions. The latter is a mouse myeloma commonly employed in generating hybridoma lines after fusion with mouse splenocytes. These cells grow to high density in suspension culture.

Cells are broken either by using nitrogen cavitation in a Parr Bomb (Parr Instruments, Moline, IL) or by means of a fluid pump disruption device [Baumbach et al, 1981; Horst et al, 1980b]. Both methods leave the nucleus and much of the cytoplasm intact. After removing unbroken cells, nuclei with attached cytoplasm and mitochondria are collected by low-speed centrifugation, and large microsomes are collected by centrifugation at 30,000g. The latter pellet is homogenized under hypotonic conditions at slightly alkaline pH, repelleted, and then loaded onto a stepped sucrose gradient. A plasma membrane-enriched fraction bands at the interface between the 25% and 40% (w/v) sucrose layers. This material is finally pelleted at 100,000g for 1 h. The preparation consists of about 0.5–1% of the total cell protein; is enriched 30–40-fold in Na^+K^+-dependent ATPase and phosphodiesterase I, which are used as markers for plasma membrane; and has an ^{125}I-specific radioactivity about 15-fold higher than that of the cell homogenate when isolated from cells surface labeled by the lactoperoxidase technique. Electron microscopy reveals a population of smooth-surfaced vesicles, 200 to 500 nm in diameter [Horst et al, 1980b].

A wide variety of other methods have been used to break cells and to isolate fractions enriched in plasma membrane. These methods have been reviewed in detail elsewhere [Baumbach et al, 1981] and will not be discussed here. However, two points are worth stressing: First, depending upon the breakage procedure, the plasma membrane may be fragmented into particles differing markedly in size and density, and the final "purified" fraction, isolated after differential and isopycnic gradient centrifugation, may not be representative of the surface as a whole. This situation is likely to be exacerbated in epithelial or other cell types where basal, lateral, and upper surfaces, as well as domains within each of these regions, are structurally differentiated and presumably distinct in their properties. Second, plasma membrane preparations, even those enriched 30–40-fold or more in "marker" enzymes, are only a part of a dynamic continuum of membrane systems and cytoplasmic vacuoles originating in the endoplasmic reticulum. The membrane is also connected directly or indirectly to various components of the cell's cytoskeletal system. For these reasons and also because cytoplasmic materials become entrapped or adsorbed during breakage, the preparations we and others have isolated contain a large number of proteins that are clearly not a part of the structure we normally conceptualize as plasma membrane.

SOLUBILIZATION OF MEMBRANE POLYPEPTIDES

Background

The procedure of O'Farrell [1975], which employs a solubilization buffer containing 9 M urea and 2% (v/v) Nonidet P-40 (NP-40), has not been rigorously tested for its ability to solubilize membrane material. Since membrane proteins are considered to be notoriously difficult to dissociate and solubilize, some workers have preferred to employ extracting solutions that contain the ionic detergent sodium dodecyl sulfate (SDS) [Imada et al, 1978; Tuszynski et al, 1979]. For example, Ames and Nikaido [1976] treated membrane with a dilute SDS solution at 70°C and then subjected the extract directly to isoelectric focusing (IEF) in the presence of excess NP-40. Although ionic detergent was present, the membrane proteins appeared to migrate according to their native charge, and it was assumed that a higher concentration of nonionic detergent could quickly replace the SDS from polypeptide subunits. The SDS, being a small, acidic molecule, moved quickly to the acidic end of the IEF gel. However, SDS in high concentrations, as well as at high solubilization temperatures, does interfere with subsequent isoelectric focusing [Ames and Nikaido, 1976; Wilson et al, 1977], thereby limiting the amount of a sample that can be solubilized and hence analyzed on a single gel.

In addition we have demonstrated that nonequilibrium pH gradient electrophoresis (NEPHGE) cannot be used in the first dimension after SDS solubilization [Horst et al, 1980a]. NEPHGE analysis, which is best employed for more basic proteins, is merely electrophoresis toward the cathode in presence of ampholytes, urea and detergent. We interpret this result to mean that the solubilized proteins, which must still bind considerable amounts of SDS and therefore have a pronounced negative charge even in presence of an excess of nonionic detergent, migrate immediately into the upper anodic buffer solution when the current is applied and before all of the SDS is exchanged. During standard IEF, on the other hand, the protein-SDS complexes can enter the gel readily and begin to migrate toward the anode. Any free SDS present in the equilibrium mixture established between the two detergents and the protein will be quickly removed electrophoretically. As more SDS dissociates to reestablish the equilibrium, it too will be removed and replaced by nonionic detergent. In time, all of the SDS will be removed and the protein will assume its own characteristic charge and migrate to the point in the gel determined by its isoelectric point (pI).

Although this scenario appears to hold true for most polypeptides provided only low concentrations of SDS are employed [Ames and Nikaido, 1976; Anderson and Anderson, 1977], we have noted that occasional polypeptides either migrate anomalously or do not appear on the gels at all when SDS-solubilization procedures are employed (see Fig. 1). Possibly membrane proteins, being quite hydrophobic, bind SDS more tenaciously and are more prone to behave in this manner.

Because proteins are usually separated in the first (IEF) dimension in presence of 9.3 M urea containing the nonionic detergent NP-40, it seemed advantageous to us to employ a solubilizing mixture which contained these components and not SDS. In addition, since urea at slightly alkaline pH had been used by others [Miner and Heston, 1972; Lim and Tadayyon, 1970] to dissolve normally insoluble brain proteins prior to either polyacrylamide gel electrophoresis or IEF, we tested potassium carbonate, as initially recommended by Neville [1967], to augment the solubilizing properties of NP-40 and 9.3 M urea. When CHO cell membrane was solubilized in 9.3 M urea and 5 mM K_2CO_3 with low-energy sonication followed by the addition of 2% (v/v) NP-40 and 5 mg/ml dithiothreitol [Horst et al, 1980a], greater than 93% of the total protein remained soluble (Table I). Alkaline urea is not only efficient but can solubilize at least 26 mg/ml of membrane protein. In the following section we describe the solubilization methodology in detail.

Procedure

We have found that 10^8 cells provide approximately 2.5 mg of partially purified plasma membrane [Fitzgerald et al, 1981]. This membrane preparation is suspended in 0.3 ml of solution 1 (urea-potassium carbonate, solution 1; see Appendix I) using a plastic pipette tip, and transferred to a small (5 × 0.5 cm) plastic test tube. The original tube containing the membrane is washed with a further 0.7 ml of solution, which is also transferred to the plastic tube. The amount of solubilizing solution should be scaled up or down according to the amount of material available.

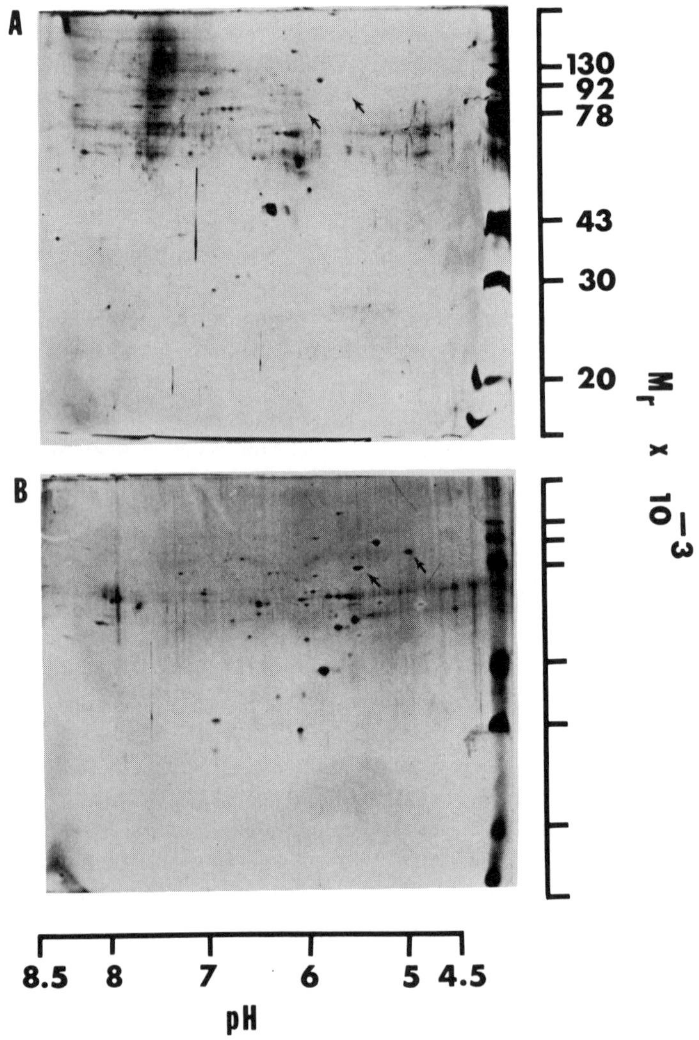

Fig. 1. *Solubilization of plasma membrane polypeptides for two-dimensional polyacrylamide gel electrophoresis using either SDS or alkaline urea. Plasma membrane was isolated by differential and isopycnic centrifugation from Sp2/0 cells disrupted with a fluid pump [see Horst et al, 1980b]. Twenty five micrograms of membrane were solubilized in 2% (w/v) SDS and 1% (v/v) 2-mercaptoethanol by boiling, then NP-40 and urea were added to final concentrations of 6% (v/v) and 9.3 M, respectively (A). An equivalent sample was solubilized in 9.3 M urea and 5 mM K_2CO_3 by sonication, then NP-40 and DTT were added to final concentrations of 2% (v/v) and 5 mg/ml, respectively (B). Both samples were analyzed by two-dimensional electrophoresis using isoelectric focusing in the first dimension followed by SDS-polyacrylamide gel electrophoresis in the second dimension. Gels were stained with silver. Note the loss of polypeptides (arrows) when membranes are solubilized in boiling SDS.*

TABLE I. Solubilization of Membrane in Alkaline Urea

Sample	Experiment	Amount solubilized (% total)	Protein concentration (mg/ml)
CHO plasma membrane	1	98.4[a]	0.96
CHO plasma membrane	2	96.6[a]	0.26
CHO plasma membrane	3	98.4[a]	0.72
CHO plasma membrane	4	98.4[a]	0.64
CHO plasma membrane	5	93.0[a]	8.26
Mitochondrial inner[b] membranes	6	95.0[c]	26.0

[a]After alkaline urea solubilization of membranes from L-[^{3}H]leucine labeled CHO cells, the remaining insoluble residue was dissolved in 0.1 M NaOH containing 1% SDS at 100°C for 5 min.
[b]Mitochondrial inner membranes were isolated from bovine liver as described by O'Brien and Kalf [1967]. The resultant "clarification pellet" was then subjected to alkaline urea solubilization as described in the text.
[c]Residual protein was solubilized as above and protein estimated colorimetrically.

The plastic tube containing the suspension is covered with a cap of aluminum foil, a hole is punched in the foil, and a narrow sonicator probe is inserted into the solution below. The suspension is then sonicated on ice (4 × 15 s at 45-s intervals) to break up large aggregates and to disperse the particles within the solubilizing solution. We normally employ a Bronwill Biosonic at 33–50% probe intensity. High-frequency sonication should be avoided since such treatment has been reported to cause cleavage of large polypeptide chains [Reisfield and Kahan, 1971].

The sonicated preparation is transferred to a 10 × 75-mm centrifuge tube. NP-40 (20 μl per ml of solubilizing solution) and 5 mg of dithiothreitol per ml are then added. The presence of dithiothreitol, which is an excellent buffer, reduces the pH from approximately 10.3 to between 7 and 8 so that exposure to alkaline conditions is not prolonged. The contents of the tube are mixed gently to avoid foaming and centrifuged at 13,000g for 30 min at 10°C. We have found that centrifugation at 100,000g for 1 h does not cause any marked increase in the amount of protein in the pellet. Therefore, we normally classify the protein in the clear 13,000g supernatant fraction as "soluble," although it should be emphasized that this term is purely operational.

Table I lists the solubilization efficiency of this procedure, employing plasma membrane preparations derived from CHO cells that had been labeled metabolically with L-[^{3}H] leucine. The appearance of trichloroacetic acid (TCA)-precipitable ^{3}H in the 13,000g supernatant fraction was used as an index of the efficiency of the method. The alkaline urea procedure also solubilized mitochondrial inner membranes with high efficiency. Note also that at least 26 mg of protein could be extracted using 1 ml of alkaline urea and that the method is potentially useful, therefore, for preparative purposes. The efficiency of solubilization is comparable to that using boiling SDS solution.

We have also considered the possibility that the very brief exposure to alkaline condition might result in loss of carbohydrate chains by β-elimination or increased rates of carbamylation, but from a series of control experiments we concluded that no such complicating side reactions occurred [see Horst et al, 1980a]. It is sometimes recommended that ampholytes be included in the initial solubilizing buffer since they then do not have to be added later. The ampholytes may also "protect" the protein from carbamylation by cyanate, a breakdown product of urea [O'Farrell, 1975]. However, unless all of the solubilized sample is to be used for IEF or NEPHGE, this habit is wasteful of expensive ampholytes.

The solution of membrane prepared in the above manner can now be subjected directly to the first-dimensional analysis, isoelectric focusing in presence of urea and NP-40. Material not to be used immediately is stored at −80°C.

TWO-DIMENSIONAL ELECTROPHORESIS

Introduction

This technique is now so widely used and so many variations on O'Farrell's original method have been published that it is difficult to recommend one procedure or one set of reagents over another. For example, ampholytes are now available from at least six different distributors, and dozens of companies sell electrophoresis equipment. For this reason, although we shall describe the method that we employ in some detail, there are numerous substitutions or variations that can be employed along the way. The reader is also referred to an excellent methodological review [Jones, 1980].

First-Dimension Separation

We shall describe two methods for first-dimension separation. The first is isoelectric focusing (IEF), which separates proteins according to their isoelectric point in a pH gradient established using a mixture of ampholytes. In general, the pH gradient runs from around 4 to around 8, although the pH is often unstable and variable at the basic end due to loss of ampholytes into the cathodic buffer. The second method, nonequilibrium pH gradient electrophoresis (NEPHGE) [see O'Farrell et al, 1977], has been briefly described earlier and allows basic proteins to be analyzed. The procedure depends upon the electrophoretic movement of polypeptides toward the cathode in an unstable pH gradient established using basic ampholytes. The separation which is eventually achieved depends not only on the buffer and ampholyte composition but also on the time of electrophoresis and the sample size. Because this technique leads to poorer resolution of components and since basic membrane proteins are relatively few, the method does not have broad utility in the context of this chapter. Nevertheless, NEPHGE should be employed at least once in any initial analysis in case highly basic proteins are present which might be missed during an IEF separation. As shown in Figure 2A, after two-dimensional separation, some polypeptides are well defined using NEPHGE with the first dimension, but are present as streaks at the basic end of an IEF gel (Fig. 2B).

However, it should be noted that the preparation of the gels for IEF and NEPHGE is identical except for the composition of the ampholytes. In both procedures (IEF and NEPHGE) diallyltartardiamide (DATD) rather than *bis*-acrylamide is used as the cross-linker. We have found that gels cross-linked in this manner permit larger proteins to be focused [Horst and Roberts, 1979a]. For example, thyroglobulin aggregates at the top of *bis*-acrylamide IEF gels but will focus at its isoelectric point in gels polymerized by our procedure. We assume that the larger pore size of DATD cross-linked gels allows less restricted movement of the proteins into the gel matrix [Horst and Roberts, 1979a; Bauman and Chrambach, 1976], thus reducing the tendency to aggregate as they become concentrated at their point of entry. The method also seems to allow larger quantities of protein to be analyzed on a single gel. In addition to this effect of the cross-linking agent, we have noted that the quality of the acrylamide influences the first-dimension separation, and we recommend the use of commercially prepared, highly purified acrylamide for IEF. Recommended brands include those distributed by BioRad or British Drug Houses. For second-dimension slab gels, acrylamide recrystallized in the laboratory is quite suitable (see Appendix II).

Preparation of IEF and NEPHGE glass gel tubes. The materials necessary and a pictorial account of preparing the first-dimension gels is shown in Figure 3. Glass tubes should be

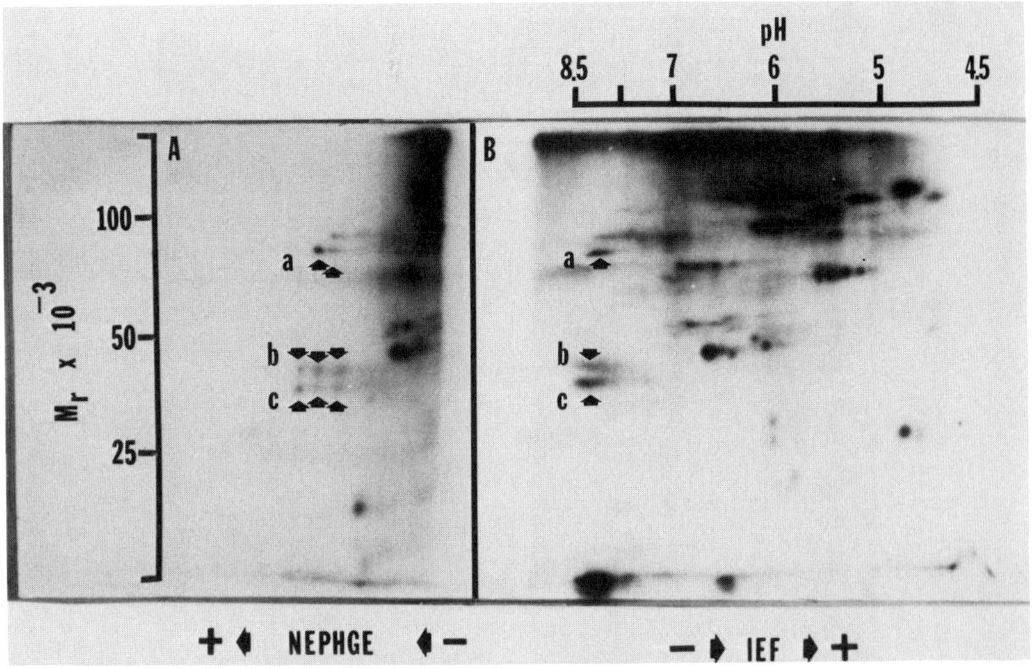

Fig. 2. *Autoradiograms of basic ^{125}I-labeled Chinese hamster ovary cell surface polypeptides separated by two-dimensional electrophoresis using in the first dimension either (A) nonequilibrium polyacrylamide gel electrophoresis (NEPHGE) or (B) isoelectric focusing (IEF). CHO cell (line 13, a wheat germ agglutinin resistant clone provided by Dr. Stuart Kornfeld [Briles et al, 1977]) surface proteins were labeled with Na^{125}I by lactoperoxidase. Cells were solubilized by the alkaline urea method as described. Samples were loaded on tubular gels and separated in the first dimension by the method indicated. A 10% acrylamide SDS slab gel was used in the second dimension. Gels were fixed, stained with Coomassie Blue, destained, dried, and autoradiograms prepared by exposure to preflashed Kodak XRP film at −70°C using an intensifying screen. Arrows highlight three highly basic polypolypeptides (a,b and c) which are separable into several isoelectric variants each by NEPHGE (A) but which streak together in an IEF gel (B).*

heavy-walled with an internal diameter (ID) of 2.5 mm and a length suitable for producing tube gels that will ultimately fit across a second-dimension slab gel apparatus. We employ tubes that are 14–16 cm long in order to provide a 12-cm first-dimension gel. The ends should be fire-polished. The tubes are cleaned thoroughly in boiling nitric acid in a fume hood. This is a dangerous procedure and care should be taken in ensuring that the glass vessel containing the acid is thick-walled and that long gloves are worn during the addition and removal of the tubes. We favor nitric acid cleaning since dichromate-H_2SO_4 mixtures do not remove acrylamide adhering to glass. After acid treatment the tubes are rinsed several times and dried. They should be marked at a point 12 cm along their length and sealed at their bottom end with several layers of parafilm, making sure that the film is wrapped around the sides of the tube several times. The tubes are then placed vertically in a leveled holder (Fig. 3A).

Gel mixture preparation and addition. Add 5.5 g of high purity urea to a clean, graduated, conical centrifuge tube. This amount will be sufficient for nine gels. Acrylamide stock solution (solution 2; 1.33 ml) and 2 ml of 10% (v/v) NP-40 are then added. Note again that this acrylamide stock solution contains DATD rather than *bis*-acrylamide as cross-linker. The reagents are then mixed on a Vortex mixer to

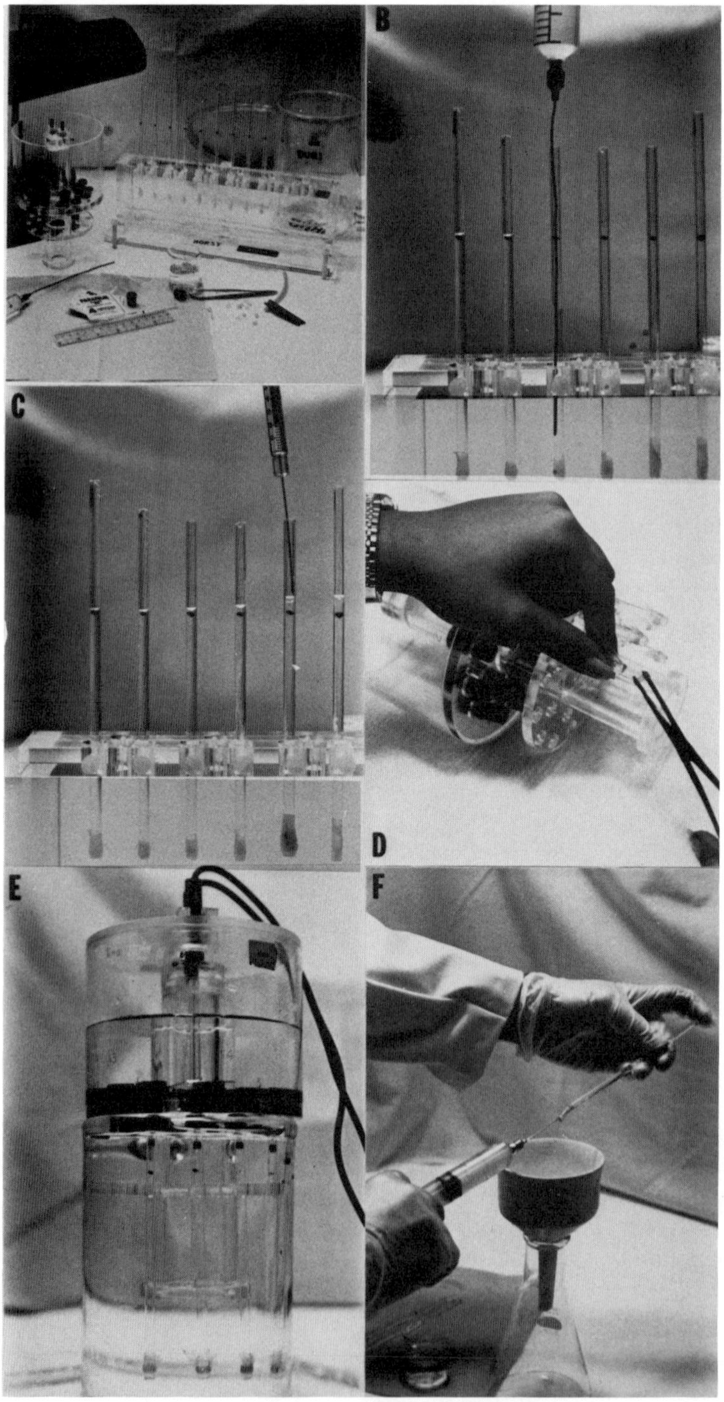

Fig. 3. *First-dimensional tube gel assembly. The necessary equipment for tube gel polymerization and electrofocusing are exhibited in A. Shown left to right in the background is a fluorescent lamp for acrylamide polymerization, the electrofusing gel tube machine with rubber stopper grommets, the tube gel polymerization rack loaded with gel tubes sealed with parafilm at one end and marked 12 cm up from the end, and the lower electrolyte cooling chamber. In the foreground are miscellaneous items including*

remove air bubbles. Water is added to give a volume of 9 ml, and the tube is immersed in warm (~50°C) water to dissolve the urea.

Ampholytes are then added to 2% (v/v) final concentration from 40% stock solutions. In our procedure we employ LKB or equivalent ampholytes in the following amounts for preparing 9 gels:

IEF			NEPHGE		
LKB Ampholines	Volume for 10 ml gel	Final % (v/v)	LKB Ampholines	Volume for 10 ml gel	Final % (v/v)
3.5 to 10	0.25 ml	1	9 to 11	0.35 ml	1.4
5 to 7	0.18 ml	0.7	8 to 9.5	0.10 ml	0.4
9 to 11	0.08 ml	0.3	6 to 8	0.05 ml	0.2

However, a single broad range or a different mixture could be substituted and pH gradients established. The ampholytes should be removed from the bottle using a sterile needle and stocks must be stored in the refrigerator.

After mixing, riboflavin-N,N,N′,N′-tetramethylethylenediamine (TEMED) solution (solution 3; 0.38 ml) is added, and the volume is adjusted to 10 ml with water. Immediately before pouring, 5 μl of an ammonium persulfate stock solution (solution 4) is added and mixed in thoroughly to increase the rate of polymerization.

The focus mix is then drawn directly into the barrel of a 10- or 20-ml syringe. An 18-gauge needle sufficiently long (6–8 in) to reach the bottom of the tube is then affixed to the syringe, and the solution is transferred carefully to each tube up to the 12-cm mark. Fill each tube from the bottom to displace air bubbles upward (Fig. 3B). It may be necessary to tap or wiggle tubes that have air bubbles trapped in the solution.

Each solution should then be overlayered to a depth of 1 mm with 8 M urea (about 25 μl) to prevent O_2 from interfering with polymerization at the top of the gel. Overlay the urea solution with an equivalent volume of water (Fig. 3C).

Place the gels directly under a fluorescent light. Polymerization should occur within 30 min. At this time remove the overlay of urea and water with a tissue paper wick and replace with 50 μl of solubilization buffer mixture (solution 5). The gels should then be allowed to stand for 1 1/2 h, then the overlay solution is removed and replaced with solution 6 (solubilizing buffer-dithiothreitol [DTT] mixture; 50 μl).

Fig. 3. *(Continued)*

a syringe fitted with a 6–8-in spinal-tap needle for dispensing the gel mixture, parafilm, rubber stopper grommets, wetted dialysis tubing, forceps, and elastic rubber bands cut from silastic tubing. Pouring the acrylamide-DATD mixture from the bottom of the tube with the aid of a spinal needle and syringe is shown in B. The tube is filled to the 12-cm mark, then overlayered with 1–2 mm of 8 M urea, then with 25 μl of water (shown in C). The gels are allowed to polymerize under the fluorescent lamp for 30 min, and then the overlayering solutions are removed and replaced with solubilization buffer (solution 5). Polymerization is allowed to proceed for another 1.5 h, then the overlay solution is removed and replaced with solution 6. The parafilm is removed, and the tubes are inserted through the rubber grommets in the electrofocusing machine. As shown in D, the tube bottoms are wrapped with a 2 × 2-cm piece of wetted dialysis membrane. Air bubbles are excluded from the tube gel bottom with a drop of glycerol and the dialysis membrane is held in place with a rubber band cut from 0.065-in diameter silastic tubing. After sample loading, the upper electrofocusing machine with gel tubes is inserted into the lower electrolyte chamber (shown in E). Note that the lower chamber is custom made to accommodate the 16-cm long tube gels so that the complete length of the acrylamide gel is surrounded by the cooled lower electrolyte solution. The upper ends of the gel tubes are barely visible in the upper electrolyte solution since the gel tube is pushed downward to the top of the rubber grommet. In F, the gel is extruded from the glass tube with a gentle stream of water as the inner surface of the tube is rimmed with a narrow-gauge needle.

It should be emphasized that many of the solutions employed in this part of the procedure, especially the acrylamide, are toxic and gloves should be worn throughout. Some workers also prefer to subject the gels to a short period of electrophoresis before loading the sample in order to remove ammonium persulfate and to establish a pH gradient. However, we have not noted that these precautions improve the final resolution.

Once the gels have completely polymerized, the parafilm must be removed from the bottom of the tubes using a razor blade and the sides of the glass tubes wetted with glycerol. They are then inserted into the grommets of the upper electrolyte reservoir rack (see Fig. 3D). We fashion grommets from No. 00 rubber stoppers.

The tubes are then sealed at their base (the anodic end) with presoaked dialysis membrane as illustrated in Figure 3D in order to prevent the acrylamide gels from sliding out. The dialysis membrane should be wrapped neatly over the end and any air bubbles squeezed out using glycerol. It can be held tightly in place by applying small rings cut from latex tubing.

A hemostat or forceps are used to hold the rings and to slip them over the end of the tube (Fig. 3D). In most instances the polyacrylamide gels will not slide out of the tubes. However, since this mishap does occasionally occur, we strongly suggest that our recommendation be followed.

Loading and running the sample. For IEF the lower chamber should be filled almost to the brim with 0.06 M H_2SO_4 (solution 7). The tubes in the removable upper reservoir rack are then placed in the acid electrolyte tank, allowing the lower end of the tubes to be completely immersed in the acid.

The solubilized sample of protein (up to 100 μl) can then be added to the top of the gels and mixed with the solubilizing buffer-DTT mixture (solution 6) already there. This dense solution is overlayered with 10 μl of 0.02% (w/v) fast green (in solution 5 diluted 1:1 with water) which acts as a marker dye and which will migrate to the anodic end of the gel during isoelectric focusing.

The cathode buffer is 0.04 M NaOH (solution 8), which is prepared well in advance and degassed thoroughly under reduced pressure prior to use. This solution is added carefully around each tube to about 1/2 in from the tube top. The tubes are then themselves filled with 0.04 M NaOH with care not to disturb the layers of sample and dye. Once filled, the tubes can be pushed down further until they are beneath the surface of the upper tank solution. Try to ensure that as much of the gel as possible is immersed in the lower chamber solution which in turn, is cooled by a circulating coolant at 18°C (Fig. 3E).

In order to minimize deleterious effects of overheating the gel, the amperage is kept low throughout electrofocusing by stepwise increases in the applied voltage. With the power supply in a "constant voltage" mode, electrofusing is initiated at 75 V for 30 min. Subsequently the voltage is increased to 150 V for 2 h and then to 300 V for a 15-h overnight run. Finally a 450–600-V run of 1 to 2 h is employed to "sharpen up" the bands and to reduce streaking at the basic end of the gel.

When NEPHGE is being carried out, the basic electrode solution (0.04 M NaOH; solution 8) is placed in the lower chamber and the acidic solution 7 at the top. The attachment of the terminals is also reversed (ie, the lower electrode is the cathode). The voltage regimen is as follows: 75 V for 30 min to allow stacking, and 400 V for 3 1/2 h to allow separation. The latter interval may be extended if sample volumes larger than 50 μl are to be employed, to increase migration toward the cathode.

Removal of gels. At the end of the first-dimension separation the tubes should be pushed out of their grommets and the dialysis membrane removed from the end. Use a syringe with a narrow-gauge spinal-tap needle to "rim" around the gels with a steady stream of distilled water (Fig. 3F). Slide the needle gently along the gel with its bevel facing away from the glass wall, squirting water as you proceed. Both ends may require this treatment. Continue until the gel slides out into a Buchner funnel. Nick the one end of the gel with the needle for future identification. The acidic end should have a green mark (IEF only) resulting from migration of the fast green dye, but it has a tendency to leach out quickly.

As an alternative to the above procedure for removing gels, place a few drops of SDS buffer in the gel tube and attach a rubber teat on the upper end. The gel may then be extruded by squeezing the teat.

Equilibration and storage of tube gels. The extruded gels should be placed in labeled 18 × 150-mm tubes and half filled with SDS-equilibration buffer (solution 9) containing 1% (v/v) 2-mercaptoethanol. The tubes are capped and gently agitated for 15 min at room temperature. The gels are then drained individually onto a Buchner funnel and then either subjected immediately to second dimension separation or stored frozen. In the latter case, the gels are placed lengthwise in glass test tubes, avoiding kinks. They are stored at −70°C in a horizontal position until used, with the tube sealed with parafilm.

Measuring the pH gradient and effect of ampholyte source. With each run, include a tube for measuring the pH gradient. This can be sliced into segments along its full length using a set of razor blades separated with metal spacers. Each slice is placed in a test tube (10 × 75 mm) containing 1 ml of distilled water for at least 1 h and the pH of the solutions measured by means of a narrow bore pH electrode. The gradients resulting during IEF with ampholytes manufactured by three separate companies (LKB, Serva, and Pharmacia) were compared and are shown in Figure 4.

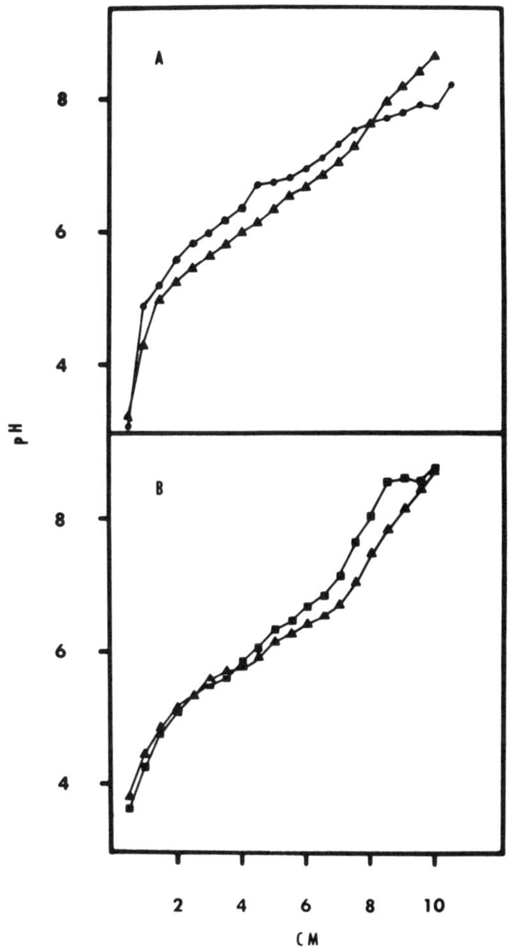

Fig. 4. *A comparison of pH gradients established with different ampholyte preparations in first-dimension polyacrylamide IEF tube gels. Equilibrium pH gradients were measured after incubating 1-cm gel slices in 1 ml distilled water for 1 h. The anodal (top) end of the gel is on the right. In A are pH gradients using (▲) LKB Ampholines 3.5–10 (1%), 5–7 (0.7%) and 9–11 (0.3%); and (■) Pharmacia Pharmalytes 3–10 (1%), 5–8 (0.7%), and 8–10.5 (0.3%). In B are gradients as described in A for (▲) LKB Ampholines and for (■) Serva Servalytes 3–10 (1%), 5–7 (0.7%), and 9–11 (0.3%).*

The pH gradients developed by isoelectric focusing (IEF) are obtained by means of specific buffering reagents called ampholytes. The theoretical properties and criteria that the ideal ampholytes must fulfill were described a number of years ago [Svensson, 1961, 1962]. These include buffering capacity, conductance at their pI, low molecular weight, pI values distributed throughout the pI range of most proteins (ie, 3–11), water solubility, and low light absorption at 280 nm. The first synthetic ampholytes were produced by Vesterberg [1969] as isomers and homologs of aliphatic polyaminopolycarboxylic acids. The synthesis of these have been further developed by commercial producers and in such cases information on their composition is limited. However, they are most likely complex mixtures of amines containing guanidic, sulfonic, and phosphonic groups supplemented with carboxylic groups. These are marketed under such trade names as Ampholine (LKB), Pharmalyte (Pharmacia), Biolyte (Bio-Rad), and Servalyt (Serva) and are available as mixtures covering broad (eg, 3–11) as well as narrow (eg, 7–8) pH ranges. The general isoelectric focusing method used in our laboratories consists of a mixture of ampholytes of the following ranges: 3.5–10, 5–7, and 9–11 [Horst and Roberts, 1979b].

We have evaluated the possibility of using different commercial sources of ampholytes since they vary considerably in cost. We have compared ampholytes using our standard technique (Fig. 4A; LKB Ampholines 3.5–10, 5–7, 9–11) to ampholytes from Pharmacia (Pharmalytes 3–10, 5–8, 8–10.5). In Figure 4B, LKB ampholytes are compared to Servalyte (3–10, 5–7, 9–11) ampholytes. Note that, except at the basic end of the gels, the pH gradients which were established are very similar. We recommend, therefore, that major criteria for use should be relative cost and availability of the product since they all appear to be of fairly similar quality. In future considerations, molecular size of the ampholytes should perhaps also be considered since these reagents elute only slowly from acrylamide gels and stain with Coomassie Blue. Smaller, more diffusible ampholytes are therefore potentially more desirable.

Second-Dimension Separation

The second-dimension separation involves polyacrylamide gel electrophoresis in the presence of SDS essentially in the manner described by O'Farrell [1975]. The tubular IEF gel is equilibrated with a buffer containing SDS and 2-mercaptoethanol and then fixed in place above a second-dimension slab gel. The latter may be made up with a standard concentration of acrylamide or else a gradient gel may be employed. We shall describe results using both methods.

Apparatus for slab gel electrophoresis. There are several commercial models that can be purchased. These are usually fairly expensive and not necessarily more convenient to use than homemade models, which can be constructed relatively easily from Plexiglass (Perspex). A detailed plan for one such apparatus has been given by Jones [1980]. Since the system we employ is also based on that described by O'Farrell [1975], we shall only describe it briefly. Essentially, the apparatus consists of a box through which coolant can be circulated. The slab gels are clamped to each side of this box and form the side walls of the upper buffer reservoir. A schematic drawing is shown in Figure 5. The sequence of assembly is shown in Figure 6. Once the slab gels are in place, the whole apparatus is placed in a tank (65 × 15 × 25 cm) containing anodic buffer deep enough to immerse the bottom of the slabs and the lower platinum electrode. Several of these set-ups can be run simultaneously.

When constructing such an apparatus a number of points should receive particular attention:

1. Since the slab gels are held to the box using large binder clips, it is important that the box be wide enough to allow these clips to be attached and removed easily. Similarly, if a four-slab apparatus (as shown in Figs. 5,6), consisting of two cooling boxes in series, is planned, there should be sufficient space between the boxes for two sets of binder clips to be attached side-by-side.

2. The silastic tubing (0.065-in outer diameter) which forms the gasket between the glass plate and the box should fit tightly into the routed groove on the box.

3. If coolant is to be circulated through the boxes, care must be taken to ensure that there are no leaks. Leaks also tend to develop with age in a poorly constructed box.

Slab gel plates. These should have dimensions compatible with the second-dimension electrophoresis apparatus and the first-dimension tube gels which they must accommodate. We cut our plates from 1/4-in glass, but others [eg, Jones, 1980] recommend 1/8-in window glass, which is cheaper, but more fragile.

In our set-up the outer plate is 18 cm × 18 cm × 1/8 in. The inner front plate (see Fig.

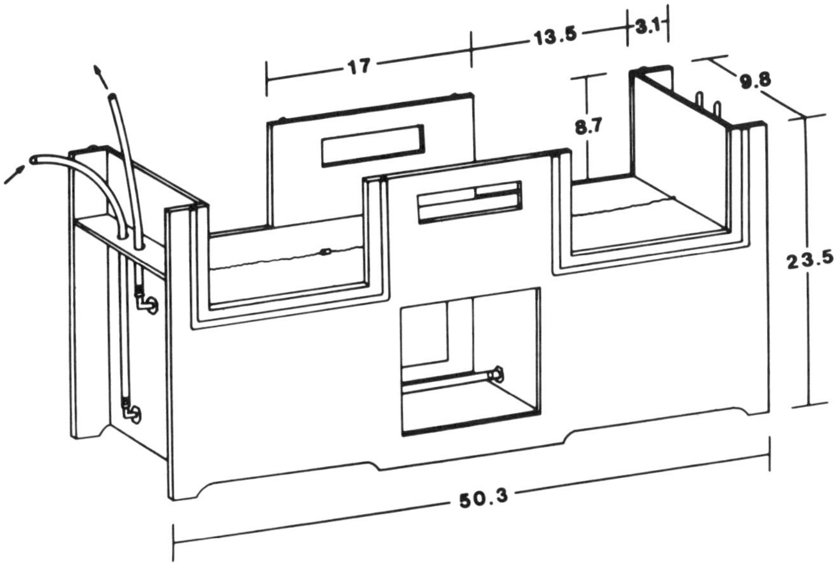

Fig. 5. *Schematic drawing of the gel box for the simultaneous electrophoresis of four polyacrylamide slab gels. The box contains electrode posts at one end (right) with platinum wire electrodes running the length of the box both on top (shown) and on the bottom (not shown). Coolant enters (arrow) passing through the left cooling chamber to fill first the right cooling chamber then returning to fill the left chamber and exiting as shown (arrow). Silicone tubing (3/16'' OD) acts as a sealing gasket to prevent the upper reservoir buffer from leaking around the attached plate. Plates are assembled (see Fig. 6) and attached by binder clips with the notched front plate flush against the box.*

7A) has the same overall dimensions but has a central notch 3.5 × 13 cm in its upper side, and, for added strength, this plate is 1/4-in thick. The bottom of this notch is beveled to a depth of about 0.5 cm (ie, at a 45° angle) to allow the easy placement of the first-dimension gel above the slab (Fig. 7B).

Many individuals recommend that spacer strips between the two plates be glued to the front plate. However, because breakage is not infrequent, particularly during cleaning, we prefer to use removable Teflon strips 1.5 mm thick.

The glass plates should be scrubbed with a laboratory detergent using a soft brush and washed thoroughly. The polyacrylamide has a tendency to stick to the glass, so immerse the plates in concentrated nitric acid for at least 20 min and rinse them thoroughly to remove the acid. Try to avoid scratching the glass, since these marks will be molded onto the gels and result in staining artifacts.

Place the dried plates in pairs flat on a table or bench with the bevels on the front plate facing upward (see Fig. 6A). A length of silastic tubing (0.065-in outer diameter) forms a mold for the acrylamide between the plates. The plates are bound together using one binder clip along the bottom and one more clip along each side. The best method for assembling the plates is first to sandwich the silastic tubing between the bottoms of the plates, then clamp the base, insert the Teflon spacers (1 × 14 × 1.5 cm) down each side (Fig. 6B), run the tubing along the side of the spacers, and clamp the plates together along their lateral edges (Fig. 6C).

Preparing slab gels. Gels can be poured up to 24 h before needed. Freshly poured gels should stand at least 2 h before they are used in order for polymerization to be completed. The gels may be poured prior to clamping to the slab gel box, in which case a small piece of Teflon spacer is placed between the bases

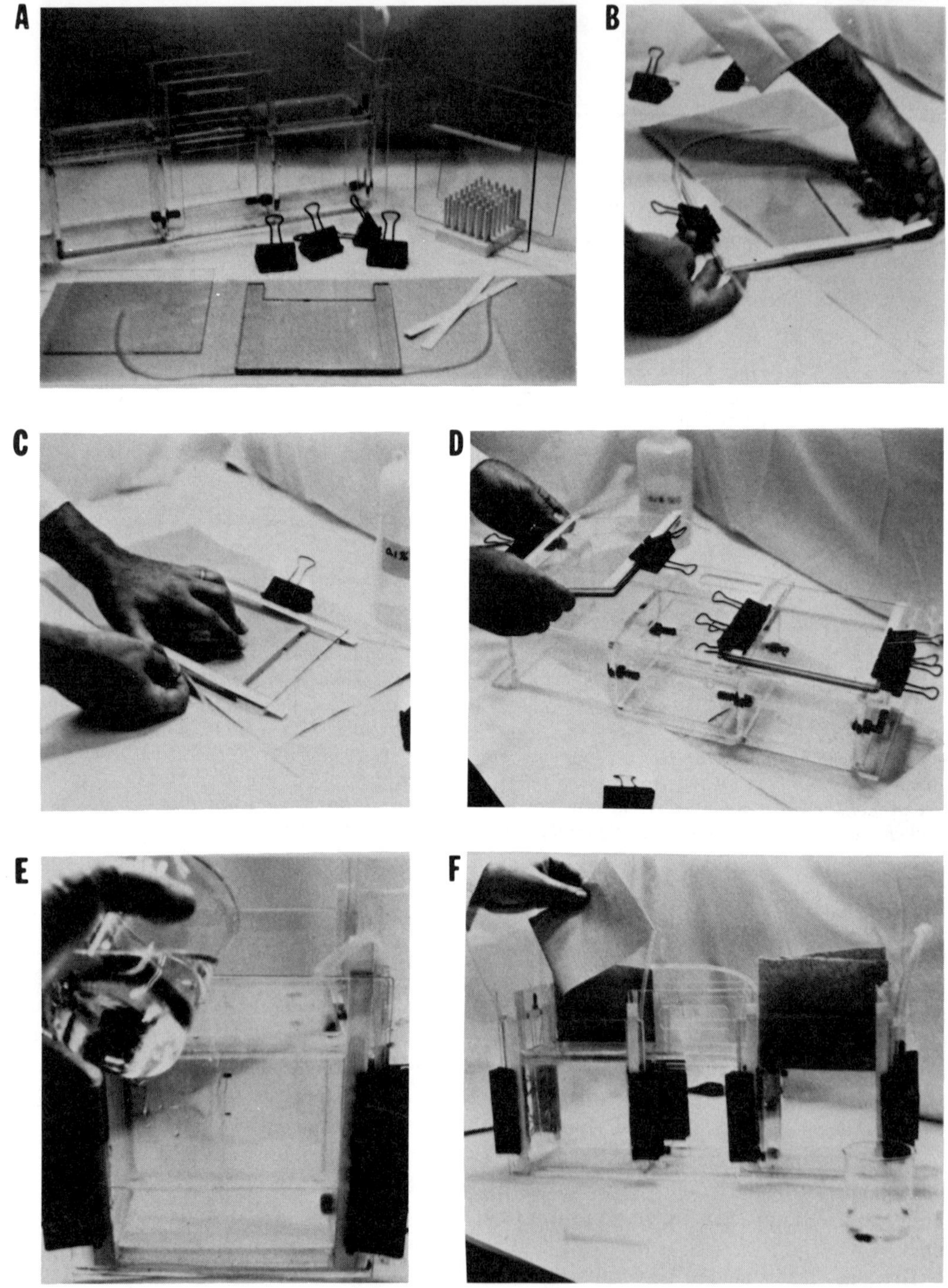

Fig. 6. *Assembly of the equipment for the simultaneous electrophoresis of four polyacrylamide slab gels. Displayed in A is the required equipment for the assembling and electrophoresis gel box, square and beveled-notched glass plates, binder clips, 1.5-mm Teflon gel plate mold spacers and the silastic tubing gasket. The glass plate gel molds are assembled first by placing the tubing gasket across the bottom edge of the inner notched plate as shown in A (foreground, center). The square outer plate is placed over the bottom plate, secured with a binder clip (see B), and then uniformly separated on the lateral edges with 1 × 16 × 1.5-mm Teflon spacers. The sides are sealed by pinching the gasket tubing between the thumb and forefinger of one hand and inserting the tubing between the plates with the thumb of the second hand in a smooth continuous motion (C). Wetting the tubing and plate edges with 0.1% (w/v) SDS aids in its insertion. Next, the assembled plates are clamped to the electrophoresis box with*

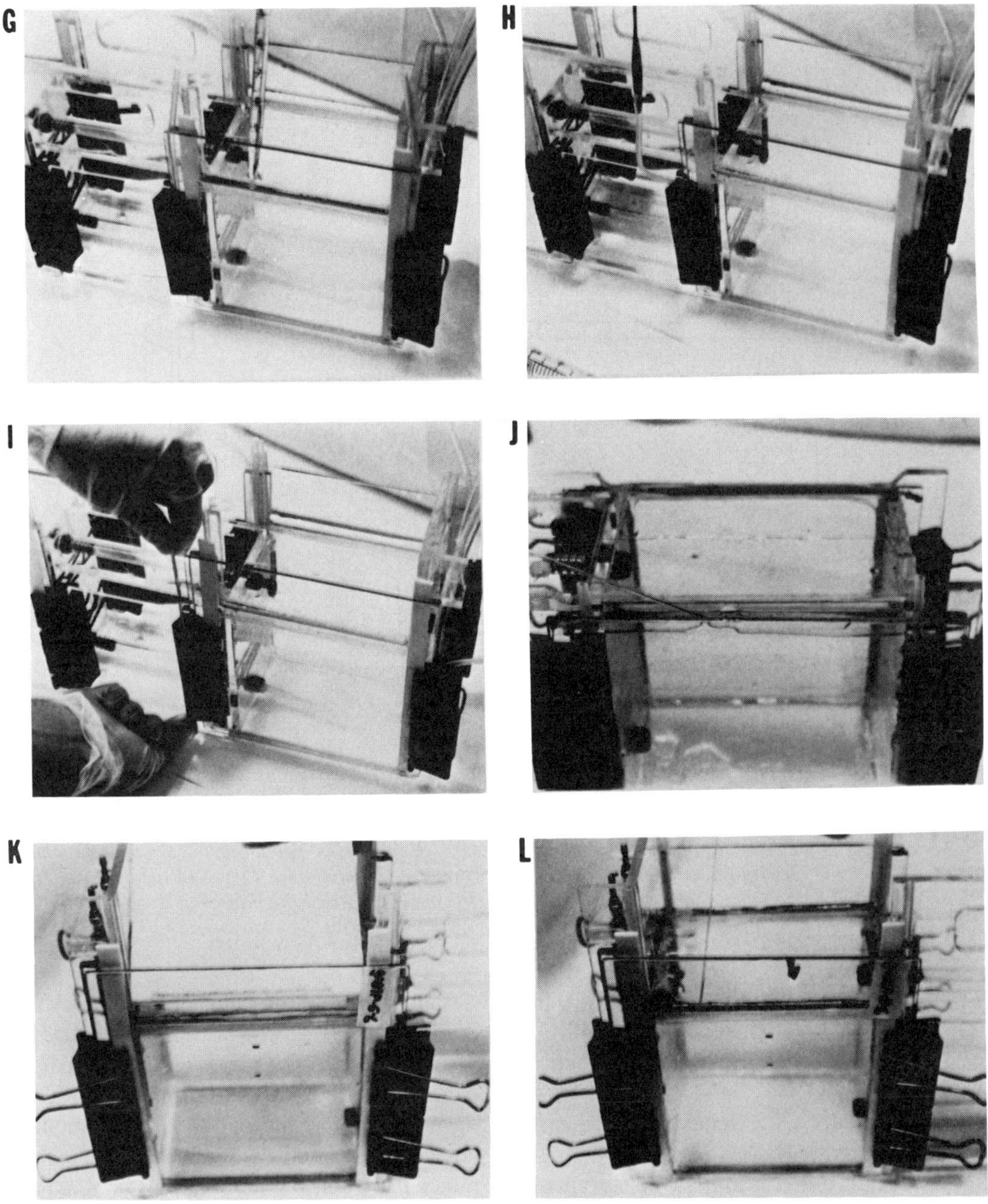

four binder clips, as shown in D. The acrylamide gel mixture is poured into each mold to a point 1 cm below the beveled notch bottom (E). The acrylamide is overlayered with 1–2 mm of 0.1% (w/v) SDS and allowed to polymerize for 30 min. Then the SDS overlay and unpolymerized acrylamide solutions are removed from the top of the slab gel with paper toweling (F). The stacking gel mixture is pipetted directly into the mold, filling it up to a point that is 0.5 cm from the top of the beveled notch (see G); the stacking gel is overlayered with 0.1% (w/v) SDS and allowed to polymerize for 30 min. The gasket tubing is removed from between the plates with a spatula tip, as seen in H. The tubing is pushed through the center of the binder clips (I) and then peeled from the bottom edge. The electrophoresis box is inverted (J) and 1-cm pieces of tubing are inserted between the glass plates over the Teflon spacers to form a mold. The acrylamide slab gel bottom is filled with molten 1% (w/v) agarose (solution 15) and allowed to gel. The box is then turned right-side-up. The first dimension tube gel is placed on a glass plate and positioned on top of the stacking gel, as seen in K. The tube gel is secured in the notch with molten 1% (w/v) agarose (solution 16) containing bromphenol blue tracking dye (L). Ready for electrophoresis, the four-slab gel box is then placed in the lower chamber filled with electrolyte buffer. The upper reservoir is also filled with the same electrolyte buffer, coolant is circulated through the electrophoresis box, and then the power supply is connected and electrophoresis begun.

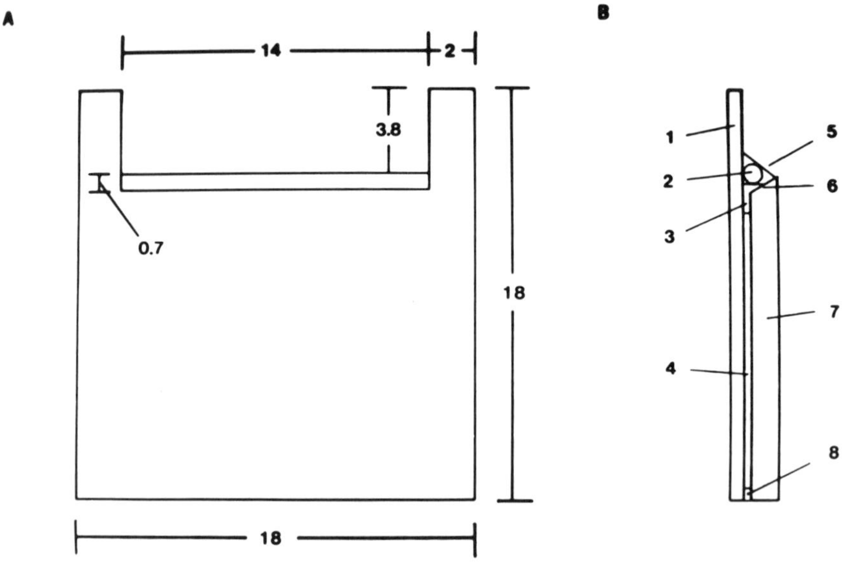

Fig. 7. *Schematic drawing of the beveled notched glass plate and a cross-sectional diagram of the assembled acrylamide gel molding plates (dimensions shown are in centimeters). A) Front view of 1/4-in thick "rabbit-ear" type glass plate with a beveled notch for holding the first dimension tube gel. B) Cross-sectional view of an assembled slab gel in its mold together with the first dimension acrylamide tube gel in place above the slab gel. Numbers indicate (1) the outer glass plate, (2) the IEF or NEPHGE tube gel, (3) the 4.5% acrylamide "stacking" gel (4) SDS-acrylamide slab gel, (5) solidified 1% agarose to anchor the tube gels, (6) middle of bevel on inner "rabbit-ear" glass plate, (7) inner glass plate (for clarity, the "rabbit-ears" have been omitted from the drawing; see Fig. 6A), and (8) solidified 1% agarose to seal the bottom of the slab mold.*

of the plates equidistant between the clips. This precaution prevents compression and maintains an even thickness along the bottoms as well as the sides of the gels. The plates can be rested on the lower binder clips when the gels are poured.

However, we prefer to pour the acrylamide after the plates are attached to the box since this avoids any possibility of damaging the gels during clamping. In this case the clips on the bottom of the plates are removed and the side clips are used to clamp the mold onto the gel box (Fig. 6D). If necessary, two molds can be clamped together on the same side of a box ("piggybacking"), thus allowing double the usual number of gels to be run on the same apparatus. If this is done, we recommend that low currents be used during the electrophoresis (ie, 5 mA/gel overnight) to avoid overheating.

Table II lists the formulations for polymerizing gels of several different acrylamide concentration. The first four ingredients (acrylamide solution 10, buffer 12, SDS, and water) are mixed thoroughly. The ammonium persulfate (solution 13) solution is then quickly mixed in by pouring the solutions between two vessels several times. Avoid shaking, which causes foaming.

The mixture is poured between the two plates up to a line 0.5 cm from the bottom of the bevel on the front plate (Fig. 6E). When all the gels have been poured, gently bump the chamber on the bench to dislodge any air bubbles that might be trapped below the solution.

Using a glass syringe with a narrow-gauge needle, cover the acrylamide with a layer (~2 mm) of 0.1% (w/v) SDS solution. Spread this SDS solution along the walls of the glass above

TABLE II. Formulations for Preparing SDS-Acrylamide Gels Sufficient for Four Slabs*

	Acrylamide monomer % (w/v)				
Stock solution[a]	5% (75,000 to 300,000)[b]	7.5% (30,000 to 200,000)	10% (20,000 to 100,000)	12.5% (10,000 to 90,000)	15% (5,000 to 60,000)
Acrylamide, solution 10 (ml)	25	37.5	50	62.5	75
Running gel buffer, solution 12, pH 8.8 (ml)	37.5	37.5	37.5	37.5	37.5
20% (w/v) SDS (ml)	0.75	0.75	0.75	0.75	0.75
Distilled H_2O (ml)	56.75	44.25	31.75	19.25	6.75
Ammonium persulfate, solution 13 (ml)	30.0	30.0	30.0	30.0	30.0
Total volume (ml)	150	150	150	150	150

*Modified from Laemmli [1970].
[a]Solution formulations are detailed in Appendix I.
[b]Molecular weight separation range: Ranges given are only approximations for globular proteins boiled in SDS with a reducing agent. Apparent molecular weights (Mr) for glycoproteins containing a large amount of carbohydrate may give anomalous results.

the acrylamide, thus allowing it to flow gently onto the denser layer below.

Allow the gels to polymerize for 30 min. As a rule we do not proceed further with gels that polymerize in less than 10 min or in longer than 30 min.

The overlay solution may then be poured off along with any fluid "squeezed" from the slab during polymerization. The upper surface of each gel must then be blotted dry with paper towels (Fig. 6F).

Gradient gels. We normally prepare gradient gels one at a time, although it is possible to pour several at once using an appropriate manifold. A two-chambered gradient mixer in which both chambers can be stirred is required. Constant agitation prevents the acrylamide from polymerizing prematurely.

The gradient maker is set up with the highest concentration of acrylamide in the first chamber (effluent side) and the lowest concentration in the second. Except for the concentration of acrylamide and *bis*-acrylamide, both solutions are identical and contain the same amount of other reagents. Formulations for several linear gradient acrylamide gels are listed in Table III. A total of 16 ml of solution is placed in each chamber, the apparatus is leveled, the port between the two chambers is opened, and the solution is allowed to flow out of the effluent chamber by gravity at a rate of about 2 ml/min directly into the mold for the second-dimension slab gel. Once the gradient maker has been emptied, the acrylamide is overlaid with 0.1% SDS solution and allowed to polymerize. The stacking gel for gradient and constant percentage acrylamide gels is the same and is described below.

Stacking gel preparation. The stacking gel mixture is prepared according to the following recipe which is sufficient for four stacking gels: acrylamide stock solution (solution 10), 4.8 ml; stacking gel buffer (pH 6.8) (solution 14), 4.0 ml; 20% (w/v) SDS solution, 0.16 ml; water, 12.6 ml; ammonium persulfate (solution 11), 10.6 ml; and TEMED, 10 μl.

Thoroughly mix the first four ingredients and then mix in the final two polymerizing agents. Working quickly, fill each mold amost to the top of the bevel (Fig. 6G). Overlay the stacking gel with 0.1% (w/v) SDS solution as before and leave 30 min for polymerization to occur. Although the catalyst, TEMED, is included in the stacking gel buffer stock solution (solution 14), we add an extra 10 μl of fresh TEMED to promote a more rapid, reliable, and uniform polymerization in these gels of low-acrylamide concentration (4.5%).

TABLE III. Linear Gradient Acrylamide Gel Formulations Sufficient for One Slab Gel

	% (w/v) Acrylamide monomer				
Stock solutions[a]	5%	10%	15%	17%	20%
Acrylamide stock, solution 11 (ml)	1.6	3.2	4.8	5.5	6.4
Running gel buffer, solution 12 (ml)	4.0	4.0	4.0	4.0	4.0
20% (w/v) SDS (ml)	0.08	0.08	0.08	0.08	0.08
Distilled H_2O (ml)	6.9	5.3	3.7	3.0	2.1
Ammonium persulfate, solution 13 (ml)	3.5	3.5	3.5	3.5	3.5

[a]Solution formulations are given in Appendix I.

After the stacking gel has polymerized, all fluid above the gel is poured off, and the upper surface is blotted dry with tissue. The amount of shrinkage that occurs during polymerization retracts the acrylamide about halfway down the bevel and provides a space which can accommodate the 2.5-mm first-dimension gel almost exactly.

Application of the first-dimension gel. At this time the two agarose-sealing solutions (solutions 15 and 16) should be removed from the refrigerator and liquefied in a boiling water bath. The silastic tubing should be pulled from the sides and bottom of each mold without removing the clips (see Fig. 6H,I). Remember to wear gloves, as the tubing will be wet with unpolymerized acrylamide. The box should now be turned upside down and small pieces of silastic tubing inserted vertically at the edges of each set of plates to act as a gasket to seal the bottom of the gel with agarose (solution 15, Fig. 6J).

Using a Pasteur pipette, pour prewarmed bottom gel sealer (solution 15) into the space between the bottom of the gel and the bottom of the glass plates (Fig. 6J). The inserts of silastic tubing prevents the solution running out to the side. It is important to fill this space completely with sealant. Once the agarose has gelled, the silastic tubing inserts can be removed and the box turned right-side-up.

The first-dimension gels are then placed on a glass plate, and any foam is removed with a Pasteur pipette. The fast green dye marks the acidic end. Each gel is allowed to slide onto the top of the second-dimension gel (Fig. 6K). A strip (1 cm) of agarose containing protein standards (see below) is placed alongside the gel at its acidic end. Once the gels and standards have been placed evenly atop the stacking gel, seal them into place with top gel sealer (solution 16, Fig. 6L). The solution must be warm, not hot. Carefully layer this agarose solution using a Pasteur pipette along the length of the tube gel to form a complete and even seal. The agarose should cover the tube gel completely to prevent it becoming dislodged during electrophoresis. This sealant solution should not be overheated or artifactual Coomassie Blue-positive bands will appear on the gel. It is also advisable not to include mercaptoethanol in this solution, since it is unnecessary and its presence also seems to cause additional staining artifacts on the second-dimension gel.

Preparation of protein standards. A choice of standards should be based on the percentage of acrylamide used in the second-dimension slab gel and the molecular weight range of the polypeptides under analysis. We use a mixture of the following standards, which can all be purchased readily:

Protein standard	~Mr
Thyroglobulin	335,000
β-Galactosidase	130,000
Phosphorylase b	92,500
Transferrin	78,000
Ovalbumin	43,000
Carbonic anhydrase	30,000
Soybean trypsin inhibitor	20,000
Cytochrome c	12,500

Standards are prepared individually at a concentration of 2 mg/ml in a solution of 10% (v/v) β-mercaptoethanol and 2% (w/v) SDS in 0.01 M Tris-HCl pH 6.8, boiled for 5 min, and stored frozen.

The protein standards are cast in agarose using 2.5-mm (ID) tube gels (identical to those used for IEF and NEPHGE) as molds. For each gel, add 10 mg of agarose to 0.8 ml of SDS-dissociation buffer (see above) and 25 μl of each protein standard in a clean tube. Warm to dissolve the agarose, mix and transfer to a gel tube sealed at its base with parafilm. After this solution has set, the gel is forced out of the tubing as described earlier. It is placed on a piece of parafilm, frozen, and cut into 1-cm pieces. These slices (containing ~ 10 μg of each standard) are conveniently placed next to the IEF or NEPHGE worm atop the second-dimension slab gel.

Electrophoresis. Electrophoresis buffer (solution 17) should be freshly prepared. The electrophoresis box with attached gels is placed in the lower electrolyte tank, coolant (18°C) is pumped through the chamber, and the apparatus is tested for leaks. Buffer is poured into the lower tank until the bottom edge of the slabs are submerged (glass marbles can be used to displace fluid and reduce the required volume of buffer). The reservoir holding the electrophoresis machine is a 65 × 15 × 25-cm Plexiglass tank. The electrodes are hooked up appropriately with the positive terminal connected to the lower (anodic) electrode.

Electrophoresis is carried out at 15 milliamps (mA) per slab for about 2 h using the "constant current" mode on the power supply. The current is then increased to 20 mA/slab and electrophoresis continued until the bromphenol blue mark reaches the bottom of the gel (~4 h). If no cooling is employed, we recommend that an overnight run of about 5 mA/gel be employed. High current and no cooling can sometimes cause vertical streaking of polypeptide spots.

After completing the run, turn off the power, lift out the chamber from the reservoir, remove the binder clips, and lift each set of two plates from the electrophoresis box. Separate the plates by first prying one corner to admit air. The spacers should be removed and the gel transferred to fixation fluid (ethanol:acetic acid:H_2O = 40:7:53). This is best achieved by wetting the gel with fixer and peeling it from the plate with a flat spatula. It is then rolled gently into a plastic tray or box containing fixative.

We have compared the separation of an identical group of plasma membrane polypeptides from Sp2/0 cells using either a uniform 10% acrylamide second-dimension gel (Fig. 8A) or a linear gradient gel (in this case 5–10% Fig. 8B). Note the flattening of the polypeptide spots on the gradient gel. This shape probably results from the leading edge of the polypeptides being unable to migrate further in the acrylamide matrix so that the spots become compressed with continued electrophoresis.

STAINING TECHNIQUES

Background

Two-dimensional polyacrylamide gel electrophoresis allows very large numbers of proteins to be resolved from complex mixtures [O'Farrell, 1975; Anderson and Anderson, 1977]. In order to detect such proteins, particularly those present in only trace quantities, highly sensitive techniques have been developed, including staining and photographic film methods. The most sensitive of all methods are the latter, which measure the prior incorporation of radioactive percursors into protein. Using autoradiography, for example, O'Farrell [1975] was able to detect several thousand polypeptides in extracts of *Escherichia coli*. Some of these contained less than 0.0001% of the total cellular protein. Such techniques are discussed in detail in a later section.

However, in many cases it is impractical to use autoradiography as a means of detection, and direct-staining procedures must be employed. Among the dyes that have been employed as stains are fast greens, amido black E, bromphenol blue and Coomassie Brilliant Blues. Of these, Coomassie Brilliant Blue R-250 has become the most popular because of its general utility, economy, and its sensitivity

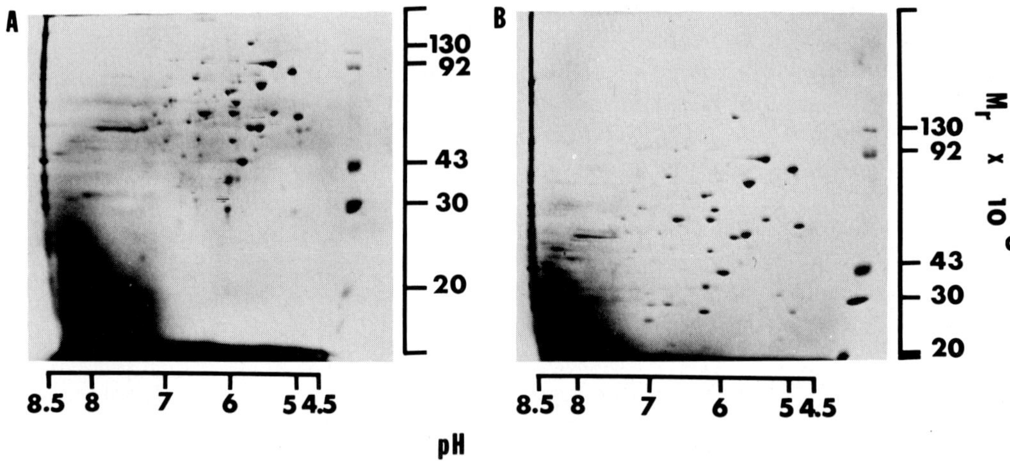

Fig. 8. *Second-dimension separation by SDS polyacrylamide gel electrophoresis using either a constant concentration or a linear gradient of acrylamide. Sp2/0 cell plasma membranes were solubilized by the alkaline urea method and subjected to tube gel isoelectric focusing in the first dimension. In A, the membrane polypeptides were separated on a 10% acrylamide SDS slab gel. In B, an equivalent sample was separated on a 5–10% linear acrylamide gradient SDS slab gel. Polypeptides were stained with Coomassie Blue.*

to low amounts of protein. An example of a gel stained by means of Coomassie Blue is shown in Figure 9A. However, an example where Coomassie Blue is not sufficiently sensitive occurs in attempts to detect trace components in a mixture which is dominated by just a few major proteins. In order to visualize minor components, large quantities of the mixture must be loaded, a situation which can lead to artifacts in both first and second dimension separation. In addition, large spots and streaks mask and distort extensive areas of the gel.

Recently, a series of staining methods, many being simple improvements on earlier procedures, have been developed in which silver is utilized for staining proteins separated on polyacrylamide gels [Switzer et al, 1979; Oakley et al, 1980; Merril et al, 1981a,b; Wray et al, 1981; Sammons et al, 1981; Morrissey et al, 1981]. These silver staining methods are based on histological staining procedures and depend upon the selective reduction of protein-bound ionic silver by photographic developers. The first histologically derived silver stains for polyacrylamide gels were shown to be highly sensitive, giving a 100-fold increase in sensitivity over the organic dye Coomassie Blue [Merril et al, 1979; Switzer et al, 1979]. However, these early procedures suffered serious drawbacks not encountered with Coomassie Blue. For example, the multiple-step methodologies were time-consuming and complex, the reagents were expensive, and the procedures gave variable results. Modifications [Oakley et al, 1980; Merril et al, 1981a,c] removed some of these objections, but nonspecific background and surface staining still often resulted and there was a loss of sensitivity with thick gels (> 1 mm). More recently, Wray et al [1981] modified and simplified the silver staining procedure still further, such that it was rapid, easily controllable, and gave a stable end point. It could be employed to detect protein at the nanogram level. In addition to monochromatic silver stains, Sammons et al [1981] recently introduced a color development system for distinguishing different classes of proteins on two-dimensional gels. This system is sensitive, reproducible, and inexpensive, although it has time-con-

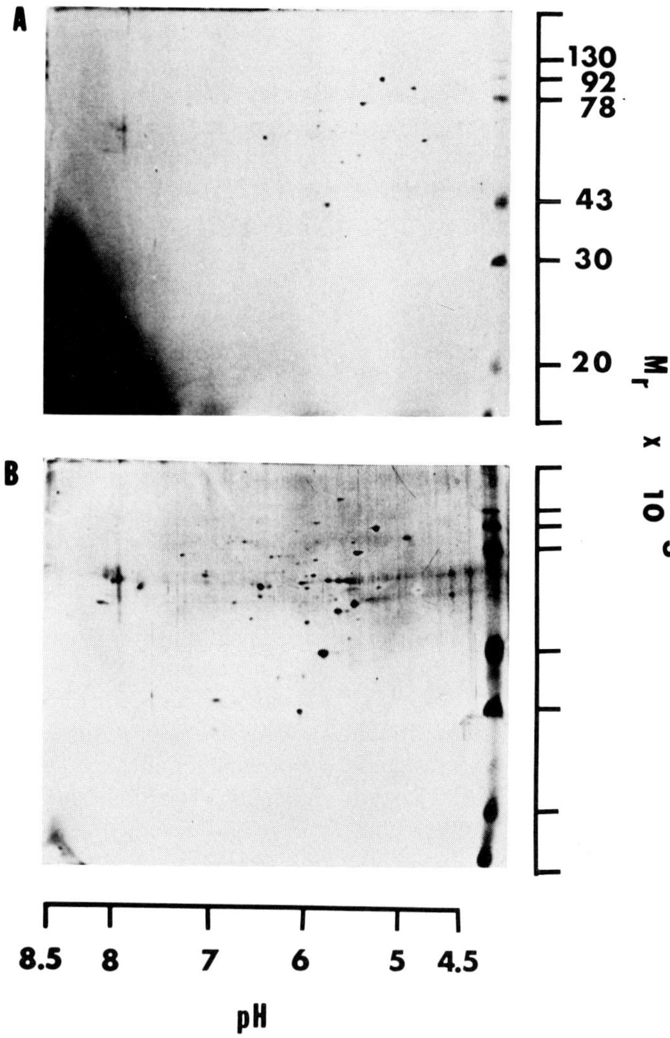

Fig. 9. *Silver staining of two-dimensional polyacrylamide gels reveals many more membrane polypeptides than does Coomassie Blue staining. Sp 2/0 plasma membrane (25 μg) was solubilized by the alkaline urea method and separated by isoelectric focusing followed by SDS polyacrylamide gel electrophoresis on a 10% acrylamide slab gel. In A, the polypeptides were stained with 0.125% Coomassie Brilliant Blue R-250 in 7% (v/v) acetic acid and 40% (v/v) ethanol, then destained in 7% (v/v) acetic acid and 10% (v/v) ethanol. In B, the polypeptides were stained with silver by the method of Wray et al [1981] and reduced with Kodak Rapid Fixer.*

suming washing steps and gives a deeply colored background.

Ochs et al [1981] compared six different silver staining methods, including all of those discussed above, with regard to sensitivity, quality of stain, expense, simplicity, and reproducibility. Their results suggested that the most desirable method was the color development system of Sammons et al [1981]. However, of the monochromatic stains it appeared that that of Wray et al [1981] was superior.

Most of the difficulties that have been encountered with the silver stain have as their origins the many chemicals used in two-dimensional electrophoresis. While some of these inhibit the development of the silver, others give undesirable background staining. Many of these contaminating chemicals are removed by soaking the gel in 50% (v/v) methanol [Wray et al, 1981]. The list of known interfering substances include acetic acid, glycerol, 8 M urea, glycine, Triton X-100, and agarose.

The mechanism of silver staining is not well understood, but the methods are broadly separable into two types: those carried out at acidic pH and those performed at alkaline pH. In acid environments it is thought that silver ions complex with carboxyl groups of proteins, as described by Freeman [1973] for a series of model compounds and by Peters [1955], who employed blocking reagents for specific groups on proteins prior to histological staining. Thus, in an acid environment, proteins with high levels of dicarboxylic amino acids should stain avidly. In basic environments there is evidence [Peters, 1955] that histidine and possibly arginine residues combine with silver. Wray et al [1981] have suggested that all protonated amino groups react with silver cations. Other amino acids that are known to react with silver under basic conditions include cysteine and methionine [Wray et al, 1981; Freeman, 1973; Peters, 1955].

In our laboratories, we make use of the Wray et al [1981] method for silver staining with several slight modifications. We find it highly sensitive, reproducible, economical and, very importantly, simple and rapid. The procedure allows the gel to be first stained with Coomassie Blue (Fig. 9A) and subsequently with silver (Fig. 9B).

The only disadvantage of silver staining methods is a quenching effect on autoradiography and fluorography in the detection of radioactively labeled proteins [Van Keuren et al, 1981]. There are several ways in which this difficulty can be overcome. One way is to run a companion gel in which one gel is silver stained and the second gel used for fluorography or autoradiography. Alternatively, a gel can be stained with silver, photographed, and destained with a photographic reducing agent prior to autoradiography or fluorography. This procedure does not eliminate quenching, but it is considerably reduced, provided that the method of Merril et al [1981c] is followed to stain and reduce the gel [Van Keuren et al, 1981].

Methods

Coomassie Blue. There are numerous ways to stain gels with Coomassie Blue. We use the following procedure, although there is no evidence for its superiority over the other methods:

Slab gels are fixed and SDS and ampholytes are removed by soaking in a solution (500 ml) of 7% (v/v) acetic acid and 40% (v/v) ethanol. This solution is changed at least twice during a 24-h period. Subsequently, gels are stained with 0.125% (w/v) Coomassie Brilliant Blue R-250 (solution 18) at room temperature in plastic containers for a minimum of 2 h on a rocking or reciprocating platform. For maximum staining, the incubation should be left overnight. The stain solution can be filtered and reused several times. Gels are destained by soaking in a solution of ethanol:acetic acid:water (10:7:83). Destaining is hastened if the destaining vessels are rocked or held at 40°C in a water bath. Several changes are required for complete destaining over 24–72 h. Used destain solution can be regenerated by passing it through a column of activated charcoal.

Silver stain. Due to the very high sensitivity of silver stain, the following precautions will ensure quality results. First, never touch the acrylamide gel with bare hands, or fingerprints will leave an indelible mark. Wear plastic gloves and wash off any talcum powder on them. Second, for molding the gel, use acid-washed glass plates that have no visible scratches. Last, never place a gel in a staining box or tray which has abrasions, signs of pitting, or raised and jagged surfaces.

Gels are soaked in 50% (v/v) methanol (reagent grade) after Coomassie Blue staining and destaining. For high-percentage acrylamide gels, thick gels (> 1.0 mm), or gradient gels an overnight soak is recommended with several changes of methanol-water.

The slab gel is placed in a glass tray which has been acid-washed and thoroughly rinsed in distilled water. The gel is then completely immersed in ammoniacal silver solution (solution 19) for 30 min with slow but constant agitation. This silver solution must be freshly prepared. Prepare no more than is immediately needed (silver nitrate is expensive).

Wash the gel in the glass tray with three changes of distilled water with gentle agitation for a total of 5 min and then soak the gel in developer (solution 20) with agitation until the bands or spots appear. This procedure normally takes 5 to 10 min and the gels are usually completely developed by 15 min. If the solution turns dark, immediately discard it, rinse the gel with more water, and add more developer. If the spots appear rapidly, the development time can be controlled by adding methanol (to a final concentration of no more than 10% by volume) to the developer. The more methanol added, the slower the development.

The reaction should be terminated when a brown-yellow color begins to appear around the outer edges of the gel, which is then quickly rinsed with distilled water. The reaction is stopped by adding 50% (v/v) methanol (500 ml) or a solution of 50% (v/v) methanol containing 10% (v/v) acetic acid. This solution should be changed at least one time after several hours. The gel, which has shrunk to about half its normal size, may be photographed at this stage. However, if a full facsimile is wanted or if the gel is distorted (gradient gels shrink unequally along the first dimension), it should be reswollen in either water or ethanol-acetic acid-water (10:7:83). The gel is best stored in food-storage bags in 50% (v/v) methanol and 10% (v/v) acetic acid, which maintains the stain. If the silver deposits are too heavy or the background is too dark, the gel may be destained with full-strength Kodak Rapid Fix (Fig. 10A–D). Note, however, that this reagent will bleach the gel completely if left in contact with it too long. Therefore, the reaction must be followed carefully and stopped by adding methanol (to about 10% by volume) when the desired level of destaining has been achieved. In Figure 10 gels are stained with silver (B–D) and the background diminished with Rapid Fixer for about 30 min (C) or 2 h (D). The fixative is removed by washing twice with approximately 500 ml of water. The gel is then irrigated with freshly prepared hypoclearing agent (25 g/L) and stored in 50% (v/v) methanol to stabilize the stain.

If the gel is to be impregnated with a scintillant fluor (see below), Rapid Fixer must be allowed to completely bleach the silver reaction. This is especially important for detecting tritium, as quenching will still be considerable [Van Keuren et al, 1981]. Autoradiographic detection of ^{14}C or ^{35}S will also be quenched, albeit to a lesser extent than for tritium. Even so, complete bleaching is highly recommended.

IDENTIFICATION OF GLYCOPROTEINS AND ANTIGENS IN GELS

Lectin Binding to Glycoproteins

Many of the polypeptide components of plasma membranes are glycoproteins, which are thought to fulfill a wide variety of important functions at the cell surface. Because glycoproteins can form highly specific associations with carbohydrate-binding proteins, usually called lectins, the latter molecules have assumed great importance as a means for probing the structure of membranes [Nicolson, 1974] and for purifying specific membrane glycoproteins [Lotan and Nicolson, 1979; Hedo, this series]. Since lectins can be iodinated [Burridge, 1976; Tanner and Anstee, 1976], coupled to horseradish peroxidase [Olden and Yamada, 1977] or conjugated with fluorescent dyes [Hoffman and McMahon, 1978], they can be used to detect glycoproteins on polyacrylamide gels. Their binding to glycoproteins does not depend upon the configuration of the polypeptide portion of the molecule, and so lectins bind as well to proteins that have been denatured in SDS as

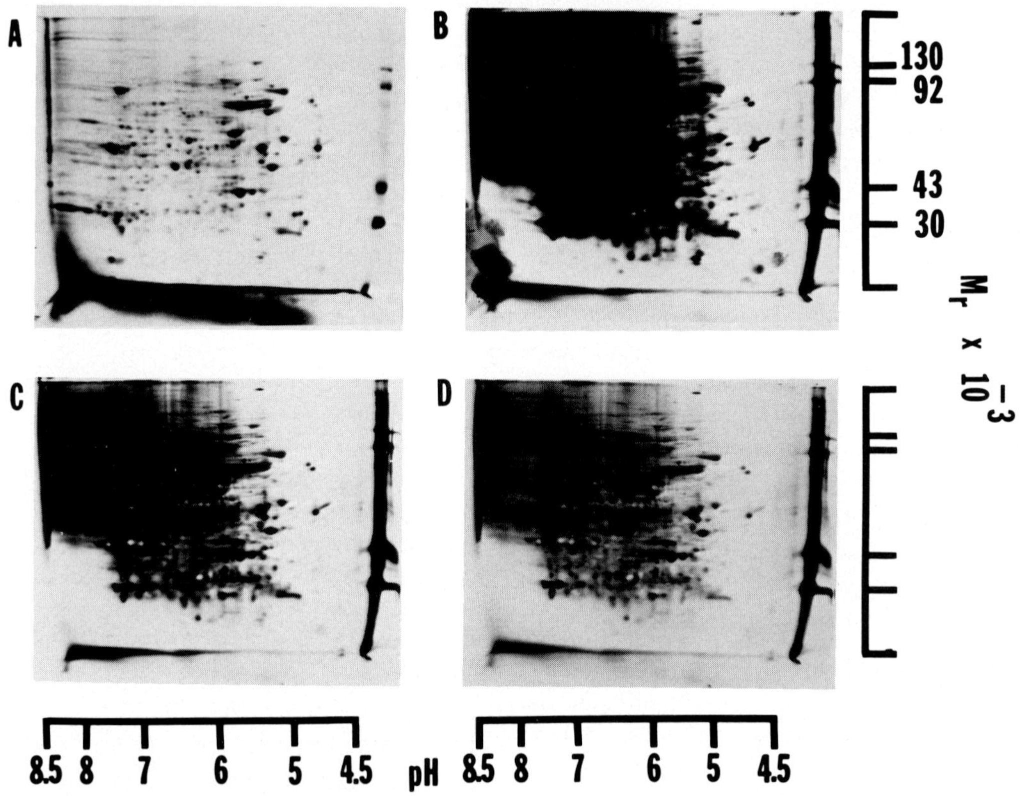

Fig. 10. *Photochemical reduction of silver stained CHO cell polypeptides. Cells were lysed in 2% (v/v) NP-40 and the 100,000g soluble fraction was separated by IEF followed by SDS-PAGE in an 8% acrylamide slab. The gel was stained with Coomassie Blue (A), soaked in 50% (v/v) methanol, and stained with silver according to the method of Wray et al [1981] for 15 min (B). The background has been reduced for 30 min (C) or 2 h (D) in Kodak Rapid Fixer.*

those in their native configuration. Therefore no major attempts have to be made to "renature" the proteins in the polyacrylamide gel as is usually required before attempting to bind antibodies.

The lectins that have been purified from different species of plant seed (particularly the legumes) have been shown to exhibit remarkable degrees of specificity for different types of oligosaccharides commonly found in membrane glocyproteins. Concanavalin A, for example, exhibits high affinity for only certain N-linked "high mannose" type oligosaccharide chains [Baenziger and Fiete, 1979], whereas wheat germ agglutinin, although showing binding to chains containing β-linked N-acetylglucosamine residues, shows highest affinity for "complex-type" oligosaccharide chains terminating in N-acetylneuraminic acid [Montsigny et al, 1980]. Because of this range of specificities, lectins not only have a value in identifying glycoproteins but can also be used to obtain information about the structures of their carbohydrate groups. For example, in CHO cells, there are large numbers of plasma membrane proteins that bind concanavalin A [Horst et al, 1979] but only one major group that bind wheat germ agglutinin [Horst et al, 1980b]. The latter are a group of acidic, high molecular weight molecules that are the major components of the cell surface that are iodinatable by means of lactoperoxidase catalyzed ^{125}I-iodination [Horst et al, 1980b; Fitzgerald et al, 1981].

These wheat germ agglutinin-positive proteins also bind a number of other lectins with specificities more or less directed towards complex oligosaccharides of the N-linked type [Horst et al, 1980b]. From these and a variety of other experiments we have concluded that these high molecular weight glycoproteins are highly glycosylated and bear a variety of different oligosaccharide chains which are mainly of the N-linked, complex type.

Lectin Staining of Glycoproteins in Acrylamide Gels

Iodination of lectins. Lectins with a wide variety of specificities can be purchased commercially from many suppliers such as Vector Laboratories (Novato, CA), Miles-Yeda (Elkhart, IN), and Sigma Chemicals (St Louis, MO). Affinity-purified lectins assure quality results. We use two methods for iodination: The first employs lactoperoxidase as a catalyst. Lectins (10 mg) are dissolved in 2 ml of 0.2 M sodium acetate buffer (pH 5.6) containing 1 mg lactoperoxidase (Sigma). After addition of Na[^{125}I] (0.5 mCi carrier-free), the reaction is initiated by adding 0.06% (v/v) H_2O_2 (10 μl). Subsequent additions are made every minute for 15 min, and the reaction is finally terminated by addition of ~1 mg of sodium azide. Lectins can also be iodinated using the immobilized reagent "Iodogen" (1, 3, 4, 6-tetrachloro-3α, 6α-diphenylglucoluril), as described by Fraker and Speck [1978].

Following labeling, the lectins are separated from salt and free iodine by gel filtration on a Biogel P-2 column (1 × 10 cm) equilibrated with Tris-buffered saline (TBS; 0.05 M Tris HCl, 0.14 M NaCl, 0.02% (w/v) NaN_3, pH 7.5). The peak of radioactivity that elutes at the void volume of the column contains most of the ^{125}I-labeled lectin and is pooled and stored frozen at −20°C in convenient aliquots of ~ 10^7 cpm. It is advisable to place the tubes in containers shielded with lead to reduce hazardous radiation. The specific activity of the ^{125}I-lectins prepared as above is somewhat variable, but is usually around 10^8 cpm/mg.

Certain of the lectins can be purified further by affinity chromatography to ensure that the labeled material used has retained its carbohydrate-binding properties [see Horst et al, 1980a,b]. For example, concanavalin A (as well as *Lens culinaris* and pea lectins) can be applied to a short column (1 × 5 cm) of Sephadex G-100, washed with TBS, and eluted with 0.5 M D-glucose. Ricin I binds to agarose beads (Sepharose-6B) and can be eluted with 0.5 M D-galactose. Finally, wheat germ agglutinin can be purified on N-acetyl D-glucosamine-Sepharose (supplied as "Affigel" by Pierce Chemical Co) and eluted with N-acetyl D-glucosamine [Shaper et al, 1973]. The excess haptens are removed by dialysis.

Lectin binding procedure. Coomassie Blue stained slab gels are equilibrated with TBS. Use five changes of 250 ml in flat plastic pans over a period of 1 to 2 days. The inclusion of azide is essential to all stages to reduce bacterial contamination during the relatively long procedure.

The gels are then transferred to plastic, heat-seal bags (eg, Kapak; 3 M Co, Minneapolis, MN). Labeled lectin (~2 × 10^7 cpm; 0.2 mg) is diluted with 20 ml of filtered TBS containing 0.1 mg/ml hemoglobin and, when appropriate, saccharide inhibitor. When using concanavalin A, include 0.5 mM Mn^{2+} and Ca^{2+} (as their chloride salts) in the buffer as this lectin requires both these divalent cations for carbohydrate binding. The radioactive lectin solution is then added to the plastic bag with the gel and heat-sealed. The bag is placed for 12 to 24 h at room temperature on a tilting or rotating platform to gently bathe the whole gel in overlay solution. A corner of the bag can then be slit open, the overlay solution drained out, and replaced with 20 ml TBS.

The gel is rinsed several times in this manner. All washings and the original overlay solution should be contained in a radioactive waste vessel and disposed of appropriately. The gel may then be removed from the bag and washed in ~8 changes (250 ml each) of

TBS over a period of 3 to 4 days until radioactivity in the buffer has reached background. The gel is then dried or placed in a fresh plastic bag and subjected to autoradiography.

In Figure 11A, the mannose-containing glycoproteins of the Sp2/0 cell plasma membrane have been detected using ^{125}I-labeled concanavalin A. Both negative- and positive-binding reactions must be performed as controls. First, a companion gel was incubated with the lectin in the presence of α-methylmannoside (0.1 M) and washed in the presence of 0.5 M glucose. No binding to membrane proteins was evident (data not shown). Second, the molecular weight standards serve as internal controls. Note that of the standards used, only ovalbumin (Mr = 43,000) contains high mannose oligosaccharides and, therefore, is bound by ^{125}I-labeled concanavalin A (Figure 14B). Transferrin (Mr = 78,000) has complex type chains and is bound by concanavalin A only slightly. Carbonic anhydrase (Mr = 30,000) and soybean trypsin inhibitor (Mr = 20,000) are not glycoproteins and are not stained in Figure 14B. When employing other lectins, the appropriate saccharide inhibitor must be chosen. The competing hapten for ricin I is 0.5 M D-galactose during exposure to the lectin and 0.1 M during washing. Because of its expense, N-acetyl D-glucosamine is not employed as a hapten for wheat germ agglutinin. Instead we employ a low concentration (1 mg/ml) of glycopeptides from pronase-digested fetuin during staining (for details of this procedure see Horst et al, [1980a]).

Technical difficulties observed with ^{125}I-lectin staining. Lectin staining may present difficulties ranging from no binding (a blank autoradiogram) to a completely blackened autoradiogram. The former condition usually results from using an unreactive lectin; the latter is usually due to incomplete washing. Between these two extremes, there are a variety of intermediate conditions. One which is commonly observed is a low labeling of almost all of the proteins on the gel. It is advisable here to note whether actin or other major cytoskeletal proteins have bound the lectin. As these are not glycoproteins, their association with the lectin is probably due to their ability to bind denatured lectin. It is particularly important to avoid denaturation of the lectins during iodination and subsequent handling. Burridge [1976] recommends keeping the lectin in the presence of a dilute solution of hemoglobin at all times. The hemoglobin must be completely dissolved, then deaggregated by filtration or centrifugation; otherwise the autoradiogram will be replete with nonspecific "blotches." For certain lectins, such as concanavalin A, appropriate concentrations of divalent cations (Ca^{2+} and Mn^{2+}) should be maintained in the buffer (see Lis and Sharon [1973] for metal requirements of lectins). We also recommend that only short autoradiographic exposures be employed (4 days or less). Longer exposures may reveal additional positive spots, but in our view these often represent nonspecific binding.

Identification of Antigens in Gels

This procedure has been described in detail by Burridge [1976] and will not be discussed at length here. It depends upon the ability of antibodies to bind to their corresponding antigen in polyacrylamide gels. Because the proteins have been dissociated first in urea and mercaptoethanol and subsequently in SDS, and since the gels themselves have been fixed, it has to be assumed that some renaturation occurs during the washing procedure prior to provision of the antibody solution. The primary antibody, bound to its antigen, is then recognized with a second radioactive macromolecule (either a second antibody or staphylococcal protein A which have been labeled with ^{125}I). The overall procedure is extremely time-consuming and has a number of other associated difficulties. For these reasons we recommend using the electroblotting procedure. Nevertheless the method has been successful in recognizing fibronectin and a large number of other proteins [Burridge, 1976; Olden and Yamada, 1977; Adair et al, 1978; Tjian et al, 1974] in one-dimensional polyacrylamide gels.

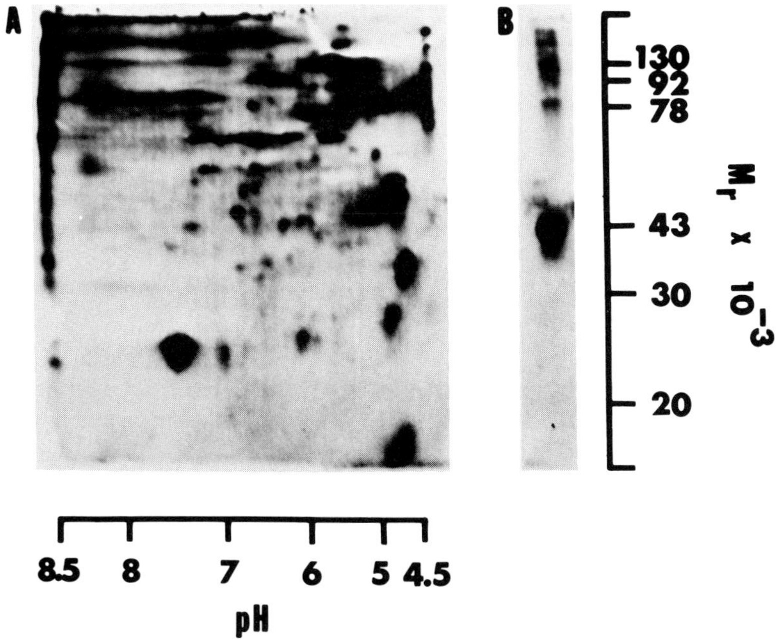

Fig. 11. *Autoradiogram of glycoproteins of Sp2/0 cell plasma membranes as detected by* ^{125}*I-labeled concanavalin A. Plasma membrane from fluid pump-disrupted Sp2/0 cells were isolated by differential and isopycnic sucrose gradient centrifugation, solubilized by the alkaline urea method, and separated in two dimensions using isoelectric focusing followed by SDS polyacrylamide gel electrophoresis (SDS-PAGE) on a 10% acrylamide slab gel. After Coomassie Blue staining, the gel slab was incubated with* ~ 10^7 *cpm of affinity-purified,* ^{125}*I-labeled concanavalin A for 24 h. The gel was extensively washed, then dried, and an autoradiogram prepared. In A, the mannose-containing polypeptides are revealed. As with CHO cells [Horst et al, 1979], there appear to be many concanavalin A-binding polypeptides on the surface of the Sp2/0 cell. Shown in B are the molecular weight marker proteins also subjected to* ^{125}*I-labeled concanavalin A binding. Note that the high mannose glycoprotein, ovalbumin (*Mr = *43,000) is bound strongly by concanavalin A.*

Electroblotting

Introduction. In the previous section we described how certain classes of polypeptide components of membranes could be detected on two-dimensional polyacrylamide gels by their ability to bind plant lectins. However, there are undoubtedly difficulties associated with the above techniques. These include the following: (1) Lengthy incubation and washing times are required to remove unbound ligands necessary to reduce the background for autoradiography; (2) ligands may fail to penetrate the gel, particularly if high-percentage acrylamide gels are used; (3) certain proteins may be eluted from the gel matrix during washing or renaturation, particularly if the buffer employed (eg, TBS) allows the gel to swell; and (4) gels are frequently torn or cracked during handling and drying. In addition, some refolding probably has to occur in order to present sufficient antigenic sites to antibody to allow binding. Such refolding (or renaturation) may occur slowly or not all within the acrylamide matrix, even if pains are taken to remove the bound SDS.

Particularly since the development of methods for raising large numbers of monoclonal antibodies against cell surface molecules, there has been a need for a simpler, more rapid, and more reliable procedure for matching these antibodies with their particular membrane antigens. Although immunoprecipitation, followed by electrophoretic identification of the polypeptide, is often the preferred technique, this approach is not always successful with minor polypeptide components (eg, those contributing less than 0.1% of the cell protein), since the background of contaminating species trapped in the immune complex can be considerable [Ivarie and Jones, 1979].

Recently, methods have been developed to transfer polypeptides from electrophoretic gels to either derivatized paper or nitrocellulose. Such supports are easier to handle than acrylamide gels; they are quickly equilibrated with binding ligand, washed, and dried. The polypeptides, once transferred, are tightly bound yet can be immediately exposed to antibody or lectin. Finally, electrophoretic elution may allow considerable renaturation to occur before the protein is trapped. Renart et al [1979] transferred polypeptides by diffusion from one-dimensional gels of simian virus 40 (SV40)-transformed cells to diazobenzyloxymethyl (DBM) paper and identified specific virion proteins using appropriate antisera. However, this method also involves (1) preparation of polyacrylamide gels using a periodate-sensitive cross-linker, DATD; (2) using agarose as a supporting matrix during capillary blotting of protein to DBM paper; and (3) exacting and time-consuming preparation of the derivitized paper and its conversion to DBM form immediately prior to use. Derivatized papers can be purchased commercially (Bio-Rad), but not in the final DBM form. This protein transfer method did have an apparent advantage over the nitrocellulose procedure (see below), namely, that the paper blot could be reused several times since the proteins are covalently attached. Bound ligands can be removed by treatment with 10 M urea, 0.1 M β-mercaptoethanol [Renart et al, 1979].

A more convenient method of protein transfer from polyacrylamide gels has been introduced by Towbin et al [1979]. This procedure involves the electrophoretic transfer of proteins to nitrocellulose and is useful for either urea or SDS gels. For ribosomal proteins the efficiency of transfer was near 100%. In addition, the nitrocellulose blots can be readily stained using amido black. In this method, standard *bis*-acrylamide cross-linked gels can be used, and the nitrocellulose is used directly as supplied commercially. In addition, the nitrocellulose paper is less expensive than the derivatized paper. Burnette [1981] has modified the Towbin method slightly (so-called Western blotting) and has provided a more complete description of the transfer process. More recently, refinements of the "Western" or electroblotting method have appeared (see below), and specifically designed transfer cells have become commercially available (Bio-Rad).

Method for electrophoretic blotting of polypeptides onto nitrocellulose. We have used the electroblotting method to determine the antigenic specificity and cross-reactivity of a number of monoclonal antibodies produced against CHO cell surface proteins. For determining antibody specificity either a whole cell lysate or a more purified plasma membrane preparation can be employed. Additionally, this method can be used to identify glycoproteins in membranes by binding iodinated lectins to polypeptides transferred from gels to nitrocellulose.

Standard two-dimensional polyacrylamide gels (using 8% rather than 10% polyacrylamide in the second dimension) are then prepared. The lower concentration of acrylamide allows larger polypeptides to be transferred more readily. With 10% acrylamide gels we have noted that proteins above Mr = 90,000 often remain in the slab gel.

Following electrophoresis, the gel plates are removed and the exact gel dimensions are measured (ie, distance of the bromophenol blue tracking dye to the top of the gel). A corresponding piece of nitrocellulose (Schleicher and Schuell, BA85) is cut. Transfer is performed with a Bio-Rad Trans-Blot™ cell or its equivalent using 0.025 M Tris·HCl, 0.193 mM glycine, pH 8.3, and 20% (v/v) methanol (degassed before use). This buffer is readily prepared without methanol as a 5×

concentrate (0.125 M Tris·HCl; 0.96 M glycine).

Place in order between two Scotch-Brite pads (supplied by Bio-Rad or purchased locally) (1) a sheet of wetted Whatman 3 MM filter paper, 6 × 8 in; (2) the polyacrylamide gel; (3) the nitrocellulose sheet prewetted in transfer buffer for 1 to 2 min before use; and (4) another sheet of 3 MM paper. Care is taken to squeeze out all air bubbles between any elements of the above set-up using wetted gloves. The gel-nitrocellulose "sandwich" with the nitrocellulose on the anodic side is placed between the plastic holders of the Trans-Blot apparatus and slowly lowered into the transfer cell filled with degassed buffer. Electrophoretic transfer is performed overnight at 25–30 V (~200 mA) with current supplied from a standard electrophoresis power supply.

The nitrocellulose blot is removed and either stained with amido black or left unstained for antibody binding. Blots are stained for 2–3 min in 1% (w/v) amido black in 45% (v/v) methanol, 10% (v/v) acetic acid, and destained in 45% (v/v) methanol and 10% (v/v) acetic acid for 4–5 changes of approximately 15–30-min duration. The anodal side (ie, the one furthest away from the gel) of the blot is stained much more faintly than the cathodal side (ie, the one facing the gel). Some major spots, such as actin, will appear on the anodal side, which may indicate a local overloading of the binding capacity of the nitrocellulose. This should be no cause for concern, since membrane proteins are present in much smaller amounts than cytoskeletal elements.

Unstained blots are placed briefly in water to remove electrophoresis buffer and the remaining binding sites of the nitrocellulose are "blocked" by incubation using 10 mg/ml bovine serum albumin (BSA, Sigma fraction V) in phosphate buffered saline (PBS) plus 0.1% (w/v) sodium azide for 1–2 h at 37°C on a rocking platform. The nitrocellulose blots are always handled using plastic gloves, or fingerprints will result. Care is taken not to introduce cracks or tears into the blots.

Incubation with primary antibody (1:100 dilution of whole ascites fluid or monoclonal antibody) in PBS containing 1% (w/v) BSA, and 0.1% (w/v) sodium azide is performed in plastic heat-seal bags in a volume of 20 ml. The blot is incubated overnight at 4°C on a rocking platform. The transfer is removed and washed for five changes of 5 min each in 100–200 ml volumes of PBS with 0.1% (w/v) BSA at room temperature in plastic trays. The blot is placed in sealable plastic bags and incubated for 6 h in presence of 20×10^6 cpm of the affinity-purified IgG fraction of ^{125}I-labeled antibody specific for mouse IgG (made in sheep) in 20 ml of PBS and 0.1% (w/v) BSA at 4°C. The paper is again washed as described above and dried between paper towels for several hours. Finally, the blot is taped to paper and autoradiography performed.

It is often more convenient to screen new antisera in nitrocellulose blots from one-dimensional gels, since many sera can then be processed together. A 7.5% acrylamide Laemmli [1970] slab gel is cast without a comb (see Table II). Approximately 3 mg of protein can be layered directly onto this gel and subjected to electrophoresis. Transfer is carried out as above, but the nitrocellulose is then cut into a series of ~1 cm strips parallel to the direction of electrophoresis such that each strip will contain an identical complement of cell proteins. Incubations with primary antibody can be performed in plastic tubes using smaller volumes of solution. Washing and binding of the second antibody can be performed similarly. Those antibodies which give a positive binding reaction can then be examined further using two-dimensional gels. Those antibodies that are negative (ie, do not bind to the one-dimensional blots) can be laid aside and their specificity examined by some other technique (eg, immunoprecipitation).

Results. A comparison of Coomassie Blue stained CHO cell polypeptides in a two-dimensional polyacrylamide gel (Fig. 12A) and after electrophoretic transfer to nitrocellulose are shown (Fig. 12B). Most polypeptides appear to transfer with little or no loss in reso-

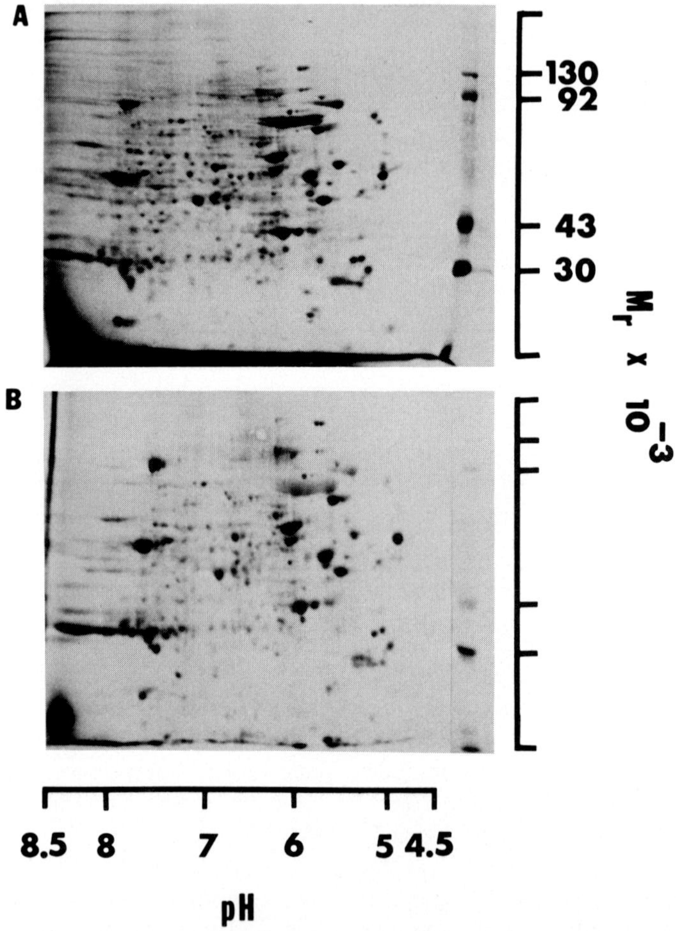

Fig. 12. *Transfer of polypeptides from a two-dimensional polyacrylamide gel to nitrocellulose. CHO cells were lysed in 0.5% (v/v) NP-40, and the 100,000g soluble fraction was separated in two-dimensions by IEF followed by SDS-PAGE in an 8% acrylamide gel. The Coomassie Blue stained polypeptides in the acrylamide gel are shown in A. After electrophoretic transfer from a companion gel to nitrocellulose, the amido black stained polypeptides are shown in B. Transfer appears quantitative even for high molecular weight polypeptides.*

lution. Differences in staining intensity of certain polypeptides are probably due to the different stains used on each (Coomassie Blue versus amido black). Autoradiographs of CHO cell surface polypeptides transferred to nitrocellulose, incubated with a monoclonal antibody, followed by a radiolabeled second antibody are shown in Figure 13. Antigenicity has been retained following solubilization in urea and reducing agents and electrophoresis in presence of SDS. The identical antigen has also been identified by means of immunoprecipitation protocols.

Applications and recent improvements in the electroblot method. Electrophoretic transfer of proteins from polyacrylamide gels to nitrocellulose is a simple and convenient method for the immunological detection of certain antigens including membrane receptors. For example, Anderson et al [1982] used standard electrophoretic blotting of bovine adrenal cortex proteins to nitrocellulose to iden-

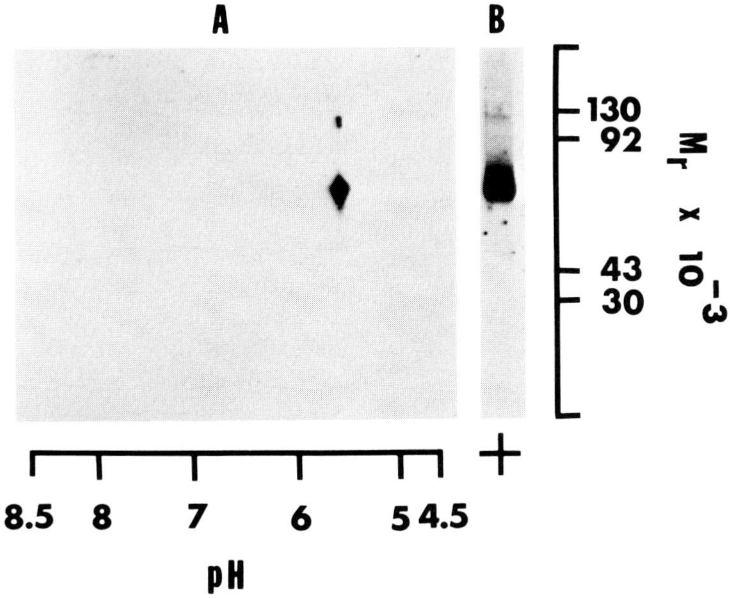

Fig. 13. *Immunological detection of cell surface antigens transferred from polyacrylamide gels to nitrocellulose. Monoclonal antibodies were produced by hybridization of spleen cells from mice injected with CHO plasma membrane glycoproteins which bind to wheat germ agglutinin [Fitzgerald et al, 1981]. A single hybridoma selected for cell surface binding using glutaraldehyde fixed CHO cells was incubated with nitrocellulose electroblots of CHO polypeptides separated by (A) two-dimensional polyacrylamide gel electrophoresis or by (B) one-dimensional SDS-PAGE. Bound antibody was visualized by a subsequent incubation with ^{125}I-labeled antibody to mouse IgG followed by autoradiography. A single polypeptide antigen of* Mr = *70,000 with an apparent pI of 5.5 is revealed.*

tify low density lipoprotein (LDL) receptors using both monoclonal and polyclonal antibodies. For detection of primary antibody either ^{125}I-labeled antiimmunoglobin or ^{125}I-labeled protein A have been used. In addition, Towbin et al [1979] have described the use of peroxidase and fluorescein conjugated antibodies to detect antibody-antigen complexes on nitrocellulose transfers.

Modifications of the basic electroblotting method are continually being reported. Erickson et al [1982] found that the presence of 0.1% (w/v) SDS in the transfer buffer of Towbin et al [1979] improved the efficiency of transfer, especially of higher molecular weight proteins. In addition, it has recently been demonstrated that nitrocellulose blots can be reused following incubation in 8 M urea and 0.1 M 2-mercaptoethanol for 1 h at 60°C, a procedure which removes antibody-protein A complexes. Erickson et al [1982] also performed transfer at 3 V/cm for 21 h, which is a considerably lower voltage than previously reported. Legocki and Verma [1981] reported that antibody-protein A complex could be removed from nitrocellulose blots using a buffer containing 0.1 M glycine, pH 2.2, for 1.5 h at room temperature. Howe and Hershey [1981] performed antibody incubations of nitrocellulose blots with a buffer containing 0.5% (v/v) Triton X-100 and 0.2% (w/v) SDS to reduce nonspecific absorption of antibody or protein A. Symington et al [1981] reported an increase in autoradiographic sensitivity compared to heat-dried polyacrylamide gels when [^{35}S]-methionine-labeled proteins were transferred to DBM paper. Both Bio-Rad and Schleicher and Schuell provide current references and a variety of electroblotting methodologies.

IDENTIFICATION OF SURFACE-EXPOSED PROTEINS ON TWO-DIMENSIONAL POLYACRYLAMIDE GELS

Introduction

It has long been realized that even the most carefully isolated preparation of plasma membrane is probably contaminated with proteins that are not truly a part of the membrane. In addition, there will be other components that may be associated only superficially with the membrane inner face. As well, there are proteins which do not penetrate through to the outside but are a part of the bilayer. This volume is concerned mainly with membrane receptors which, by our definition, are molecules which are exposed at the exterior of the cell to their binding ligands. In this section we shall deal briefly with a few methods that are useful in defining such molecules that are accessible to chemical modification by certain reagents which themselves penetrate the membrane only poorly. A large number of such procedures have been developed, particularly for use with red blood cells. However we shall discuss only four: lactoperoxidase catalyzed iodination, galactose oxidase-$NaB[^3H]_4$ labeling, periodate-$NaB[^3H]_4$ labeling, and the introduction of immunoreactive groups by trinitrobenzenesulfonate and related compounds.

Our purpose is less to provide detailed methodologies and more to suggest approaches that might be taken in conjunction with two-dimensional electrophoresis to define those molecules that are chemically reactive on the surface of a cell.

Lactoperoxidase-Catalyzed Iodination

This method employs the enzyme lactoperoxidase to catalyze the iodination of cell surface polypeptides. Iodine is incorporated mainly into tyrosine and, to some extent, histidine residues [Morrison, 1974].

The method we employ is based on that originally described by Phillips and Morrison [1971]. Confluent cells in monolayer in 75 cm^2 flasks (15×10^6 CHO cells/flask) are covered with 5 ml of Hank's balanced salt solution (HBSS), which contains 0.5% (w/v) glucose, 10^{-8} M KI, 0.2 mg/ml lactoperoxidase (Sigma), and 0.125 mCi/ml carrier-free $Na[^{125}I]$ (Amersham). Hydrogen peroxide (12 μl of a freshly prepared 0.06% [v/v] solution) is added every minute for 20 min. The vessels are kept on ice throughout this procedure. The reaction can also be carried out on suspended Sp2/0 cells. Cold temperatures are recommended in order to minimize the internalization of the labeling reagents into vacuoles. The solution is decanted, and the cells are washed twice with 10 ml HBSS containing 10^{-3} M KI, and three times with 10 ml HBSS. Under these conditions, 1.5×10^7 cells incorporate 1.5×10^6 cpm of acid-insoluble radioactivity.

Morrison [1974] has studied the mechanism of the reaction and has proposed the following sequence of events: The enzyme first reacts with hydrogen peroxide, and is only then capable of reacting with the next substrate, which can be either I^- or I_2. The enzyme next forms an enzyme-substrate complex with an accessible tyrosine residue. This scheme is supported by the fact that the reaction using L-tyrosine as substrate follows ordered ping-pong (multi-step) kinetics [Morrison and Bayse, 1970] as well as by other evidence [Bayse and Morrison, 1971]. The labeling in this case is restricted to the cell surface because the enzyme (Mr = 77,500) is too large to penetrate the plasma membrane.

However, in the absence of iodinatable substrate, the enzyme will oxidize I^- to I_2, and free I_2 reacts readily with tyrosine at pH 7.4. The reaction of I_2 with tyrosine is entirely distinct from the enzyme-mediated reaction, and because I_2 is membrane-permeable, there is a possibility that internal proteins can become labeled as well as those of the plasma membrane. The most effective means of minimizing this problem is to use a very low concentration of iodide. In the above protocol, the total concentration of iodide is 10^{-8} M. If the enzyme does oxidize I^- to I_2, the concentration of I_2 will be low enough such that its rate of penetration across the plasma membrane will be slow. Some oxidation of I^- to I_2 probably does occur, since iodination of cell surface lipids has been reported [Mersel et al, 1976; Schlager, 1979] using conditions similar to those described above, and it seems more likely that this labeling results from the sequestration of I_2 in the lipid bilayer than from the formation of an enzyme-substrate complex between the lipids and lactoperoxidase. Under our conditions of labeling, any I_2 sequestered in the bilayer does not enter the cytoplasm, since actin and other intracellular proteins do not become labeled. Therefore, reaction is restricted to the cell surface.

Two types of control should be run early in any investigation: The first will involve a complete reaction mixture but without addition of hydrogen peroxide, the second would exclude both lactoperoxidase and peroxide [Morrison, 1974]. ^{125}I incorporation into protein should in each case be low.

Hubbard and Cohn [1972] have described a modification of the above method in which the enzyme glucose oxidase is used to provide a continuous source of hydrogen peroxide throughout the reaction. This procedure does not appear to provide any major advantage over the one described by Phillips and Morrison [1971], although it is more expensive and includes an additional soluble protein that can become iodinated. Indeed, the presence of soluble or adsorbed proteins not derived from the cells under study can give very misleading information, for not only will these components act as competing substrates in the reaction, they will also appear on the two-dimensional gels used for analysis. Clearly, cells whose membranes are leaky can also provide such soluble protein substrates and allow the labeling reagent to reach cytoplasmic components which are normally protected. Gels should always be checked, therefore, to determine whether any serum proteins, such as albumin, α-fetoprotein, and transferrin, or cytoskeletal proteins, such as actin and tubulin, are labeled.

One final point stressed by Morrison [1974] is that a considerable number of low molecular weight compounds, including several amino acids and antimicrobial compounds, such as azide, very effectively inhibit the lactoperoxidase reaction. Care should be taken to exclude such compounds from the buffers used in the procedure.

Autoradiograms of ^{125}I-labeled Sp2/0 and CHO cell surface polypeptides separated by two-dimensional electrophoresis are shown in Figure 14 (A and B, respectively). Sp2/0 cells contained a more diverse group of iodinatable polypeptides, and the pattern was distinct from that seen with CHO cells (Fig. 14A). When CHO cells are labeled by the lactoperoxidase procedure, the major components that become labeled are a group of high molecular weight acidic glycoproteins (Fig. 14B). These are identical to the plasma membrane components that bind wheat germ agglutinin [Horst et al, 1980b].

Labeling With $NaB[^3H]_4$ After Chemical Modification of Carbohydrate Groups

Cell surface carbohydrate can be oxidized using either galactose oxidase or sodium metaperiodate and then reduced with $NaB[^3H]_4$. Galactose oxidase converts D-galactosyl and N-acetyl-D-galactosaminyl residues at nonreducing terminals of glycoproteins and glycolipids to carbon-6 aldehydes [Gahmberg, 1978]. The aldehydes are then reduced back to galactose or N-acetyl-galactosamine with $NaB[^3H]_4$. Removal of sialic acid with neuraminidase exposes more galactose residues and therefore enhances the incorporation of label. Gahmberg [1978] provides a detailed methodology and it will not be discussed further here. One disadvantage of this procedure is that in order to obtain efficient labeling the glycoproteins are desialylated, which modifies their charge, and hence position on two-dimensional polyacrylamide gels (Fig. 15A).

Cell surface carbohydrate may also be oxidized using sodium metaperiodate [Gahmberg and Andersson, 1977]. Periodate oxidatively cleaves sialic acids between carbons 7 and 8

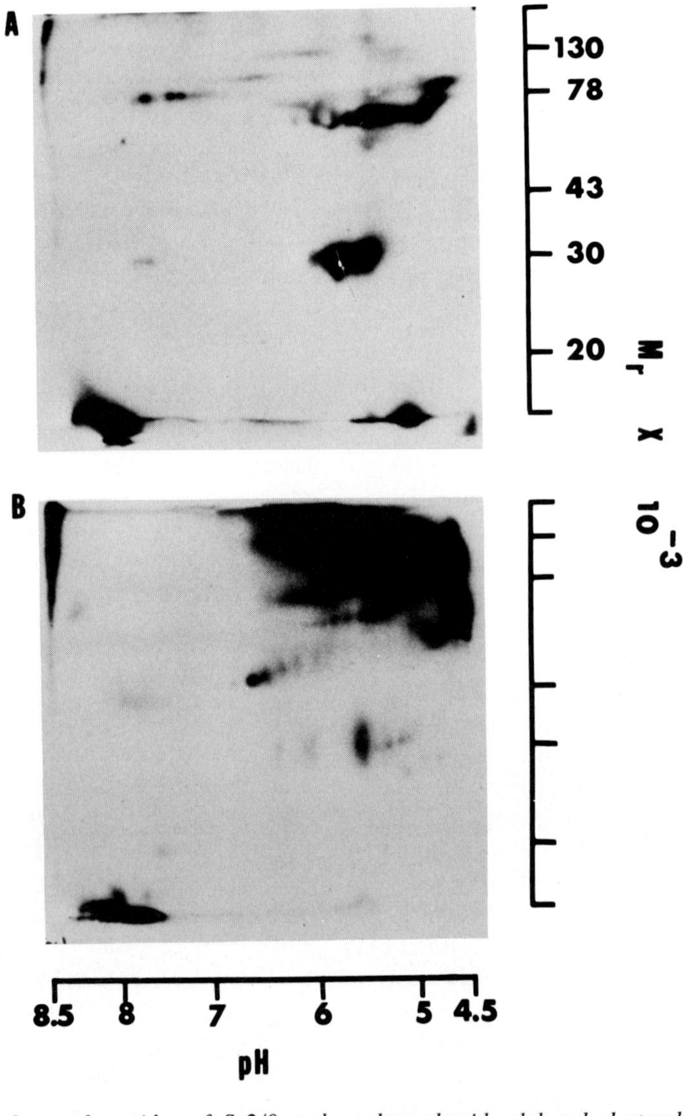

Fig. 14. *Cell surface polypeptides of Sp2/0 and Chinese hamster ovary (CHO) cells labeled with Na^{125}I by lactoperoxidase. Sp2/0 and CHO cell surface proteins were vectorially labeled with Na[^{125}I] using lactoperoxidase and 0.06% (v/v) H_2O_2. Both cells were then disrupted in a fluid pump and plasma membrane isolated by differential and isopycnic sucrose gradient centrifugation. Membrane polypeptides were solubilized with alkaline urea and separated by isoelectric focusing followed by SDS polyacrylamide slab gel electrophoresis on a 10% acrylamide slab gel. Gels were stained with Coomassie Blue, dried, and autoradiograms prepared. A) The surface exposed polypeptides of the Sp2/0 cell labeled with ^{125}I. B) The iodinated surface proteins of the CHO cell. These polypeptides are the major wheat germ agglutinin receptors of the CHO cell [Horst et al, 1980b]. They are characterized further in Figure 15.*

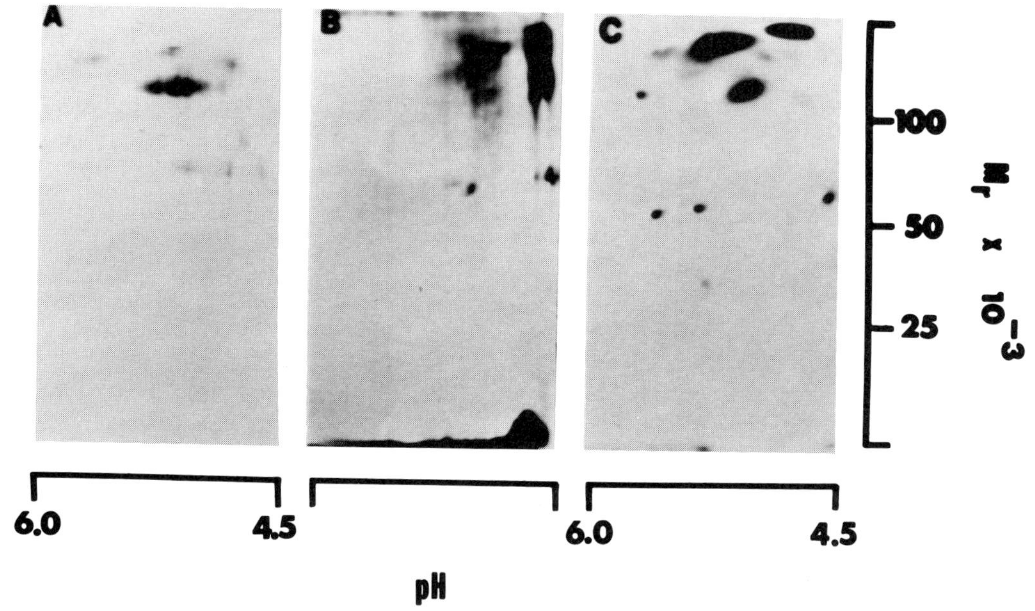

Fig. 15. *Cell surface labeling of plasma membrane glycoproteins by galactose oxidase, periodate, and metabolic labeling with radiolabeled fucose. Plasma membrane was isolated by differential and isopycnic centrifugation of fluid pump-disrupted CHO cells. Polypeptides were separated by two-dimensional electrophoresis and only the acidic portion is shown. A) Galactose and galactosamine residues were labeled with NaB[^{3}H]$_4$ after treatment of cells with neuraminidase and galactose oxidase prior to disruption. B) The result of labeling sialic acid-containing polypeptides by periodate treatment and NaB[^{3}H]$_4$ on whole CHO cells prior to disruption is shown. Note that the high molecular weight acidic glycoproteins are preferentially labeled; these are the major receptors for wheat germ agglutinin on the CHO cell surface. C) The high molecular weight acidic glycoproteins from CHO cells can also be labeled metabolically with L-[^{3}H]-fucose (100 μCi/ml) for 24 h in culture prior to membrane isolation. This confirms the presence of oligosaccharide on these proteins.*

or between carbons 8 and 9, and the resulting aldehydes are then reduced using NaB[^{3}H]$_4$. The labeling is restricted to the cell surface because low concentrations (1 mM) of periodate are used and the periodate anion, IO_4^-, penetrates the plasma membrane very slowly at 0°C. A detailed procedure is provided by Gahmberg and Andersson [1977]. Two-dimensional electrophoresis of CHO cell polypeptides after labeling with periodate-NaB[^{3}H]$_4$ is shown in Figure 15B. Note that labeling is confined to a group of high molecular weight, acidic components, possibly identical to those labeled after binding ^{125}I-labeled wheat germ agglutinin [Horst et al, 1980b] or after lactoperoxidase iodination (Fig. 14B). These same molecules incorporate ^{3}H after growing the cells in presence of L-[^{3}H]-fucose (Fig. 15C), confirming the glycosylation of these polypeptides.

Immunoreactive Reagents

Trinitrobenzenesulfonic acid (TNBS) reacts readily with protein amino groups above neutral pH. It is specific for cell surface proteins because the charged sulfonate group prevents penetration of the compound across the plasma membrane. If the cells are first metabolically labeled with [^{35}S]-methionine or [^{3}H]-glucosamine, the reacted proteins can be isolated using antidinitrophenyl antibody (anti-DNP) and visualized after electrophoresis and fluorography. This technique has been used successfully to identify the externally disposed proteins of mouse macrophages [Kaplan et al, 1981].

In the protocol described by Kaplan et al [1981], macrophages in monolayer are labeled for 3 h at 37°C with methionine-free medium containing 30–50 μCi/ml of [^{35}S]-methionine (600–1,300 Ci/mmol). The radiolabeled cells are then labeled on ice with 10 mM TNBS at pH 7.8 for 30 min, followed by washing and incubation with 0.5 mg/ml rabbit anti-DNP IgG on ice for 20 min. Anti-DNP antiserum is available commercially. We have applied exactly the same procedure to CHO cells in monolayer except that the labelings with TNBS and antibody are carried out in a cold room at 3°C. The cells are then washed, scraped from the dishes, collected by centrifugation, and lysed with 200 μl of 0.5% NP-40 in calcium- and magnesium-free PBS containing 1 mM phenylmethanesulfonyl fluoride. The lysate is centrifuged and the 45,000g soluble fraction is mixed with 100μl of a 10% (v/v) suspension of fixed *Staphylococcus aureus* (Bethesda Research Labs, Inc, Gaithersburg, MD). The *S. aureus* is collected by centrifugation and washed several times with 0.1% SDS, 0.05% NP-40, 0.6 M NaCl, 10 mM Tris HCl (pH 8.6), and finally with 1 ml water. We then incubate the washed *S. aureus* with 200 μl of alkaline urea solution containing 9.3 M urea, 5 mM K_2CO_3, 2% (v/v) NP-40, and 5 mg/ml dithiothreitol for 15 min at room temperature. The *S. aureus* is removed by centrifugation and the supernatants are applied to two-dimensional gels. Complications may arise in the analysis of the resulting trinitrophenyl-labeled proteins by two-dimensional electrophoresis, since trinitrophenylation neutralizes amino groups and hence decreases the isoelectric point of proteins. For this reason we have developed the membrane-impermeable reagent isethionyl 3-(N-2,4-di-nitrophenyl)-aminopropioimidate (IDNPAP) [Denny and Roberts, 1982], which has immunochemical properties similar to those of TNBS but reacts with proteins without altering their isoelectric points.

RADIOGRAPHIC METHODS FOR DETECTION OF RADIOACTIVITY IN ELECTROPHORETIC GELS

Introduction and Choice of Methods

Radiolabeled proteins separated by polyacrylamide gel electrophoresis are visualized on medical x-ray film by one of three basic methods: (1) direct autoradiography, (2) indirect autoradiography, and (3) fluorography [Laskey, 1980]. In choosing a procedure the particular isotope and its activity in the sample are the critical factors. In this section, it is assumed that maximal sensitivity using the minimal amount of radioactivity is the goal of the researcher.

Direct autoradiography is used with isotopes emitting either x-rays or β-particles of sufficiently high energy, such as ^{14}C, ^{35}S, ^{32}P, or ^{125}I, to directly expose the film. Silver bromide in the film is directly activated to a silver atom by the charged particles released from the radiolabeled proteins in the electropherogram. Tritium emits a particle of such low energy (18.5 keV) that it cannot penetrate through the gel, and therefore the signal is quenched [Rogers, 1979]. Direct autoradiography is of value only when the amount of radioactivity is sufficiently high to produce an effective latent image. For example, to detect an image from a single band in 24 h on an acrylamide gel, the minimal amounts of radioactivity necessary are: 6,000 dpm ^{14}C or ^{35}S, 575 dpm of ^{32}P, and 1,600 dpm of ^{125}I (calculated from Laskey [1980]).

Enhancement of autoradiographic images is possible using indirect autoradiography [Laskey and Mills, 1977] whereby emitted particles of sufficiently high energy (^{32}P and ^{125}I) not only expose silver bromide in the film directly, but more important, the particles also penetrate through the x-ray film. Placed behind the film is a fluorescent surface (intensifying screen) which emits many photons of light when excited by the β-particles or x-rays. The fluorescent light will also expose the film, thus amplifying the original signal. Further enhancement is possible by stabilizing the exposed silver atoms in the film by two separate methods in tandem. First, the film is preflashed (fogged) with a brief (< 1 ms) burst of visible light to establish stable nucleation sites within silver halide crystals [Laskey and Mills, 1975; Rogers, 1979]. Preflashing has another advantage in that the density of the exposed image becomes linearly related to the amount of radioactivity in the source. Second, the back reaction of a silver atom to silver bromide is temperature dependent and totally inhibited at ultralow temperatures (−70°C; Koren et al [1970]). By using intensifying screens, preflashed film, and exposure at −70°C, a 10.5- and 16-fold enhancement of sensitivity for ^{32}P and ^{125}I, respectively, is possible when compared to direct autoradiography [Laskey and Mills, 1977].

The last method, fluorography, is most useful for ^{3}H, ^{35}S, and ^{14}C when the amount of radioactivity is very low (minimum of 400 dpm per source for a 24-h exposure). Fluorography is essential for the detection of tritium in polyacrylamide gels [Bonner and Laskey, 1974], since the β-particles travel only a few microns [Rogers, 1979]. The polyacrylamide support is impregnated with a scintillant which then fluoresces when excited by β-particle emissions. Maximal sensitivity and linearity of film response to radioisotope dose is obtainable using preflashed film and exposure at −70°C [Laskey and Mills, 1975].

These methods are detailed below and the methods of choice are summarized in Table IV.

Choice of film. Screen-type films (Kodak XRP and XAR) are essential when image production depends on fluorescence, either from an intensifying screen or from an impregnated fluor. These films are also used for direct autoradiography. Although the direct-type films (Kodak NS) are somewhat more sensitive for direct autoradiography, they are not applicable to indirect autoradiography or fluorography. Therefore, if only one type of film is to be stocked, Kodak XAR is the film of choice for all applications. However, Kodak XAR film is approximately twice the cost of Kodak XRP. We routinely stock two types of film: (1) XRP, since it offers reasonable sensitivity for most applications, and (2) Kodak

TABLE IV. Optimal Detection Methods and Exposure Conditions for Isotopically Labeled Macromolecules in Polyacrylamide Gels

Isotope	Relative amount of radioactivity[a]	Direct autoradiography[b]	Indirect autoradiography[c] (intensifying screen)	Fluorography[c] (gel impregnation)
^{3}H	All amounts	Never	Never	Essential
^{14}C	Low (400 dpm)	No	No value	Optimal
	High	Yes	No value	Best, but not essential
^{35}S	Low (400 dpm)	No	No value	Optimal
	High	Yes	No value	Best, but not essential
^{32}P	Low (50 dpm)	No	Optimal	No
	High	Yes	Optimal	No
^{125}I	Low (100 dpm)	No	Optimal	Best, but not essential
	High	Yes	Optimal	No

[a]For an optimal 24-h exposure giving a detectable image (dpm/cm^2) [Laskey and Mills, 1975, 1977; Laskey, 1980].
[b]There is no major enhancement of sensitivity using preflashed film or ultralow-temperature exposure.
[c]Maximum sensitivity depends on the use of preflashed film and ultralow-temperatures for exposure.

XAR for samples of very low activity (^{3}H especially). If access to an automatic film processor is available, Kodak XRP and XAR films may be conveniently developed. However Kodak NS must be hand-processed [Kodak, 1980].

Drying of polyacrylamide gels. Increased sensitivity for detecting labeled proteins routinely requires the drying of slab gels to a firm support for autoradiography or after fluor impregnation for fluorography.

Drying gels is routinely performed by heating the gel under vacuum in a commercial apparatus (eg, Bio-Rad). However, similar home-made apparatuses [Lim et al, 1969; Fairbanks et al, 1965] work equally well. For the economy-minded, a gel placed on a gelatin-coated glass plate can be easily dried under a hairdryer or infrared lamp. To minimize distortions and cracking, the gel may be placed for a brief equilibration in 2% (v/v) aqueous glycerol in 7% (v/v) acetic acid. However, glycerol should be omitted if the gel is impregnated with a fluor. Polyacryalmide gels adhere firmly to paper while drying, and although the use of Whatman No. 1 or 3 MM papers are frequently cited in the literature, we use desk-blotter paper, which is cheaper and more resilient, purchased from a local office supply store.

The following procedure is applicable to commercial drying machines:

1. Label blotter paper with an indelible ink to identify the sample gel.
2. Wet the paper with water and place it on the drying machine support.
3. Place the gel on the paper, being careful not to entrap air between the gel and paper.
4. Carefully place a plastic food wrap (eg, Saran Wrap) over the gel, avoiding both air bubbles and wrinkles in the wrap. Smooth out any wrinkles.
5. Cover with a firm sheet of mylar plastic.
6. Dry the gel under vacuum for 1–2 h using heat according to the manufacturer's instructions.

Caution: Dried gels may curl up immediately after drying. To avoid this, place the dried electrophoretogram between two glass plates and clamp them together with binder clips. Gels impregnated with a fluor do not curl as readily.

House vacuum and water aspirators are not recommended. Efficient drying depends on the high vacuum provided by a mechanical pump (eg, Duo-Seal Vacuum Pump, W.W. Welsch Inc). To protect the pump, a vapor trap and a cold trap should be placed in the vacuum line between the pump and the drying equipment. The cold trap contents are to be discarded after drying a gel according to normal radioisotope-handling procedures. Be forewarned that interruption of the vacuum before the gel is completely dried (either manually or by an overfilled frozen cold trap) results in the immediate cracking of the gel and possibly its total destruction.

Cracks in gels are not due only to vacuum failures. If the gel is stretched as it is placed on the wet blotter paper, cracks will almost definitely appear. Gradient acrylamide gels tend to distort upon drying. If extra care is taken not to stretch the gel as it is placed on the wet blotter paper, the risks of the gel cracking are reduced. This is accomplished by submerging the blotter paper in a shallow tray filled with 7% (v/v) acetic acid. The gradient slab gel is placed on the tray over the paper. The fluid is then aspirated. In this manner, handling of the gel is avoided. Alternatively, distortions or cracking can be eliminated by using a hydrophobic gel support (GelBond, Marine Colloids), although at an increased cost.

Preflashing of film. X-ray film can be easily fogged to increase its sensitivity to the visible light emitted by fluorescent intensifying screens or the impregnated fluors. Fogging also results in a linear response of image density on developed film to the amount of radioactivity in the gel when using indirect autoradiography and fluorography [Laskey and Mills, 1975]. Preflashed film does not increase the sensitivity of autoradiography [Laskey and Mills, 1975], and the relationship between the amount of radioactivity and image density is linear [Randerath, 1970]. Low temperatures are also not necessary for autoradiography.

The light for fogging is supplied by an accessory electronic flash unit to a 35-mm camera. To reduce the light output the window is covered with a Kodak Wratten No. 22 (Deep Orange) filter over which is placed a piece of Whatman No. 1 filter paper [Laskey and Mills, 1975]. Alternatively, a photographic printing enlarger can be used [Schleif and Wensink, 1981]. The flash of light should last no longer than 1 ms.

The amount of fogging depends on the light intensity of the flash unit and its distance from the film. Adjustment in fogging density is controlled by empirical determination of the optimal distance, usually between 70 cm and 1 m. To avoid uneven fogging, do not flash closer than 50 cm [Laskey, 1980]. In the darkroom, a piece of yellow paper (the interleaf sheets in Kodak X-ray film is convenient) is placed under the film. A safelight (Kodak GBX, 15-W bulb) may be used, but avoid prolonged or close exposure of x-ray films. Flash several sheets at different distances. The electronic flash unit can be mounted on a tall ring stand for added stability, and the stand rod should be marked with measured gradations for convenience and reproducibility. Optimal prefogging is determined after film development by measuring the optical density of the film at 540 nm [Laskey and Mills, 1975] on a densitometer or on a spectrophotometer. Maximal sensitivity is achieved at absorbances of 0.1 to 0.15 units above the absorbance of unflashed film [Laskey and Mills, 1975; Laskey, 1980]. Higher absorbances result in further image enhancement but with a concomitant increase in background [Laskey and Mills, 1975]. Storage of prefogged film is not recommended as background intensity increase as the emulsion deteriorates [Laskey, 1980].

When using preflashed film, the film side nearest to the light flash source is also the side placed nearest to the fluorescing medium (intensifying screen for indirect autoradiography, or the fluor impregnated gel for fluorography) during exposure [Laskey and Mills, 1975; Laskey, 1980].

Direct Autoradiographic Exposure Method

When there are several thousand disintegrations per minute in a single protein on an acrylamide gel, such as with ^{125}I-lectin or ^{125}I antibody overlays, direct autoradiography is the most convenient method. Simply stated, the x-ray film is placed in direct contact with the dried electropherogram and held firmly in place with glass plates and binder clips or in a commercial x-ray film cassette. Obviously, the film-gel sandwich must be protected from light, either in an x-ray cassette, wrapped in foil or sealed in a Cronex Lo Dose Mammography bag (Dupont, with the glass plates and clips on the outside). Low temperatures and fogged film are of no value.

As an alternative and particularly if some loss of sensitivity and resolution can be tolerated, hydrated gels can be exposed to film (direct autoradiography of ^{35}S, ^{32}P, or ^{125}I. The gels are most conveniently handled by placing them in common boilable cooking pouches and closing the bag with a kitchen-type plastic-bag sealer [Yu and Spring, 1978]. In the darkroom, the bagged gel is placed on a sheet of x-ray film between two glass plates. Contact is maintained by taping the plates together. The gel-film-glass sandwich is either wrapped in aluminum foil or placed in a black opaque film pouch (Cronex Lo Dose Mammography Bag, Dupont) and heat sealed. Exposure does not require low temperatures or preflashed film.

Indirect Autoradiographic Exposure Method

Intensifying screens can be used to enhance the detection of ^{32}P and ^{125}I (and other high-energy emitters such as ^{59}Fe or ^{45}Ca), but not ^{14}C, ^{3}H, or ^{35}S [Swanstrom and Shank, 1978]. For optimal results, the film must be preflashed and exposure performed at −70°C [Laskey and Mills, 1977].

The use of intensifying screens does lower the resolution of images, but not seriously. If resolution loss is not a major problem, then the use of two intensifying screens will enhance the sensitivity for ^{32}P [Laskey and Mills, 1977]. The film must be matched to intensifying screens sensitive to blue light. XAR and XRP are most sensitive when used with Cronex Lighting Plus (Dupont) or Mach II (Fuji) screens [Laskey and Mills, 1977]. Screens are, however, very expensive, and in practice we use a variety of screens obtained at no cost as discards from the Radiology Department. These should be carefully cleaned before use, using a screen cleaner. Most blue-light-sensitive screens will enhance the detection of ^{32}P or ^{125}I [Laskey and Mills, 1977].

We recommend the following procedure:

1. The fogged side of the film is placed directly next to the intensifying screen.

2. The dried gel is then placed on the other side of the film (a piece of Saran Wrap may be placed between the gel and the film if the gel is the least bit sticky. The set-up is thus: gel:film:screen (toward fogged side of the film).

3. For two screens the set-up is screen: gel:film:screen.

4. The use of an x-ray cassette in which the screen is an integral part of the cassette is advantageous since it is also light-tight. If a cassette is not used, the set-up must be held together with glass plates and binder clips inside a light-tight mammography bag or an aluminum foil pouch.

Fluorographic Exposure Methods

Choice of fluors. For detection of ^{3}H and low levels of ^{14}C, ^{35}S, and ^{125}I, several fluorographic methods are available which differ only according to the fluor employed. Presently 2,5-diphenyloxazole (PPO) in dimethylsulfoxide (DMSO), sodium salicylate, and commercially available EN^{3}HANCE™ and ENLIGHTNING™ (New England Nuclear) are the fluors of choice.

PPO/DMSO [Bonner and Laskey, 1974] offers the greatest resolution, since fluorography using the other above-mentioned fluors results in broadening of the radioactive image. However, preparation time is approximately 5 h and requires constant handling of gels in potentially hazardous solutions (PPO is a carcinogen and DMSO is rapidly absorbed by the skin). Extreme caution must be observed when PPO/DMSO is used. Alternately, PPO in glacial acetic acid can be used to avoid DMSO [Pulleyblank and Booth, 1981].

Sodium salicylate [Chamberlain, 1979] is equally as sensitive in detecting β-emissions as PPO/DMSO; however, some loss of resolution occurs. The main advantages of sodium salicylate are (1) rapid preparation time (1 h), (2) relatively safety, and (3) high cost-efficiency. We use sodium salicylate in our laboratories exclusively.

EN^{3}HANCE™ has slightly greater sensitivity than PPO/DMSO or sodium salicylate with little loss of resolution. Preparation time is approximately 2 h. It is reputedly not hazardous. A new product, ENLIGHTNING™ has all of these advantages, including shorter preparation times of 15–30 min. Both EN^{3}HANCE™ and ENLIGHTNING™ are expensive, but may prove to be very cost-efficient as compared to PPO/DMSO when the expense of technical personnel is taken into consideration.

Fluor impregnation methods

PPO/DMSO [Laskey, 1980]. 1. After fixation and staining, the gel is immersed in 20 gel vol of DMSO for 30 min. A second equilibration in fresh DMSO allows for the complete removal of water, since PPO is insoluble in aqueous solution. The DMSO baths may be reused several times. The gel will shrink in DMSO.

2. The gel is impregnated by equilibration in 4 gel vol of 22% (w/v) PPO in DMSO for 3 h. Band broadening occurs with longer equilibration times and therefore should be avoided.

3. The PPO is precipitated in the gel and the DMSO removed by soaking in water for 1 h (0.75–1.5-mm thick gels). The DMSO must

be completely removed in order for the gel to dry efficiently. Change the water bath several times. The gel will reswell.

4. Dry the gel.

5. PPO may be retrieved from used PPO/DMSO solutions by precipitation with 3 vol of 10% (v/v) ethanol in water. PPO crystals are collected on filter paper and washed with 20 vol of water. Crystals are dried at room temperature.

PPO/acetic acid [Pulleyblank and Booth, 1981]. 1. Equilibrate the gel in 20% (w/v) PPO in glacial acetic acid. Upon immersion, PPO will initially precipitate but subsequently redissolves. Terminate the equilibration when all of the PPO has redissolved. The gel will swell in acetic acid.

2. Wash gel in water for one hour to precipitate PPO in situ.

3. Dry the gel.

Sodium salicylate [Chamberlain, 1979]. 1. Gels which contain acids from staining or fixation must be washed in tap water for 30 min.

2. Equilibrate the gel in 1 M sodium salicylate (pH 5–7) using approximately 20 gel vol for 30 min. Protein bands rapidly fade with longer equilibration times.

3. Dry the gel.

4. The salicylate solution may be reused up to three times but should not be stored for extended periods.

EN^3HANCE^{TM} [New England Nuclear, 1981]. 1. Use 3–4 gel vol of EN^3HANCE^{TM} in order to cover the gel. Use a glass dish or polyethylene tray. EN^3HANCE^{TM} is not compatible with polystyrene or acrylic materials.

2. Precipitate the EN^3HANCE^{TM} with cold water and gentle agitation for one hour. Longer times are to be avoided.

3. Dry the gel.

4. EN^3HANCE^{TM} will evaporate from the gel at room temperature. Gels should be stored at −70°C and exposed to x-ray film without delay.

ENLIGHTNING™ [New England Nuclear, 1982]. 1. Equilibrate the gel in 5 gel vol of ENLIGHTNING™ for 15–30 min with gentle agitation. The gel should be free-floating.

2. Dry the gel.

Exposure method. The use of low temperature (−70°C) and preflashed film is essential for optimal sensitivity. If low temperature is not used, there is no advantage to preflashing the film.

The dried gel is placed next to the preflashed side of the film. Intimate contact is necessary, and therefore glass plates and binder clips are placed on the outside of a Cronex Lo Dose Mammography bag (Dupont) which contains the gel and x-ray film. Alternately, an x-ray film cassette can be used.

Miscellaneous Techniques

Dual isotopes. On occasion, an experiment is performed to compare gel patterns of samples prepared under different conditions using different isotopic tags, eg, detecting membrane protein differences between normal and transformed cells. Direct comparison of several two-dimensional gels can be ambiguous, since no two gels are exactly superimposable for every region of the gel, and most laboratories do not have sophisticated scanners available to them with programs capable of handling gel variability. However, if each sample is labeled separately, one (transformed cells) with ^3H amino acids and the other (normal cells) with either ^{14}C (or ^{35}S) amino acids, the resulting membrane samples can be mixed and electrophoresed together on one gel. Techniques are outlined below to discriminate between those polypeptides containing ^3H, ^{14}C (or ^{35}S), and those containing both isotopes [McConkey, 1979].

Optimal results require that the specific activity of the ^3H-labeled sample be at least eightfold higher than ^{14}C. Also thinner acrylamide slabs (0.4 mm) are necessary. After electrophoresis and staining, the gel is impregnated with a fluor, dried, and exposed to Kodak XAR film at −70°C. Both ^3H and ^{14}C will be detected. Next, the gel is exposed at room temperature to Kodak NS (No Screen) x-ray film which is insensitive to photons emitted by the fluor. Alternately, Kodak XAR film can be used if a piece of carbon paper is placed between the gel and the film to block the fluor emissions [Choo et al, 1980]. Thus the second exposure of the gel is by autoradiography and allows the detection of ^{14}C (or ^{35}S) only. However, this second exposure may have to be relatively long.

Polypeptides labeled only with tritium can be identified by a "subtraction" procedure [McConkey, 1979] whereby the fluorograph (first exposure, $^{3}H + {}^{14}C$) is exposed to a new piece of film to make a contact negative. The negative is then placed on a light box and the autoradiogram (^{14}C only) is placed on top of the negative. Polypeptides in common to both isotopes are lost in the dark background of the negative. However, polypeptides that contain only tritium appear white on a dark background, thus revealing polypeptides specific to the one set of membranes labeled with ^{3}H only.

Silver stained gels. Silver staining of proteins in polyacrylamide gels quenches ^{14}C and ^{3}H detection by autoradiography for fluorography, [Van Keuren et al, 1981]. The quenching is especially severe in the case of ^{3}H.

The problem is partially alleviated by using the photochemical silver stain method of Merril and co-workers [1981c] which employs low concentrations of silver nitrate. Following photographic recording of the stained gel, the silver must be removed with a photographic reducer. Subsequent autoradiography of ^{14}C is minimally quenched. Tritium detection by fluorography is quenched approximately 60%.

FUTURE TRENDS: NATIVE TWO-DIMENSIONAL ELECTROPHORESIS

Introduction

Surface receptors appear to be integral membrane proteins and are frequently composed of more than one subunit. The dissociating conditions used for their analysis in two-dimensional gels will, therefore, most likely destroy their ability to bind their specific ligands. Recently there have been successful attempts to renature a number of enzymes composed of only a single subunit after electrophoresis in SDS gels [Lacks and Springhorn, 1980; Scheele et al, 1981]. One possibility is that two-dimensional electrophoresis could be carried out without using conditions that cause polypeptide subunits of multimeric complexes to dissociate. In this manner whole receptors might remain intact and retain some of their biological properties. One restriction imposed on this system would, of course, be size, since large particles might not easily enter the polyacrylamide matrix of either the first or second dimension.

The technique of two-dimensional electrophoresis under nondenaturing conditions is still a developing procedure, however, although it might clearly have considerable implication with regard to studies on membrane receptors. It is likely to be particularly useful for screening complex mixtures of polypeptides for specific components, which retain their biological activity after electrophoretic separation. For example, it may be possible to detect membrane receptors directly on two-dimensional gels or electrophoretic "blots" using fluorescent or radiolabeled ligands. Already certain classes of enzymes may be visualized by specific staining procedures. For example, dehydrogenases can be localized utilizing a coupled, nitroblue tetrazolium dye reaction.

Although several procedures for native, two-dimensional electrophoresis have been published for soluble polypeptides [Manabe et al, 1979, 1980; Felgenhauer, 1979], none of these has been optimized for analysis of membrane proteins. The following method may offer the researcher a reliable prototype for analysis of either class of proteins.

Procedure

Preparation of IEF tube gels. Nitric acid-cleaned IEF tubes are marked and closed at one end with parafilm as described in an earlier section. Due to the omission of urea, the total acrylamide concentration is slightly higher to provide a more resilient gel. The liquid gel mixture for 10 gels contains acrylamide-DATD solution 2, 1.5 ml; 10% (v/v) NP-40, 0.05 ml; riboflavin-TEMED solution 3, 0.5 ml; ampholytes (2% (v/v) final concentration from 40% (w/v) commercial stocks), 0.5 ml; and water, 7.55 ml.

This mixture is degassed and then 15 μl of ammonium persulfate stock (solution 4) is added. The gel tubes are then filled with the mixture which is then overlayed with 25 μl of 0.05% (v/v) NP-40 containing 2% (v/v) ampholytes (pH 3.5–10) and allowed to polymerize under a fluorescent lamp for 2 h.

The tubes containing the polymerized gels are then prepared for isoelectric focusing as described earlier.

Sample preparation. Soluble proteins are dialyzed against water or a low ionic strength neutral buffer. Sucrose, 3.5–10 ampholytes, and NP-40 are added to final concentrations of 20% (w/v), 2% (v/v), and 0.5% (v/v), respectively. Dithiothreitol may be added to stabilize enzyme activity.

Membrane proteins are solubilized in 5 mM K_2CO_3 containing 1% (v/v) NP-40 (0.05 to 1%). Solubilization is initiated at 4°C by gentle sonication; urea (0.5 to 2 M) and dithiothreitol (1 mM) may be added to aid in solubilization. Before application to the IEF gels, samples are clarified by centrifugation (30,000g/45min) and sucrose (0.1 g/ml) is added.

Isoelectric focusing. Samples (50–500 μg protein in 100 μl) are applied to the top of the IEF gels and overlayered with 0.05% (v/v) NP-40 containing 0.01% (w/v) fast green. The tubes and the upper reservoir are filled with 0.04 M NaOH (solution 8). The lower (anodal) reservoir contains 0.06 N H_2SO_4. (solution 7). Electrofocusing is carried out at 4°C as follows: 75 V, 30 min; 150 V, 1 h; 300 V, 15 h; 45 V, 1 h. Tubes are then removed and the gels extruded. They may be either applied directly to the second-dimension slab gel or equilibrated for 15 to 30 min in 0.05% (v/v) NP-40 containing 10 mM Tris-HCl, pH 7.0. Tubular IEF gels are then sealed to the top of the second-dimension slab gel using 1% (w/v) agarose dissolved in the same buffer.

Nondenaturing gradient slab gel electrophoresis. The second-dimension slab gel consists of a 5 to 15% acrylamide gradient containing 0.05% (v/v) NP-40. The two solutions required for the gradient are prepared by using a stock acrylamide solution containing 45.7% (w/v) acrylamide monomer and 2.4% (w/v) *bis*-acrylamide:

	5%	15%
Acrylamide: *bis*-acrylamide stock (45.7%:2.4%)	1.03 ml	3.1 ml
1.5 M Tris·HCl, pH 8.9, containing 0.32% TEMED	3.9 ml	3.9 ml
10% (v/v) NP-40	0.77 ml	0.77 ml
Ammonium persulfate (solution 4)	25 μl	25 μl
Water	10.8 ml	7.7 ml

The slab gel is poured over a 10-min period using a standard gradient maker (eg, PGM-1; Isolabs, Inc, Akron, OH), overlayered with 0.05% (v/v) NP-40 and allowed to polymerize for 1 h. A stacking gel (containing 0.05% [v/v] NP-40 instead of SDS) is prepared and applied as described in an earlier section. After polymerization of the stacking gel, the IEF gel is affixed to the slab gel as described above, and the electrophoresis chamber is loaded into the tank. Reservoir buffer is added (3 g Tris-HCl, 14.4 g glycine, 5 ml of 10% [v/v] NP-40 made to a final volume of 1,000 ml with distilled water). Bromphenol blue tracking dye is placed in the upper reservoir and electrophoresis is initiated using a current of 15 mA/slab gel. Coolant is circulated through the chamber at 4°C. After 1 h, the current is adjusted to 30 mA per gel and is continued for about 8 h or until the tracking dye reaches the bottom of the slab gel. Alternatively, electrophoresis may be carried out overnight using a current of 8 mA per slab throughout the run. If needed, dansylated protein standards may be added to preformed wells at either edge of the slab gel; using a long-wave ultraviolet lamp, the migration of these polypeptides in the second dimension may be monitored. Other fluorescent derivatives may also be used, eg, fluorescamine or fluorescein. Dansylation of the sample is not recommended, since it modifies the isoelectric point of constituent polypeptides. If desired, basic proteins can be separated by NEPHGE as described in an earlier section except that urea is omitted. The NEPHGE gel is then affixed to a gradient slab gel which contains the acidic buffer system of Reisfield et al [1963], containing 0.05% (v/v) NP-40. By reversing the polarity on the electrophoresis apparatus, these basic polypeptides will migrate toward the cathode and be separated according to their molecular weight.

Analysis of the Two-Dimensional Separation

Depending upon the experimental requirements, two (or more) procedures may be used to analyze the slab gel after electrophoresis: (1) detection of total protein components and (2) detection of specific enzymatic or binding activity. There are at least three approaches to detection of proteins: first, the gel may be fixed and the denatured polypeptides visualized by either Coomassie Blue or silver stain procedures as described in an earlier section. The second method involves using a sample radiolabeled with a high-energy β- or γ-emitting isotope such as ^{125}I, ^{32}P, ^{35}S, ^{14}C, or ^{59}Fe. After two-dimensional separation the polypeptides are localized on the wet gel by autoradiography as previously described. Presuming that the sample is labeled to a high specific activity, this is a sensitive method for detection of native proteins.

Finally, procedures for protein detection utilizing fluorescent dyes such as the magnesium salt of 1-anilino-8-naphthalene sulfonate (Eastman-Kodak) under nondenaturing conditions are available which may detect as little as 10 μg protein.

Numerous methods exist for detection of specific enzymes in polyacrylamide gels [Gabriel, 1971] which need not be discussed here. Polypeptides may be transferred to derivatized paper by electroblot procedures described in an earlier section. Some denaturation of polypeptides may occur during coupling and washing procedures. Identification of a protein of interest can be made using specific antibody as outlined earlier.

Nondenaturing fixation of proteins in gels. Since detection procedures are often lengthy, considerable diffusion of protein spots in the acrylamide slab gel may occur. Although the pores of the acrylamide gradient limit diffusion and loss to some extent, it may be desirable to fix the proteins with retention of at least some of their biological activity. This can be accomplished by first soaking the gel in 50 mM potassium phosphate, pH 7.0, and then in the same buffer containing 0.5% (v/v) glutaraldehyde for 1 h at 4°C. The gel is washed with phosphate buffer and then reduced with 0.1% (w/v) sodium borohydride dissolved in the buffer. After 15 min, the gel is removed, washed, and reequilibrated with the buffer of choice. The proteins in such gels are reasonably stable and may be used for a battery of detection procedures.

Preliminary Findings

The method described above is a prototype for the analysis of biologically active membrane receptors in polyacrylamide gels. Although we cannot as yet report on its application to this problem, the methodology has been developed using a purified soluble enzyme from beef kidney, NADP-L-hexonate dehydrogenase [Horst, 1975].

NADP-L-hexonate dehydrogenase was dialyzed against water, and approximately 50 μg was applied to a native IEF tube gel. A pH gradient of approximately 8.5 to 4.5 (with a linear slope between 7.0 and 4.5) was established using the same composition of ampholytes routinely used for denaturing gels, ie, 1% (v/v) 3.5 to 10; 0.7% (v/v) 5 to 7; and 0.3% (v/v) 9 to 11 ampholytes (Ampholines, LKB). The enzyme migrated as a single species with an apparent pI of ~5.9. No aggregation of the protein was evident either upon entering the gel or during isoelectric focusing. Due to the acrylamide matrix selected, even large molecular weight polypeptide aggregates (up to Mr = 10^6) would enter the IEF gel [Anker, 1970; Ziegler and Hengartner, 1977].

The second-dimension separation is based on molecular weight due to the linear gradient of diminishing pore sizes in the acrylamide gel [Rodbard et al, 1971; Hearing et al, 1976]. Calibration of the 5–15% slab gel with standard proteins indicates a linear relationship between log Mr of the native protein and the distance migrated into the second dimension gel (Horst, unpublished results.) The second-dimension slab gel stacks at pH 8.9 and runs at pH 9.5 [Davis, 1974]. Thus, any polypeptide within the IEF gel will migrate toward the anode when an electric field is applied. NADP-L-hexonate dehydrogenase migrated in the gradient acrylamide slab gel as two species with apparent molecular weights of approximately 35,000.

NADP-L-hexonate dehydrogenase retained biological activity following native two-dimensional electrophoresis. In Figure 16A, two enzymatically active forms of this enzyme are revealed using L-gulonolactone as the substrate in the presence of $NADP^+$, phenazene methosulfate, and nitroblue tetrazolium dye [Gabriel, 1971]. No other polypeptides were revealed by subsequent Coomassie Blue staining, as seen in Figure 16B.

A comment concerning the separation of NADP-L-hexonate dehydrogenase into two distinct species during gradient acrylamide gel electrophoresis is deserving of mention. Native gradient acrylamide gel electrophoresis is subject to the effects of protein conformation or shape. Previous work with this enzyme [Horst, 1975] determined that it exists as a single molecular weight species of Mr ~35,000 on a SDS polyacrylamide gel, yet separates as multiple forms on a native one dimensional acrylamide gel. These multiple species differ in the amount of disulfide bond reduction and hence, differ subtly in shape. Isoelectric focusing (Fig. 16A,B) confirms that these differences are not due solely to charge, since only one isoelectric species is revealed. Therefore, the second-dimension NP-40 gradient polyacrylamide gel separated NADP-L-hexonate dehydrogenase into two distinct polypeptides which probably differ in shape, not in mass. We chose to perform a "velocity" separation [Rodbard et al, 1971] whereby electrophoresis in the slab gel was stopped when the tracking dye reached the gel bottom. The distance of migration depended on the interaction of the gel matrix and the protein's mass and shape. It may be possible to eliminate shape effects by allowing electrophoresis to continue well after the tracking dye has run off the end of the gel. The protein's migration will halt when the polyacrylamide pore size prohibits passage through it ("approach to equilibrium" separation) not merely retarding migration ("velocity" separation).

SUMMARY

Two-dimensional polyacrylamide gel electrophoresis [O'Farrell, 1975; O'Farrell et al, 1977] is a high-resolution analytical technique easily applicable to studies on plasma membrane proteins [Horst and Roberts, 1979a,b]. It may be particularly useful for studying receptor molecules, which are often only minor components of the total cell surface protein. In this chapter we first described a highly efficient alkaline urea solubilization procedure developed for, but not limited to, plasma membrane polypeptides [Horst et al, 1980a]. Alkaline urea solubilization eliminates the artifactual loss of polypeptides on two-dimensional gels which results from SDS solubilization [Ames and Nikaido, 1976] and is also compatible with subsequent two-dimensional electrophoresis of polypeptides carrying a net basic charge (NEPHGE analysis; O'Farrell et al [1977]; Horst et al [1980a]). This method takes advantage of several separate solubilizing components: (1) an alkaline pH of ~10, (2) disruption of noncovalent interactions with 9.3 M urea, (3) low-

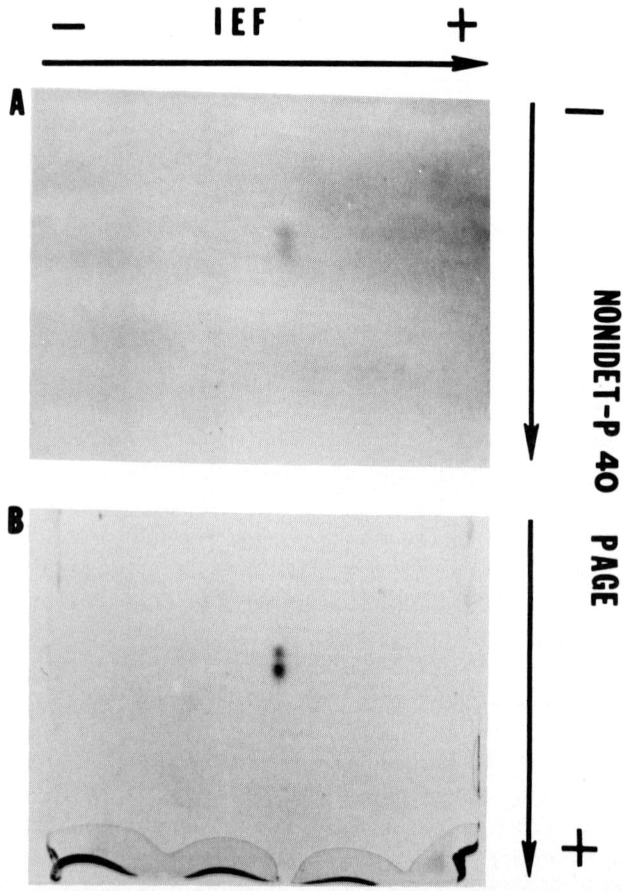

Fig. 16. *Separation of NADP-L-hexonate dehydrogenase by nondenaturing two-dimensional polyacrylamide gel electrophoresis. Purified bovine kidney enzyme was separated by isoelectric focusing in a pH gradient of 8.5 to 4.5. Isoelectric focusing was performed in a DATD cross-linked acrylamide tube gel containing 0.05% (v/v) NP-40. The second-dimension separation was performed in a 5–15% linear acrylamide slab gel containing 0.38 M Tris-HCl, pH 8.9, and 0.05% (v/v) NP-40. Following electrophoresis, the enzyme was fixed in situ with 0.5% (v/v) glutaraldehyde and reduced with 0.1% sodium borohydride. A) Oxidoreductase activity was detected by the nitroblue tetrazolium technique [Gabriel, 1971] in the presence of L-gluconolactone as substrate. B) The same slab gel stained with 0.125% (w/v) Coomassie Blue.*

frequency sonication, (4) dissolution of the lipid bilayer by a nonionic detergent and the subsequent incorporation of amphiphilic membrane proteins into detergent micelles, and (5) the reduction of covalent cystines with dithiothreitol. The solubilization protocol has broad applications, and our laboratories have used alkaline urea for the solubilization of polypeptides from isolated plasma membrane [Baumbach et al, 1981], whole cell preparation [Denny and Roberts, 1981], endometrial tissue [Basha et al, 1979], cotelydonary tissue of plant seeds [Horst et al, 1980a], exoskeleton of the brine shrimp, *Artemia* [Horst, 1981], uterine secretions [Basha et al, 1980], developing trophoblasts [Godkin et al, 1982], and bacterial exotoxins (Buhi, unpublished results). Mammalian virus proteins have also been solubilized in alkaline urea as shown by Van Dyke and Flanegan [1981].

In addition, we have described detailed protocols for the electrophoretic separation of once-solubilized polypeptides in two dimensions. For acidic polypeptides, isoelectric focusing either under native or totally denaturing conditions is the method of choice in the first dimension. Alternatively, basic polypeptides are separated in a nonequilibrium pH gradient electrophoresis (NEPHGE). The second dimension separates proteins according to their size, using either SDS-polyacrylamide gel electrophoresis in a denaturing system or NP-40 gradient polyacrylamide gel electrophoresis under nondenaturing conditions. We have provided a description of an apparatus which is capable of handling four second-dimension slab gels simultaneously. Other laboratories have published blueprints of apparatuses for one slab gel [Jones, 1980] or for many slab gels [Anderson and Anderson, 1978]. However, our equipment is more compatible with workloads requiring moderate numbers of two-dimensional electropherograms, and a single trained technician is capable of assembling three of these slab gel set-ups (a total of 12 gels) in less than 3 h.

Once two-dimensional separation has been achieved a variety of methods can be used for detecting proteins and glycoproteins in the gels. We have described two standard staining techniques (Coomassie Blue and silver stain), procedures for autoradiography and fluorography, and the "staining" of specific glycoprotein components by means of iodinated lectins. Wherever possible we have emphasized possible pitfalls and various approaches that might be taken to control the sensitivities of the procedures. Proteins, after two-dimensional electrophoresis, may also be transferred electrophoretically to nitrocellulose papers and then stained or else identified by means of specific antibodies. We have also included a short section on methods for identifying cell surface "exposed" polypeptides on two-dimensional gels and a new approach for separating proteins under nondenaturing conditions. The latter procedure may have long-term applicability for studying membrane receptors, since it is hoped that such structures will remain intact and maintain some of their biological properties even after two-dimensional electrophoretic separation.

ACKNOWLEDGMENTS

The authors thank Mr. John Berceann for typing this manuscript and Ms. E. Jodi Skeen for the photographic preparation of the figures.

This work was supported over several years by grants from the University of Florida, Division of Sponsored Research (to M.N.H. and R.M.R.); the National Science Foundation (#PCM 77–18016, PCM 79–03280, and PCM 81–0471 to R.M.R.); the National Institutes of Health (#HD–08560 and HD–10436 to R.M.R.); and the American Cancer Society, Florida Division (#F80UF-1 to R.M.R. and M.N.H.). During the early development of this work, M.N.H. was supported by the National Cancer Institute on a postdoctoral fellowship; G.A.B. and J.B.D. by predoctoral fellowships (Training Grant #NIH-1-T32-CA09126); and R.M.R. by a Career Development Award (K4-AM-70389) from the National Institutes of Health.

REFERENCES

Adair WS, Jurivich D, Goodenough UW (1978): Localization of cellular antigens in sodium dodecylsulfate-polyacrylamide gels. J Cell Biol 79:281–285.

Ames GF-L, Nikaido K (1976): Two dimensional electrophoresis of membrane proteins. Biochemistry 15:616–623.

Anderson NL, Anderson NG (1977): High resolution two dimensional electrophoresis of human plasma proteins. Proc Natl Acad Sci USA 74:5421–5425.

Anderson NL, Anderson NG (1978): Analytical techniques for cell fractions. XII. Two dimensional analysis of serum and tissue proteins: Multiple gradient-slab electrophoresis. Anal Biochem 85:341–354.

Anderson RG, Brown MS, Beisiegel V, Goldstein JL (1982): Surface distribution and recycling of the LDL receptor as visualized with anti-receptor antibodies. J Cell Biol 93:523–532.

Anker H (1970): A solubilizable acrylamide gel for electrophoresis. FEBS Lett 7:293.

Baenziger JU, Fiete D (1979): Structural determinants of Concanavalin A specificity for oligosaccharides. J Biol Chem 254:2400–2407.

Basha SMM, Bazer FW, Roberts RM (1979): The secretion of a uterine specific, purple phosphatase by cultured explants of porcine endometrium. Dependency upon the state of pregnancy of the donor animal. Biol Reprod 20:431–441.

Basha SMM, Bazer FW, Geisert RD, Roberts RM (1980): Progesterone-induced uterine secretions in pigs. Recovery from pseudopregnant and unilaterally pregnant gilts. J Anim Sci 50:113–123.

Bauman G, Chrambach A (1976): A highly crosslinked, transparent polyacrylamide gel with improved mechanical stability for use in isoelectric focusing and isotachophoresis. Anal Biochem 70:32–38.

Baumbach GA, Horst MN, Olympio MA, Aldrich HC, Roberts RM (1981): Isolation and characterization of plasma membranes from cultured cells. In Reid E, Cook GMW, Morre DJ (eds): "Cancer Cell Organelles: Methodological Surveys in Biochemistry." Chichester, England: Ellis-Horwood Ltd., Vol 10, pp 351–370.

Bayse GS, Morrison M (1971): Peroxidase catalyzed reactions of iodide at low pH. Arch Biochem Biophys 145:143–148.

Bonner WM, Laskey RA (1974): A film detection method for tritium-labeled proteins and nucleic acids in polyacrylamide gels. Eur J Biochem 46:83–88.

Briles EB, Li E, Kornfeld S (1977): Isolation of wheat germ agglutinin-resistant clones of Chinese hamster ovary cells deficient in membrane sialic acid and galactose. J Biol Chem 252:1107–1116.

Burnette WN (1981): "Western blotting:" Electrophoretic transfer of proteins from sodium dodecyl sulfate-polyacrylamide gels to unmodified nitrocellulose and radiographic detection with antibody and radioiodinated protein A. Anal Biochem 112:195–203.

Burridge K (1976): Direct identification of specific glycoproteins and antigens in sodium dodecyl sulfate gels. Methods Enzymol 50 (Part C):55–64.

Chamberlain JP (1979): Fluorographic detection of radioactivity in polyacrylamide gels with the water soluble fluor sodium salicylate. Anal Biochem 98:132–135.

Choo KH, Cotton RGH, Danks DM (1980): Double-labeling and high precision comparison of complex protein patterns on two-dimensional polyacrylamide gels. Anal Biochem 103:33–38.

Davis BJ (1974): Disc electrophoresis. II. Method and application to human serum proteins. Ann NY Acad Sci 121:404–427.

Denny JB, Roberts RM (1981): Turnover of plasma membrane polypeptides in transformed and normal cells using $^{125}I/^{131}I$ double label techniques. In Reid E, Cook GMW, Morre DJ (eds): "Cancer Cell Organelles: Methodological Surveys In Biochemistry." Chichester, England: Ellis-Horwood Ltd., Vol 10, pp 388–395.

Denny JB, Roberts RM (1982): A new immunoreactive probe for the isolation and analysis of plasma membrane polypeptides. Synthesis and properties of isethionyl 3-(N-2,4-dinitrophenyl)-aminopropioimidate. J Biol Chem 257:2460–2468.

Erickson PF, Minier LN, Lasher RS (1982): Quantitative electrophoretic transfer of polypeptides from SDS polyacrylamide gels to nitrocellulose sheets: A method for their re-use in immunoautoradiographic detection of antigens. J Immunol Methods 51:241–249.

Fairbanks G, Levinthal C, Reeder RH (1965): Analysis of ^{14}C-labeled proteins by disc electrophoresis. Biochem Biophys Res Commun 20:393–399.

Felgenhauer K (1979): Characterization of native multicomponent protein mixtures by one- and two-dimensional gradient electrophoresis. J Chromatogr 173:299–311.

Fitzgerald LA, Baumbach GA, Horst MN, Noonan KD, Roberts RM (1981): The purification and immunocytochemical localization of the major iodinatable cell surface glycoproteins of Chinese hamster ovary cells. J Cell Sci 52:405–424.

Fraker PJ, Speck JC Jr (1978): Protein and cell membrane iodinations with a sparingly soluble chloramine, 1,4,4,6-tetrachloro-3α,6α-diphenylglucoluril. Biochem Biophys Res Commun 80:849–857.

Freeman HC (1973): Metal complexes of amino acids and peptides. In Eichhorn GL (ed): "Inorganic Biochemistry." New York: Elsevier, Vol I, pp 121–161.

Gabriel O (1971): Locating enzymes in gels. Methods Enzymol 22:578–604.

Gahmberg CG (1978): Tritium labeling of cell-surface glycoproteins and glycolipids using galactose oxidase. Methods Enzymol 50 (Part C): 204–206.

Ghamberg CG, Andersson LC (1977): Selective radioactive labeling of cell surface sialoglycoproteins by periodate-tritiated borohydride. J Biol Chem 252: 5888–5893.

Godkin JD, Bazer FW, Moffatt J, Sessions F, Roberts RM (1982): Purification and properties of a major, low molecular weight protein released by the trophoblast of sheep blastocysts at Day 13–21. J Reprod Fertil 65:141–150.

Harell D, Robinson M (1979): Two-dimensional separation of erythrocyte membrane proteins. Arch Biochem Biophys 193:158–168.

Hearing V, Klingler W, Ekel T, Montague P (1976): Molecular weight estimation of Triton X-100 solubilized proteins by polyacrylamide gel electrophoresis. Anal Biochem 72:113–122.

Hoffman S, McMahon D (1978): Defective glycoproteins in the plasma membrane of an aggregation minus mutant of dictyostelium discoideum with abnormal cellular reactions. J Biol Chem 253:278–287.

Horst MN (1975): Purification and properties of NADP-L-hexonate dehydrogenase and its use in the fluorometric assay of uronic acids. PhD dissertation, The Florida State University.

Horst MN (1981): The biosynthesis of crustacean chitin by a microsomal enzyme from larval brine shrimp. J Biol Chem 256:1412–1419.

Horst MN, Basha SMM, Baumbach GA, Mansfield EH, Roberts RM (1980a): Alkaline urea solubilization, two dimensional electrophoresis and lectin staining of mammalian cell plasma membrane and plant seed proteins. Anal Biochem 102:399–408.

Horst MN, Baumbach G, Roberts RM (1979): Two-dimensional electrophoretic analysis of concanavalin A binding components in the plasma membranes of Chinese hamster fibroblasts. FEBS Lett 100:385–388.

Horst MN, Baumbach GA, Olympio MA, Roberts RM (1980b): Isolation of a domain of the plasma membrane in Chinese hamster ovary cells which contains iodinatable, acidic glycoproteins of high molecular weight. Biochim Biophys Acta 600:48–61.

Horst MN, Roberts RM (1979a): Solubilization, electrofocusing and two-dimensional electrophoresis of plasma membrane polypeptides from Chinese hamster ovary cells. In Haglund H, Westerfeld JG, Ball JT Jr (eds): "Electrofocus/78." New York: Elsevier/North-Holland Press, pp 99–118.

Horst MN, Roberts RM (1979b): Analysis of polypeptide turnover rates in Chinese hamster ovary cells plasma membranes by two-dimensional electrophoresis. J Biol Chem 254:5000–5007.

Howe JG, Hershey JW (1981): A sensitive immunoblotting method for measuring protein synthesis initiation factor levels in lysates of *Escherichia coli*. J Biol Chem 256:12836–12839.

Hubbard AL, Cohn ZA (1972): The enzymatic iodination of the red cell membrane. J Cell Biol 55:390–405.

Imada M, Hsieh P, Sueoka N (1978): Separation of mammalian cell surface proteins by a two-dimensional gel electrophoresis system. Biochim Biophys Acta 507:459–469.

Ivarie RD, Jones PP (1979): A rapid sensitive assay for specific protein synthesis in cells and cell free translations: Use of *Staphylococcus aureus* as an adsorbent for immune complexes. Anal Biochem 97:24–35.

Jones PP (1979): Analysis of H-2 and Ia molecules by two dimensional gel electrophoresis. J Exp Med 146:1261–1279.

Jones PP (1980): Analysis of radiolabeled lymphocyte proteins by one- and two-dimensional polyacrylamide gel electrophoresis. In Mishell RB, Shiigi SM (eds): "Selected Methods in Cellular Immunology." San Francisco: W.H. Freeman, pp 398–440.

Kao FT, Puck TT (1967): Genetics of somatic mammalian cells. IV. Properties of Chinese hamster cell mutants with respect to the requirement of proline. Genetics 55:513–524.

Kaplan G, Plutner H, Mellan I, Unkeless JC (1981): Studies on externally disposed plasma membrane proteins. Trinitrobenzenesulfonic acid derivatization and immune precipitation. Exp Cell Res 133:103–114.

Kodak (1980): "Biomedical Imaging Information. Autoradiography of Macroscopic Specimens." Rochester, New York: Eastman Kodak Company, Technical bulletin No. M3-508.

Koren JF, Melo TB, Prydz S (1970): Enhanced photoemulsion sensitivity at low temperatures used in radiochromatography. J Chromatogr 46:129–131.

Lacks SA, Springhorn SS (1980): Renaturation of enzymes after polyacrylamide gel electrophoresis in the presence of sodium dodecyl sulfate. J Biol Chem 255:7467–7473.

Laemmli UK (1970): Cleavage of structural proteins during the assembly of the head of bacteriophage T4. Nature 277:680–685.

Laskey RA (1980): The use of intensifying screens or organic scintillators for visualizing radioactive molecules resolved by gel electrophoresis. Methods Enzymol 65:363–371.

Laskey RA, Mills AD (1975): Quantitative film detection of ^{3}H and ^{14}C in polyacrylamide gels by fluorography. Eur J Biochem 56:335–341.

Laskey RA, Mills AD (1977): Enhanced autoradiographic detection of ^{32}P and ^{125}I using intensifying screens and hypersensitized film. FEBS Lett 82:314–316.

Legocki RP, Verma DP (1981): Multiple immunoreplica technique: Screening for specific proteins with a series of different antibodies using one polyacrylamide gel. Anal Biochem 111:385–392.

Lim R, Huang JJ, Davis GA (1969): Autoradiography with acrylamide gel slab electrophoresis. Anal Biochem 29:48–57.

Lim R, Tadayyon E (1970): Fractionation of the insoluble brain proteins with acrylamide electrophoresis. Anal Biochem 34:9–15.

Lis H, Sharon N (1973): The biochemistry of plant lectins (phytohemagglutinins). Annu Rev Riochem 42:541–575.

Lotan R, Nicolson GL (1979): Purification of cell membrane glycoproteins by lectin affinity chromatography. Biochim Biophys Acta 559:329–376.

Manabe T, Tacki K, Kojima K, Okuyama T (1979): Two-dimensional electrophoresis of plasma proteins without denaturing agents. J Biochem (Tokyo) 85:649–659.

Manabe T, Takahaski N, Kojima K, Shinoda T, Okuyama T (1980): Two-dimensional electrophoresis of immunoglobulin myeloma proteins in the absence of denaturing agents. J Biochem (Tokyo) 87:451–464.

McConkey EH (1979): Double-label autoradiography for comparation of complex protein mixtures after gel electrophoresis. Anal Biochem 96:39–44.

Meril CR, Danau ML, Goldman D (1981a): A rapid sensitive silver stain for polypeptides in polyacrylamide gels. Anal Biochem 110:201–207.

Merril CR, Goldman D, Ebert M (1981b): Protein variation associated with Lesch-Nyhan syndrome. Proc Natl Acad Sci USA 78:6471–6475.

Merril CR, Goldman D, Sedman SA, Ebert MH (1981c): Ultrasensitive stain for proteins in polyacrylamide gels shows regional variation in cerebrospinal fluid proteins. Science 211:1437–1438.

Merril CR, Switzer RD, Van Keuren ML (1979): Trace polypeptides in cellular extracts and human body fluids detected by two-dimensional electrophoresis and a highly sensitive silver stain. Proc Natl Acad Sci USA 76:4335–4339.

Mersel M, Benerson A, Doljanski F (1976): Lactoperoxidase-catalyzed iodination of surface membrane lipids. Biochem Biophys Res Commun 70:1166–1171.

Miner GD, Heston LL (1972): Method for acrylamide gel isoelectric focusing of insoluble brain proteins. Anal Biochem 50:313–316.

Montsigny M, Roche A-C, Sens C, Magel-Dana R, Delmotte F (1980): Sugar-lectin interactions: How does wheat germ agglutinin bind sialoglycoconjugates? Eur J Biochem 104:147–153.

Morrison M (1974): The determination of the exposed proteins on membranes by the use of lactoperoxidase. Methods Enzymol 32(Part B):103–109.

Morrison M, Bayse GS (1970): Catalysis of iodination by lactoperoxidase. Biochemistry 9:2995–3000.

Morrissey JH (1981): Silver stain for proteins in polyacrylamide gels: A modified procedure with enhanced uniform sensitivity. Anal Biochem 117:307–310.

Neville DM (1967): Fractionation of cell membrane protein by disc electrophoresis. Biochim Biophys Acta 133:168–170.

New England Nuclear (1981): A guide to autoradiography enhancement. Technical information compiled May 1981, No. 10M581A–1766.

New England Nuclear (1982): Technical information compiled April 1982, No. 8M482M–1985.

Nicolson GL (1974): Interaction of lectins with animal cell surfaces. Int Rev Cytol 39:89–190.

Oakley BR, Kirsch DR, Morris NR (1980): A simplified ultrasensitive silver stain for detecting proteins in polyacrylamide gels. Anal Biochem 105:361–363.

O'Brien TW, Kalf GF (1967): Ribosomes from rat liver mitochondria. II. Partial characterization. J Biol Chem 242:2180–2185.

Ochs, DC, McConkey EH, Sammons DW (1981): Silver stains for proteins in polyacrylamide gels: A comparison of six methods. Electrophoresis 2:304–307.

O'Farrell PH (1975): High resolution, two-dimensional polyacrylamide gel electrophoresis of proteins. J Biol Chem 250:4007–4021.

O'Farrell PZ, Goodman HM, O'Farrell PH (1977): High resolution two-dimensional electrophoresis of basic as well as acidic proteins. Cell 12:1133–1142.

Olden K, Yamada KM (1977): Direct detection of antigens in sodium dodecyl sulfate-polyacrylamide gels. Anal Biochem 78:483–490.

Peters A (1955): Experiments on the mechanism of silver staining. Part I. Impregnation. Q J Microsc Sci 96:84–102.

Phillips DR, Morrison M (1971): Exposed protein on the intact human erythrocyte. Biochemistry 10:1766–1771.

Pulleyblank DE, Booth GM (1981): Improved methods for the fluorographic detection of weak β-emitting radioisotopes in agarose and acrylamide electrophoresis media. J Biomed Biophys Methods 4:339–346.

Randerath K (1970): An evaluation of film detection method for weak β-emitters, particularly tritium. Anal Biochem 34:188–205.

Reisfield RA, Kahan BD (1971): Extraction and purification of soluble histocompatibility antigens. Transplant Rev 6:81–112.

Reisfield R, Lewis V, Williams D (1963): Disk electrophoresis of basic proteins and peptides on polyacrylamide gels. Nature 195:281–283.

Renart J, Reisen J, Stark GR (1979): Transfer of proteins from gels to diazobenzyloxymethyl-paper and detection with antisera: A method for studying antibody specificity and structure. Proc Natl Acad Sci USA 76:3116–3120.

Rodbard D, Kapadia G, Chrambach A (1971): Pore gradient electrophoresis. Anal Biochem 40:135–157.

Rogers AW (1979): "Techniques of Autoradiography." New York: Elsevier/North-Holland Biomedical Press, pp 13–53.

Rosenblum BB, Hanash SM, Yew N, Neil JV (1982): Two dimensional electrophoretic analysis of erythrocyte membranes. Clin Chem 28:925–931.

Sammons DW, Adams LD, Nishizawa EE (1981): Ultrasensitive silver-based color staining of polypeptides in polyacrylamide gels. Electrophoresis 2:135–141.

Sato T, Ita K, Yura K (1977): Membrane proteins of *Escherichia coli* K-12: Two dimensional polyacrylamide gel electrophoresis of inner and outer membranes. Eur J Biochem 78:557–567.

Scheele G, Pash J, Bieger W (1981): Identification of proteins according to biological activity following separation by two-dimensional isoelectric focusing/sodium dodecyl sulfate gel electrophoresis: Analysis of human exocrine pancreatic proteins. Anal Biochem 112:304–313.

Schlager SI (1979): Specific ^{125}I-iodination of cell surface lipids: Plasma membrane alterations induced during humoral immune attack. J Immunol 123:2108–2113.

Schlief RF, Wensink PC (1981): "Practical Methods in Molecular Biology." New York: Springer-Verlag, pp 184–186.

Schulman M, Wilde CD, Kohler G (1978): A better cell line for making hybridoma secreting specific antibodies. Nature 276:269–270.

Shaper JH, Barker R, Hill RC (1973): Purification of wheat germ agglutinin by affinity chromatography. Anal Biochem 53:564–570.

Svensson H (1961): Isoelectric fractionation, analysis and characterization of ampholytes in natural pH gradients. I. The differential equation of solute concentrations at a steady state and its solutions for simple cases. Acta Chem Scand 15:325–341.

Svensson H (1962): Isoelectric fractionation, analysis and characterization of ampholytes in natural pH gradients. II. Buffering. Acta Chem Scand 16:456–466.

Swanstrom R, Shank PR (1978): X-ray intensifying screens greatly enhance the detection by autoradiographs of the radioactive isotopes ^{32}P and ^{125}I. Anal Biochem 86:184–192.

Switzer RC, Merril CR, Shifrin S (1979): A highly sensitive silver stain for detecting proteins and peptides in polyacrylamide gels. Anal Biochem 98:231–237.

Symington J, Green M, Brackman K (1981): Immunoautoradiographic detection of proteins after electrophoretic transfer from gels to diazo-paper: Analysis of adenovirus encoded proteins. Proc Natl Acad Sci USA 78:177–181.

Tanner MJA, Anstee DJ (1976): A method for the direct demonstration of the lectin-binding components of the human erythrocyte membrane. Biochem J 153:265–270.

Tjian R, Stinchcomb D, Losick R (1974): Antibody directed against *Bacillus subtilis* σ factor purified by sodium dodecyl sulfate slab gel electrophoresis. J Biol Chem 250:8824–8828.

Towbin H, Staehelin T, Gordon J (1979): Electrophoretic transfer of proteins from polyacrylamide gels to nitrocellulose sheets: Procedure and some applications. Proc Natl Acad Sci USA 76:4350–4354.

Tuszynski GP, Buck CA, Warren L (1979): A two-dimensional polyacrylamide gel electrophoresis (PAGE) system using sodium dodecyl sulfate-PAGE in the first dimension. Anal Biochem 93:329–338.

Van Dyke TA, Flanegan JH (1980): Identification of polio virus polypeptide P63 as a soluble RNA-dependent RNA polymerase. J Virol 35:732–740.

Van Keuren ML, Goldman D, Merril CR (1981): Detection of radioactively labeled proteins is quenched by silver staining methods: Quenching is minimal for ^{14}C and partially reversible for ^{3}H with a photochemical stain. Anal Biochem 116:248–255.

Vesterberg O (1969): Synthesis and isoelectric fractionation of carrier ampholytes. Acta Chem Scand 23:2653–2666.

Wilson DL, Hall ME, Stone GC, Rubin RW (1977): Some improvements in two-dimensional gel electrophoresis of proteins. Anal Biochem 83:33–44.

Wray W, Boulikas T, Wray VP, Hancock R (1981): Silver staining of proteins in polyacrylamide gels. Anal Biochem 118:197–203.

Yu S-H, Spring TG (1978): A convenient method for the storage of slab gels. Anal Biochem 85:287–290.

Ziegler A, Hengartner H (1977): Sodium dodecyl sulfate electrophoresis of high molecular weight proteins in N,N′-diallyltartardiamide-crosslinked polyacrylamide slab gels. Eur J Immunol 7:690.

APPENDIX I: SOLUTIONS

Solutions for Membrane Protein Solubilization, Electrophoresis, and Staining

1. Solubilizing buffer (9.3 M urea, 5 mM K_2CO_3). Dissolve 5.5 g of "high purity" urea in 5 mM K_2CO_3 to a final volume of 9.5 ml. Finally add 0.5 ml of 4 mM phenylmethanesulfonyl fluoride. After sonicating the membranes in this solution, 20 μl of Nonidet P-40 and 5 mg dithiothreitol are added per ml of solubilizing buffer.

2. Acrylamide stock solution for IEF and NEPHGE. Bring 31.96 g acrylamide and 5.64 g DATD (diallyltartardiamide) up to 100 ml in water, filter, degas and store in a dark bottle in a refrigerator. Storage life is 3 to 4 mo. Check that the acrylamide is fully in solution and warm to room temperature before use.

3. Riboflavin-TEMED solution. Make up 2 mg riboflavin and 0.4 ml TEMED (N,N,N′, N′-tetramethylenediamine) to 50 ml in water. The solution can be stored in the refrigerator in a dark bottle. It is stable for months.

4. Ammonium persulfate stock solution for IEF and NEPHGE gels. Dissolve 0.28 g of ammonium persulfate in 1.0 ml distilled water.

5. Solubilizing buffer-ampholyte mixture. Make up 5.5 g of urea, and 0.2 ml of 40% ampholytes (pH3.5–10) to a total of 10 ml, using 2% (v/v) Nonidet P-40 in 5 mM K_2CO_3. This solution should be split into small lots (each 0.5 ml–1 ml) and stored frozen at −20°C to reduce the rate of decomposition.

6. Reducing solubilization buffer mixture. Make up 5.5 g of urea and 50 mg dithiothreitol to a total of 10 ml, using 2% (v/v) Nonidet P-40 in 5 mM K_2CO_3. This solution should be stored as above.

7. 0.06 M H_2SO_4 electrolyte solution. Add 3.3 ml of concentrated H_2SO_4 to water and bring up to 2 liters.

8. 0.04 M NaOH electrolyte solution. Dissolve 3.2 g NaOH in 2.0 liters of water. Prepare in advance and store in a tightly closed vessel. Thoroughly degas before use.

9. Tube gel equilibration buffer. Dissolve 7.87 g of Tris base and 10 g of ultrapure SDS in 700 ml distilled water. Titrate with 1 M HCl to pH 6.8. Dilute to 1 liter. Immediately prior to use, add 2-mercaptoethanol to 1% (v/v) final concentration to a 100–150 ml portion.

10. Acrylamide: bis stock solution for second-dimension slab gels. Dissolve 150 g of recrystallized acrylamide monomer and 4 g of N,N-methylene-*bis*-acrylamide in 500 ml water. Filter through Whatman No. 1 paper and degas under reduced pressure. Store refrigerated in an amber colored bottle. Warm to room temperature before use.

11. Acrylamide: bis stock solution for linear gradient acrylamide slab gels. Dissolve 150 g recrystallized acrylamide monomer and 4 g of N-N-methylene-*bis*-acrylamide in 300 ml water. Filter, degas, and store as described above.

12. Running gel buffer, pH 8.8. Dissolve 90.5 g of Tris base and 1.6 ml of TEMED (N,N,N′,N′-tetramethylenediamine) in 350 ml of water. The pH is titrated to 8.8 with approximately 120 ml of 1 M HCl. The volume is adjusted to 500 ml. The solution is filtered through Whatman No. 1 paper and degassed under reduced pressure.

13. Ammonium persulfate solution for acrylamide slab gels. Dissolve 0.28 g of ammonium persulfate in 100 ml of water. Make fresh daily.

14. Stacking gel buffer pH 6.8. Dissolve 29.9 g of Tris base and 2.3 ml of TEMED (N,N,N′,N′-tetramethylenediamine) in 200 ml of water. Titrate the pH to 6.8 with approximately 240 ml of 1 M HCl. Adjust the volume to 500 ml. The solution is filtered through Whatman No. 1 paper and degassed under reduced pressure.

15. Bottom gel sealer. Add 1 g of agarose to 75 ml of water and 25 ml of running gel buffer (solution 12). Add 0.5 ml of 20% (w/v) SDS. Warm to 60°C on a hot plate with stirring until the agarose dissolves. Dispense 10 ml at a time into 16 × 150-mm test tubes; cap and store in the refrigerator.

16. Top gel sealer. Add 1 g of agarose to 100 ml of solution 9 containing no β-mercaptoethanol. This solution is warmed to 60°C with stirring until the agarose dissolves. Add 0.1 ml of bromphenol blue saturated water. Dispense and store as above.

17. SDS-electrophoresis buffer. Dissolve 4.5 g Tris base, 21.6 g glycine, and 1.5 g of SDS in 1.5 liters of deionized water. This is sufficient for four slab gels in our apparatus. It is not necessary to use ultrapure reagents for this solution.

18. Coomassie Blue stain (0.125%). Dissolve 1.25 g Coomassie Brilliant Blue R-250 in 400 ml ethanol (95%) with stirring for several hours. Add 70 ml concentrated acetic acid and 530 ml distilled water and continue stirring. Filter to remove particulate matter in the dye through Whatman No. 1 paper.

19. Silver stain solution. This must be freshly prepared and used within 5 min. Dissolve 0.8 g of silver nitrate in 4 ml of water (solution A). Prepare 21 ml of a solution of 0.36% (w/v) sodium hydroxide and 1.4 ml of 14.8 M ammonium hydroxide (solution B). Add A dropwise into solution B with constant stirring. Bring volume up to 100 ml with water.

20. Silver stain developer. Mix 2.5 ml of a solution of 1% (w/v) citric acid and 0.25 ml of 38% (v/v) formaldehyde and bring volume to 500 ml with water. This should be freshly prepared.

Miscellaneous Solutions

Other stock solutions that should be on hand include the following:

20% (w/v) sodium dodecyl sulfate (SDS). Dissolve 10 g of ultrapure SDS in 50 ml of water.

0.1% (w/v) SDS. Mix 5 ml of 20% (w/v) SDS with 995 ml of water.

5 mM potassium carbonate K_2CO_3). Dissolve 0.83 g of K_2CO_3 in 1 liter of water.

10% (v/v) Nonidet-P40 (NP-40). Mix 10 ml NP-40 into 90 ml of water.

2% (v/v) NP-40 in 5 mM K_2CO_3. Add 2.0 ml of NP-40 to 98 ml of 5 mM K_2CO_3.

8 M urea. Dissolve 24 g of ultrapure urea in a total volume of 50 ml of water.

4 mM phenylmethane sulfonyl fluoride (PMSF). Dissolve 34.8 mg of PMSF into 1.0 ml 95% ethanol. Dilute with 49 ml of water.

Fast green tracking dye. Mix together 100 μl of solution 6 (reducing solubilization buffer), 100 μl of water, and 5 μl of an aqueous 1% (w/v) stock solution of Fast green.

Bromphenol blue saturated water. Add 5 g bromphenol blue to 5 ml of water. Mix thoroughly. Let undissolved dye settle before using.

APPENDIX II: RECRYSTALLIZATION OF ACRYLAMIDE

The cost of acrylamide for casting slab gels may become burdensome when large amounts of high-purity acrylamide monomer are consumed to satisfy the needs of routinely running many slabs. We prefer to buy reagent-grade acrylamide in reasonable bulk quantities (3–10 kg) and recrystallize it ourself. Below is our protocol.

Materials

The following reagents and equipment are necessary: reagent-grade chloroform, 300 g acrylamide, heating stir plate and magnetic Teflon stir bar, 6-liter Erlenmeyer flask, thermometer, large funnel and filter paper, clean 4-liter reagent bottle, large Buchner funnel, vacuum desiccator.

Method

All operations are carried out in a fume hood away from open flames. Wear gloves. Add 300 g of reagent-grade acrylamide and 4 liters of chloroform to the 6-liter Erlenmeyer flask. Add a stir-bar and gently heat the mixture on a hot plate to about 50°C or until the acrylamide is dissolved. The hot solution is then filtered through a large funnel equipped with filter paper into a clean, dry 4-liter reagent bottle. Keep the funnel fairly full of the hot mixture to prevent cooling, which might result in the acrylamide's beginning to crystallize and clog the filter. Loosely cap the reagent bottle containing the filtrate and place it in an explosion-proof −20°C freezer overnight. Also, chill an additional liter of chloroform at −20°C.

On the following day, remove the crystals of acrylamide while still cold by vacuum filtration through a Whatman No. 1 filter paper on a Buchner funnel. Press as much of the chloroform as possible out with a spatula, then wash with 1 liter of cold (−20°C) chloroform. Place the funnel in the vacuum dessicator for ~1 day. The acrylamide can be stored dry in a capped bottle or other closed vessel at room temperature.

Molecular and Chemical Characterization of Membrane Receptors, pages 115–123

4

Peptide Mapping as a Tool to Characterize Components of Hormone Receptor Systems

Gary L. Johnson and Thomas H. Hudson

Department of Biochemistry, University of Massachusetts Medical Center, Worcester, Massachusetts 01605 (G.L.J.) and Section on Biophysical Chemistry, Laboratory of Neurochemistry, National Institute of Mental Health, Bethesda, Maryland 20205 (T.H.H.)

INTRODUCTION

Purification and characterization of membrane-associated proteins involved in mediating or regulating hormone action is an important but difficult task. Recent progress in specifically labeling or purifying membrane proteins involved in the action of several different hormone recognition systems has made structural analysis of these proteins feasible. Hormone receptor and regulatory proteins which have been purified and whose structure is beginning to be understood include those for insulin, acetylcholine, β-adrenergic and α-adrenergic agents, asialoglycoproteins, transferrin, low-density lipoproteins, and the guanine nucleotide regulatory protein of adenylate cyclase. In several instances techniques are available to specifically covalently label these proteins using either labeled ligand and photoaffinity labeling or chemical crosslinking procedures [Pilch and Czech, 1979; Jacobs et al, 1979; Rashidbaigi and Ruoho, 1981; Shorr et al 1982; Das et al, 1977; see also Pilch and Czech, 1980; Ruoho et al,this series]. For the guanine nucleotide regulatory component of adenylate cyclase it is possible to specifically covalently label subunits of the protein using cholera toxin which uses NAD to enzymatically ADP-ribosylate the protein [Johnson et al, 1978; Gill and Meren, 1978; Cassel and Pfeuffer, 1978].

Electrophoresis in gels containing sodium dodecyl sulfate (SDS) has proven a powerful tool for the separation and characterization of the size of these proteins. However, these proteins are often oligomeric, composed of heterogenous subunits by their multiple bands on SDS-acrylamide gels, and structural analysis of their relationships is desirable. Peptide mapping either by complete or partial proteolytic digestion proves to be an easy and powerful technique to characterize structural relationships between subunits of proteins involved in binding and transmitting appropriate signals to the cell. In addition, if the receptor or regulatory proteins can be specifically labeled in intact cells, membranes, or in a functional solubilized form, partial proteolytic digestion and SDS-acrylamide gel analysis may yield useful information relating to structural changes during hormone binding or interaction of these proteins with other cellular components. In this chapter, we will outline several of the techniques which have proven useful in our structural analysis of the guanine nucleotide regulatory component of adenylate cyclase and how they are generally applicable to many other systems.

METHODS OF PEPTIDE MAPPING

Conventional peptide mapping usually involves a two-dimensional system where electrophoresis and chromatography are used sequentially after spotting the digest on thin-layer plates or paper [Gracy, 1976; Dahmus, 1981]. A more recent advance in peptide mapping that is generally more applicable to hormone receptor systems, if the desired protein is impure, is the proteolytic digestion of protein substrates resolved by prior electrophoresis on SDS-acrylamide gels. The peptides can be digested while remaining in the gel itself, or eluted electrophoretically and digested. The digests are then run on a second-dimension SDS-gel and analyzed by staining or autoradiography if they are radiolabeled. Several reports [Nathanson and Hall, 1979; Pilch and Czech, 1980; Hudson and Johnson, 1980, 1981a] using these procedures have appeared and will be outlined below.

Cleveland et al [1977] first devised a procedure whereby proteins resolved by SDS-electrophoresis were cut from the gel and the protein eluted electrophoretically in a dialysis bag. The proteins eluted from the gel piece were then precipitated using tricholoroacetic acid at a final concentration of 20%. The precipitate was collected by centrifugation washed twice with ethyl ether, and resuspended in a sample buffer of 1.25 mM Tris/HC1 at pH 6.8, 0.5% SDS, 10% glycerol, and 0.0001% bromophenol blue. Samples were heated at 100°C for 2 min and then cooled. Proteolytic digestions were performed for 30 min at 37°C with varying amounts of different proteases. Proteolysis was stopped by the addition of 2-mercaptoethanol and SDS to final concentrations of 10% and 2%, respectively, followed by boiling of the samples for 2 min. Samples were then loaded into wells of a 15% acrylamide gel and the samples electrophoresed in a manner similar to that described by Laemmli [1970]. Using Coomassie blue staining techniques and loading 10 to 15 μg protein from each sample, up to 20 peptide bands could be detected using this procedure.

A modification of this procedure was also shown by Cleveland et al [1977] to be quite satisfactory which involved proteolytic digestion without prior elution of the protein sample from the first gel slice. The gel slice was placed in the sample wells of a second SDS gel and then overlayed with protease. Appropriate proteases are not denatured during the subsequent electrophoresis and digestion occurs directly in the stacking gel. The best results have been found to occur when the power is turned off just before the bromophenol blue tracking dye reaches the bottom of the stacking gel. The gels are allowed to sit for 30 min at room temperature, during which time the proteases digest the sample proteins. The power is then turned on again and the partial proteolytic digest products separated on a 15% acrylamide gel. This procedure is particularly well suited for radioactively labeled proteins in which enough radioactivity is found in one gel band to generate a number of labeled digest fragments. If the sample is not radioactively labeled, it has been found that at least 10 μg of protein is needed to detect digest fragments by Coomassie blue staining. Recent advances in silver staining techniques [Oakley et al, 1980; Merril et al, 1979] should reduce this to 1–2 μg of protein, although with small proteins silver staining has been found to be inefficient.

An additional modification of the procedure described by Cleveland et al [1977] involved the ability to analyze by partial proteolytic digestion a heterogeneous sample without any purification of the individual components of the mixture [Siegel et al, 1981]. Samples can be electrophoresed on either a tube or slab first dimension SDS-acrylamide gel.At the end of the first electrophoresis the gel is incubated in 125 mM Tris/HCl, pH 6.8, 0.1% SDS for 40 min at room temperature. If a slab gel has been used for the first dimension the sample lane must be excised with a razor blade. The first-dimension gel lane is then layed over a second dimension slab gel. The first- and second-dimension gels are solidified using a molten 1% agarose solution held at 42–43°C in 125 mM Tris/HCl, pH 6.8, 0.1% SDS. When the agarose solidifies, the gel is placed in the electrophoresis apparatus and overlayed with a second 1 ml of 1% agarose that has been maintained at 42°C with the appropriate concentration of a protease added to this solution immediately before layering evenly over the first-dimension gel that has been molded to the second-dimension gel. The co-stacking of both protease and samples occurs in the second-dimension stacking gel. The amount of proteolysis is controlled by the amount of protease overlayed on the sample gel and the duration of time the samples are in the stacking gel. The partial proteolytic digest products are then separated in the separating gel, which is usually 15% acrylamide.

PROCEDURES FOR PEPTIDE MAPPING

Gel Electrophoresis

Electrophoresis in either tube or slab gels containing SDS is done using the discontinuous system described by Laemmli [1977]. Gels may be stained in a solution of 1% Coomassie blue, 50% ethanol, and 10% acetic acid. To avoid possible acid hydrolysis of important peptides staining for subsequent localization of gel bands should be for no more than 30 min and destaining for 1 h or less.

Electrophoretic Elution of Proteins From Gels

Figure 1 shows a simple apparatus to elute proteins from gel slices. The gel slice containing the appropriate sample is frozen and ground using a mortar and pestle and then placed in a dialysis bag containing ten-fold diluted SDS-running buffer having a final concentration of 2.5 mM Tris base, 19.2 mM glycine, and 0.01% SDS. The dialysis bag is then immersed in the Plexiglass box containing the same buffer. The box, containing two

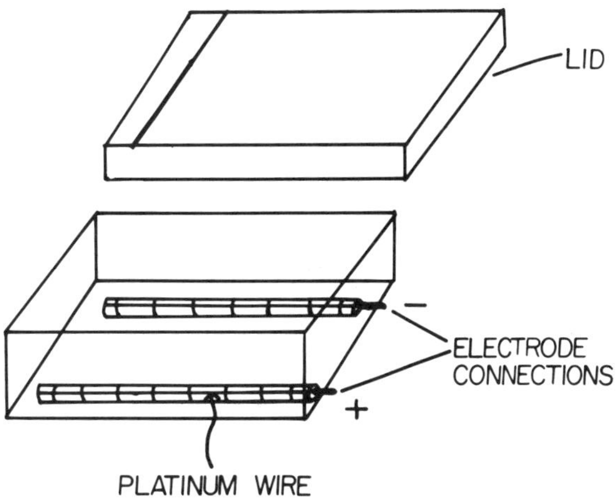

Fig. 1. *Electrophoresis apparatus to elute proteins from acrylamide gel slices contained in a dialysis bag.*

platinum electrodes, is connected to a power supply and the samples eluted from the gel using 2–10 mA for 8–12 h. The power is turned off and the contents of the dialysis bag removed using a Pasteur pipette. Gel residue is removed by centrifugation, and the sample protein, now in the diluted running buffer, is collected.

Iodination of Protein Samples in Gel Slices

In some instances it may be possible to isolate from a gel a specific protein corresponding to a subunit of a hormone receptor or regulatory protein. The protein may have been identified originally by covalently cross-linking labeled hormone, for example, to the receptor followed by first-dimension gel electrophoresis. The band may be located by autoradiography, and it is now desirable to perform some peptide mapping for analysis of subunit structure. This band can be easily iodinated by first mincing the gel slice and iodinating using a standard chloramine-T procedure. The protein in the minced gel slice is iodinated after thorough washing of the gel slices in 0.3 M sodium phosphate, pH 7.2, and the gel washed thoroughly by repeated dilution and centrifugation to remove free iodine after the iodination. The gel pieces can then be placed in a sample well of a second-dimension acrylamide gel and overlayed with protease and electrophoresed as described above.

Preparation of Proteases

The partial proteolytic digestion and peptide mapping procedures described here require stability of the protease in moderately low SDS concentrations and in some instances in temperatures of 42–43°C for short periods of time. The most appropriate proteases for these procedures appear in general to be *Staphylococcus aureus* V8 protease, elastase, chymotrypsin, and papain. Trypsin is generally unstable in even low SDS concentrations and is unsuitable for these procedures. Varying amounts of protease must be tested for each different protein sample to optimize concentrations and time of proteolysis, but a general reference point is 0.4 to 25 μg per ml of protease and digestion for 15 to 60 min.

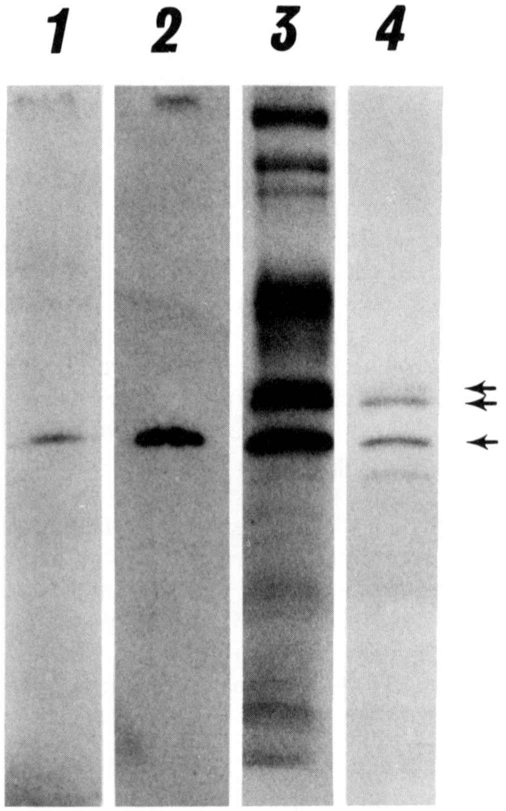

Fig. 2. *Autoradiography of SDS-polyacrylamide gel electrophoresis of detergent extracts of membranes incubated with [^{32}P]NAD and cholera toxin. Lanes represent human erythrocyte membranes (1), pigeon erythrocyte membranes (2), wild-type S49 membranes (3), and HTC4 membranes (4). Arrows indicated the Mr = 52,000/53,000 and Mr = 42,000 subunits of the guanine nucleotide regulatory protein of adenylate cyclase.*

AN EXAMPLE: PEPTIDE MAPPING OF THE Mr 42,000 SUBUNIT OF THE GUANINE NUCLEOTIDE REGULATORY PROTEIN OF ADENYLATE CYCLASE

Figure 2 shows the radiolabeled pattern of membrane extracts from pigeon and human erythrocytes, S49 mouse lymphoma, and HTC4 rat hepatoma cells after incubation with [^{32}P]NAD and cholera toxin. It has been previously shown that cholera toxin uses NAD to ADP-ribosylate the Mr = 42,000 and 52,000 peptides, which are subunits of the guanine nucleotide regulatory protein of adenylate cyclase [Cassel and Pfeuffer, 1978; Northup et al, 1980]. Since there was more than one specifically labeled peptide in S49 and HTC4 membranes as well as several non-specifically labeled peptides in S49 lymphoma membranes, proteolysis for peptide mapping of the guanine nucleotide regulatory protein subunits could not be done in membranes or detergent extracts. Therefore, a modification of the procedure described by Cleveland et al [1977] was used to generate peptide maps. Figure 3 shows two findings: (1) Peptides generated from the Mr 42,000 and 52,000 cholera toxin-labeled proteins are similar, suggesting the proteins are closely related. (2) The maps from S49- and HTC4-labeled peptides are very similar, supporting the finding that the function and structure of the guanine nucleotide regulatory protein is conserved.

Using pigeon erthrocyte membranes, where there is only the Mr 42,000 cholera toxin substrate, we have been able to extend the peptide mapping procedures to study functional changes in the guanine nucleotide regulatory protein of adenylate cyclase. Figure 4 shows that with cholera toxin-labeled membranes treatment with trypsin results in the partial digestion of the Mr = 42,000 radiolabeled peptide and production of several minor trypsin-specific fragments. Incubation with GTPγS alters the partial tryptic digestion of the Mr = 42,000 radiolabeled peptide, resulting in the quantitative generation of a Mr = 41,000 peptide. From a number of experiments we demonstrated that GTPγS causes a conformational change of the guanine nucleotide regulatory protein, exposing a new site to tryptic cleavage [Hudson and Johnson, 1981b]. GDPβS, which unlike GTPγS does not activate adenylate cyclase, did not cause the generation of the Mr = 41,000 tryptic fragment. However, GDPβS does competitively inhibit GTPγS action for both cyclase

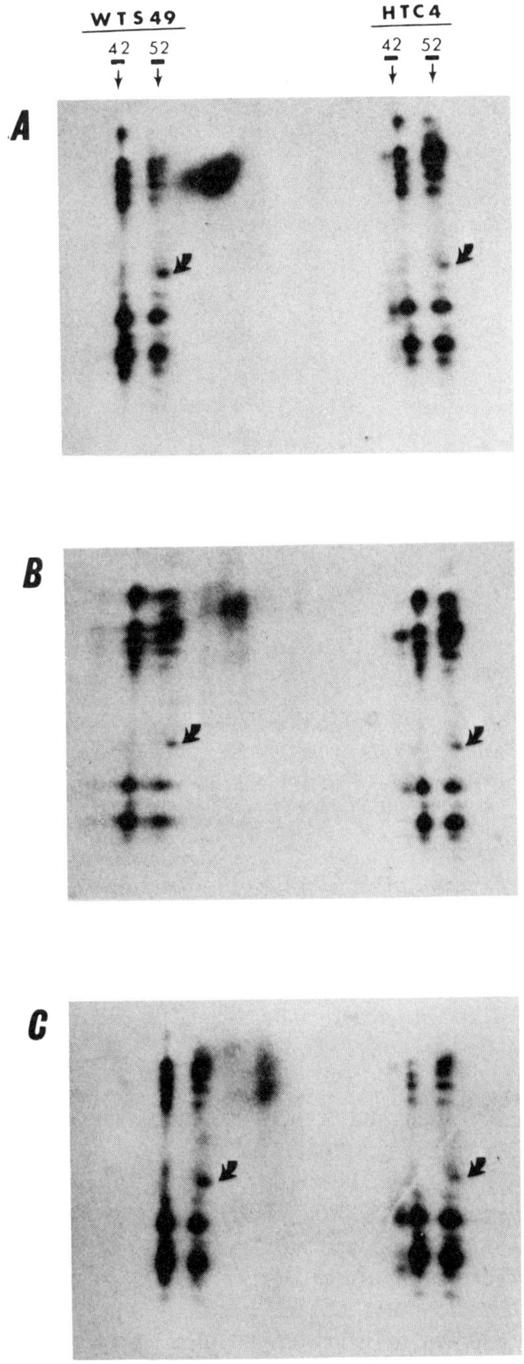

Fig. 3. *Autoradiographs of partial proteolytic digests of the Mr = 42,000 and 52,000/53,000 cholera toxin substrates of wild-type S49 and HTC4 membranes. Membranes were labeled with [^{32}P]NAD in the presence of cholera toxin. A two-dimensional modification of the procedures of Cleveland et al [1977] was used to generate partial proteolytic digests. The first dimension was a 10% polyacrylamide-SDS gel. This first-dimension SDS gel was then placed on top of a discontinuous SDS gel system containing 15% acrylamide in the separating gel. A 1% agarose gel was used to keep the first-dimension SDS gel in place and also contained the specific proteolytic enzyme. Electrophoresis was performed in the normal manner except that the current was turned off for 30 min when the bromophenol blue dye was near the bottom of the stacking gel. The peptide fragments generated were then separated on a 15% acrylamide-separating gel. A)* S. aureus *V8 protease, 12.5 μg. B) Elastase, 8 μg. C) Chymotrypsin, 50 μg.*

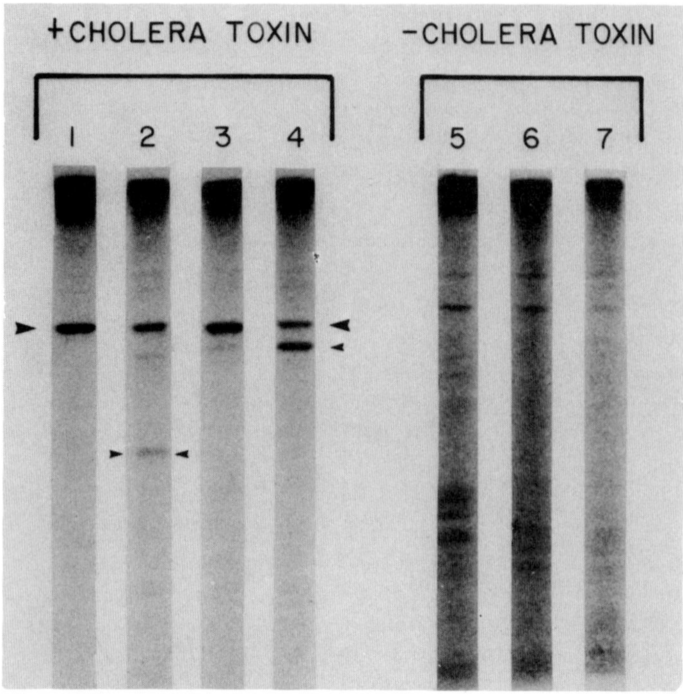

Fig. 4. *Effects of guanine nucleotides on tryptic digestion of the Mr = 42,000 cholera toxin substrate. Pigeon erythrocyte membranes were labeled with [^{32}P]NAD in the presence (lanes 1–4) or absence (lanes 5–7) of cholera toxin. Membranes at 5 mg of protein/ml were incubated for 45 min at 22°C with either no additions, or 100 μM GDPβS or GTPγS. Aliquots from these incubations were then incubated in the absence or presence of 25 μg/ml trypsin for 12 min. Digestion was stopped using aprotinin, and the samples were prepared and subjected to SDS-acrylamide electrophoresis and autoradiographed. The large arrows mark the Mr = 42,000 cholera toxin substrate, and the small arrows designate the Mr = 41,000 and 30,000 peptide fragments of the Mr = 42,000 peptide. Lanes 1 and 5, no guanine nucleotide, no trypsin; lane 2, no guanine nucleotide, plus trypsin; lanes 3 and 6, GDPβS plus trypsin; lanes 4 and 7, GTPγS plus trypsin. Autoradiographs of lanes 1 and 4 were exposed for 36 h; whereas lanes 5 and 7 were exposed for 72 h in order to bring out any low abundance fragments which might interfere with the analysis of fragments generated specifically from the Mr = 42,000 cholera toxin substrate. No difference in Coomassie blue staining patterns of high abundance proteins was observed in the presence or absence of GDPβS and GTPγS or with or without cholera toxin treatment.*

activation and tryptic digestion, suggesting that their action is mediated by a common binding site. It was also possible to demonstrate that isoproterenol diminished the lag time for adenylate cyclase activation and the generation of the Mr = 41,000 fragment. Furthermore, tryptic digests of membranes exposed to isoproterenol alone results in a profile on SDS gels containing the Mr = 42,000 radiolabeled protein and a specific Mr = 38,000 fragment generated from the Mr = 42,000 protein. Thus, cholera toxin labeling and tryptic digestion is able to measure specific states of the guanine nucleotide regulatory protein related to receptor coupling and guanine triphosphate binding.

Figure 5 shows that the action of isoproterenol and GTPγS on the sensitivity of the cholera toxin-labeled Mr = 42,000 subunit of the guanine nucleotide regulatory protein is reversible. After preloading the labeled regulatory protein with GTPγS, the addition of

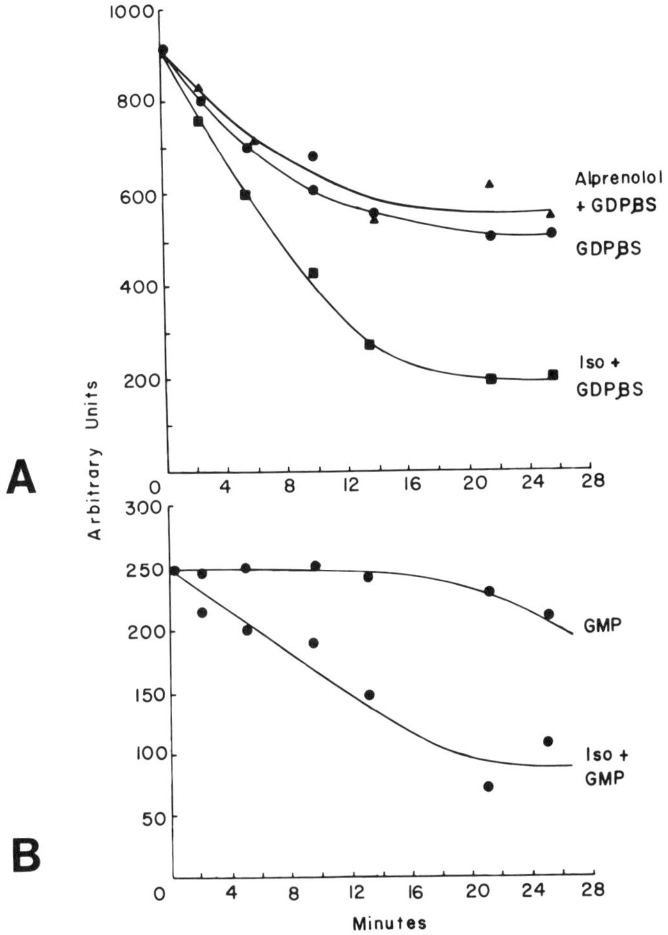

Fig. 5. *Reversal of GTPγS action on the proteolytic susceptibility of the Mr = 42,000 cholera toxin labeled pepetide. Pigeon erythrocyte membranes were labeled with [^{32}P]NAD and cholera toxin and then incubated for 45 min at 22°C in the presence of 100 μM GTPγS. This incubation maximally converts the guanine nucleotide regulatory protein of adenylate cyclase into a GTPγS-activated state. Trypsin efficiently digests the labeled Mr = 42,000 peptide to a Mr = 41,000 peptide as shown in Figure 4. The membranes were then incubated in the presence of 100 μM GDPβS (A) or GMP (B) in the presence or absence of the β-adrenergic receptor agonist isoproterenol (ISO, 1 μM) or the antagonist alprenolol (1 μM). At the indicated times aliquots were taken and the reaction stopped by dilution and centrifugation. Membranes were resuspended (5 mg/ml) and incubated with trypsin (25 μg/ml) for 12 min at 22°C. Samples were prepared and run on 10% acrylamide-SDS gels, fixed dried and autoradiographed. The Mr= 42,000 and Mr = 41,000 peptides were then quantitated using scanning densitometry. Figure shows that isoproternol in the presence of GDPβS or GMP is capable of rapidly reversing the activated state induced by GTPγS which can be measured both enzymatically and structurally.*

isoproterenol and GDPβS or GMP is capable of reversing both tryptic generation of the Mr = 41,000 fragment and adenylate cyclase activity, presumably by competitive displacement of GTPγS from the regulatory protein. Thus, the tryptic formation of the Mr = 41,000 fragment is a measure of the activated state of the guanine nucleotide regulatory protein.

This ability to correlate changes in tryptic digest patterns with functional activity of the guanine nucleotide regulatory protein has been used to measure functional alterations in components of pigeon erythrocyte adenylate cyclase following desensitization to isoproterenol [Hudson and Johnson, 1981b]. These procedures, performed using control and desensitized membranes, were used to correlate structural and functional changes in the regulatory protein following desensitization. These changes included (1) that isoproterenol has a diminished ability to alter the conformation of the G-protein in desensitized membranes, (2) that a lag in the conformational change induced by GTPγS in the presence or absence of isoproterenol is observed in desensitized preparations, and (3) that the fraction of G-proteins found in the GTPγS-specific conformation is significantly decreased in desensitized versus control membranes incubated with GTPγS with or without isoproterenol. These findings were interpreted to indicate that alterations occur during desensitization which affect coupling of hormone receptor to the regulatory protein and the GDP exchange reaction of adenylate cyclase. The results are dependent on the ability to measure adenylate cyclase. The only procedure available to do this has been by cholera toxin labeling, followed by proteolytic digestion and quantitation of peptide maps generated from different states of the hormone-sensitive adenylate cyclase system.

ACKNOWLEDGMENTS

This work was supported by USPHS grant GM30324, the Juvenile Diabetes Foundation, and the American Heart Association. G.L.J. is an Established Investigator of the American Heart Association.

REFERENCES

Cassel D, Pfeuffer T (1978); Mechanism of cholera toxin action: Covalent modification of the guanyl nucleotide-binding protein of the adenylate cyclase system. Proc Natl Acad Sci USA 75:2669–2673.

Cleveland DW, Fischer SG, Kirschner MW, Laemmli UK (1977): Peptide mapping by limited proteolysis in sodium dodecyl sulfate and analysis by gel electrophoresis. J Biol Chem 252:1102–1106.

Dahmus ME (1981): Calf thymus DNA polymerases I and II do not contain subunits structurally related to casein kinases I and II. J Biol Chem 256:1123–1124.

Das M, Miyakawa T, Fox CF, Pruss RM, Aharonov A, Herschman HR (1977): Specific radiolabeling of a cell surface receptor for epidermal growth factor. Proc Natl Acad Sci USA 74:2790–2794.

Gill DM, Meren R (1978): ADP-ribosylation of membrane proteins catalyzed by cholera toxin: Basis of the activation of adenylate cyclase. Proc Natl Acad Sci USA 75:3050–3054.

Gracy RW (1976): Two dimensional thin-layer methods. Methods Enzymol XLVII, pp 195–204.

Hudson TH, Johnson GL (1980): Peptide mapping of adenylate cyclase regulatory proteins that are cholera toxin substrates. J Biol Chem 256:7480–7486.

Hudson TH, Johnson GL (1981a): Conformational changes of adenylate cyclase regulatory proteins mediated by guanine nucleotides. J Biol Chem 256:1459–1465.

Hudson TH, Johnson GL (1981b): Functional alterations in components of pigeon erythrocyte adenylate cyclase following desensitization to isoproterenol. Mol Pharmacol 20:694–703.

Jacobs S, Hazum E, Schechter Y, Cautrecasas P (1979): Insulin receptor: Covalent labeling and identification of subunits. Proc Natl Acad Sci USA 76:4918–4921.

Johnson GL, Kaslow HR, Bourne HR (1978): Genetic evidence that cholera toxin substrates are regulatory components of adenylate cyclase. J Biol Chem 253:720–7123.

Laemmli UK (1970): Cleavage of structural proteins during the assembly of the head of bacteriophage T4. Nature 227:680–685.

Merril CR, Switzer RC, Von Keureu ML (1979): Trace polypeptides in cellular extracts and human body fluids detected by two-dimensional electrophoresis and a highly sensitive silver stain. Proc Natl Acad Sci USA 76:4335–4339.

Nathanson N, Hall ZW (1979): Subunit structure and peptide mapping of junctional and extrajunctional acetylcholine receptors from rat muscle. Biochemistry 18:3392–3401.

Northup JK, Sternweiss PC, Smigel MD, Schleifer LS, Ross EM, Gilman AG (1980): Purification of the regulatory component of adenylate cyclase. Proc Natl Acad Sci USA 77:6516–6520.

Oakley BR, Kirsch DR, Morris NR (1980): A simplified ultrasensitive silver stain for detecting proteins in polyacrylamide gels. Anal Biochem 105:361–363.

Pilch PF, Czech MP (1979): Interaction of crosslinking agents with the insulin effector system of isolated fat cells. J Biol Chem 254:3375–3381.

Pilch PF, Czech MP (1980): Hormone binding alters the conformation of the insulin receptor. Science 210:1152–1153.

Rashidbaigi A, Ruoho AE (1981): Iodoazidobenzylpindolol, a photoaffinity probe for the β-adrenergic receptor. Proc Natl Acad Sci USA 78:1609–1613.

Siegel TW, Ganguly S, Jacobs S, Rosen DM, Rubin CS (1981): Purification and properties of the human placental insulin receptor. J Biol Chem 256:9266–9273.

Shorr RGL, Heald SL, Jeffs PW, Lavin TN, Strohsacker MW, Lefkowitz RJ, Caron MG (1982): The β-adrenergic receptor: Rapid purification and covalent labeling by photoaffinity crosslinking. Proc Natl Acad Sci USA 79:2778–2782.

Molecular and Chemical Characterization of Membrane Receptors, pages 125–160

5

Determination of Carbohydrate Structures in Glycoproteins and Glycolipids

Heinz Baumann and Darrell Doyle

Department of Cell and Tumor Biology, Roswell Park Memorial Institute, Buffalo, New York 14263 (H.B.) and Department of Biological Sciences, State University of New York at Buffalo, Buffalo, New York 14214 (D.D.)

INTRODUCTION

Glycoconjugates, due to their characteristic localization within cells and tissues of eukaryotic organisms, exercise functions mainly connected with cell-cell interactions and cellular communication. In many cases, to learn the biological significance of the carbohydrates in glycoconjugates, structural properties of the sugar moieties had first to be established. A few examples can illustrate how new perspectives in our understanding of the biological role of glycoconjugates were gained by knowledge of the carbohydrate structures.

Carbohydrate moieties of glycolipids and glycoproteins which are displayed on cell surfaces can act as immunogenic determinants. Some of these determinants change during development and differentiation [Jacob, 1977; Solter and Knowles, 1979]. After elucidation of the structures responsible for the immunogenic properties, a clear definition of the antigens was possible, eg, the case of blood group antigens [Szulman, 1977; Stern et al, 1978; Kapadia et al, 1981; Muramatsu et al, 1979]. Comparisons of the hapten structures indicated functional and biogenetic relationships of different molecules, like the similarity showed by I- and i-antigens with other blood group antigens [Watanabe et al, 1979]. Moreover, the information accumulated on the structural properties of antigenic carbohydrate moieties allowed a classification of the developmentally regulated antigens [Kannagi et al, 1982].

Carbohydrate structures not only determine the type of blood group antigens but are also important in the immunogenic behavior of polysaccharides [Glaudemans, 1975]. Again, sugar sequence and arrangements of glycosidic bonds play a major factor in establishing the antigenic determinants [Pazur et al, 1976; Pazur and Forsberg, 1978].

Carbohydrate structures of plasma membrane glycoconjugates were found to be influenced by growth and malignant transformation [Hakomori et al, 1974]. Only by determining the structures of the affected carbohydrate moieties could conclusions be made about the effect of growth on biogenesis of the particular glycoproteins and glycolipids [Larraga and Edidin, 1972; Hakomori, 1975; Critchley, 1979; Hunt et al, 1981; Funakoshi and Yamashina, 1982; Basu and Basu, 1982].

Tissue culture cells, which were selected for their resistance to cytotoxic plant lectins, often have a change in the structure of the carbohydrate moieties of their externally displayed membrane glycoconjugates [Stanley and Sudo, 1981; Hunt, 1981]. The biochemical nature of the structural alterations helped not only to pinpoint functional importance of the carbohydrates [Stanley and Sudo, 1981], but also to identify the enzyme activities involved in the biosynthetic pathways of the glycoconjugate, such as a specific glycosyl transferase activity [Cummings et al, 1982] or a processing glycosidase activity [Reitman et al, 1982].

The recognition of a certain carbohydrate structure in intermediate and in end products of metabolism allows conclusions about the action of distinct modifying enzymes as contributors to the biosynthesis of the glycoconjugates [Kornfeld, 1982; Byrd et al, 1982]. Once the presence of specific enzymes, like glycosyltransferases, is predicted from carbohydrate analyses [Ando and Yu, 1979; Weinstein et al, 1982a], these activities can be characterized, and the enzyme proteins can be isolated [Yohe et al, 1982; Weinstein et al, 1982b].

The presence of carbohydrate moieties in many glycoproteins have been found to be essential in providing functionality to, and in preserving structural integrity of, the molecule [Geoghegan et al, 1980; Pierce and Parsons, 1981]. When the structure of the carbohydrate moieties is known, this can then be specifically modified, degraded, or removed; the effect of the alterations on the biological activity [Bennett and Schmid, 1980; Kalyan et al, 1982; Kalyan and Bahl, 1981, 1983] or metabolic stability [Bernhard et al, 1982] can then be assessed.

Comparison of structural features among glycoconjugates allows, to some degree, predictions about the role played by the carbohydrates [Rauvala and Finne, 1979]. For example, if the carbohydrate moiety of a molecule with unknown function is structurally similar to that of another compound whose functional importance is already established, analagous biological properties might be expected and often are detected (for instance, similarities in the case of histocompatibility antigens [Freed and Nathenson, 1977] or toxins and glycoprotein hormones [Ledley et al, 1976]).

During the last few decades, enormous progress has been made in the methodology available for analyses of carbohydrate structures. Several of the most important techniques include methylation analysis, enzymatic degradation, mass spectroscopy, and nuclear magnetic resonance spectroscopy. The result from the efforts of numerous laboratories is that many individual carbohydrate structures are now known [Montreuil, 1975; Kornfeld and Kornfeld, 1976; Macher and Sweeley, 1978; Kobata, 1977; Kobata et al, 1978; Smith et al, 1978; Lundblad, 1977, 1978; Herp et al, 1979; Ledeen and Yu, 1982]. Because of the fact that structures within a given molecular class exhibit common features [Dawson, 1978], the experimental approach in elucidating an unknown structure is facilitated or can be predicted using the wealth of available reference material and information already established.

Elucidation of the primary structures of carbohydrates in glycoconjugates requires the following steps:

1. Purification of the carbohydrate structure.
2. Determination of the identity and molecular ratio of the monosaccharide residues composing the carbohydrate structure.
3. Sequence of the carbohydrate residues.
4. Arrangement of the glycosidic linkages.
5. Anomeric configuration of the glycosidic bonds.
6. If noncarbohydrate residues such as sulfate, phosphate, or alkyl-ethers/esters are present, identity and site of attachment to the carbohydrate.

In addition to the above, two determinations are required to complete a structural characterization: absolute configuration (D or L) and ring form (pyranose or furanose) of each of the carbohydrate residues [McNeil et al, 1982]. Although such analyses were found to be essential in the cases of polysaccharides and glycosaminoglycans, they are usually not required for glycoproteins and glycolipids of animal origin. Numerous studies have confirmed that the latter glycoconjugates, which include those that are mainly associated with receptor structure and function, contain a rather restricted set of glycosides whose physicochemical properties are constant. The principal monosaccharides constituting carbohydrate moieties of glycoproteins and glycolipids are the following:

Hexoses:	D-mannose, D-glucose, D-galactose
Amino sugars:	2-acetamido-2-deoxy-D-glucose (N-acetyl-D-glucosamine); 2-acetamido-2-deoxy-D-galactose (N-acetyl-D-galactosamine)
Deoxy sugar:	L-fucose
Uronic acids:	D-glucuronic acid, D-galacturonic acid
Sialic acids:	Derivatives of neuraminic acids (5-amino-3,5-dideoxy-D-glycero-D-galacto-nonulasonic acid)
Pentose:	D-xylose

Considering the topic of these volumes, we will restrict our presentation primarily to the analysis of carbohydrates in glycoproteins and glycosphingolipids. Since all techniques essential for structure determination are described in detail by specialists and each major method has been dealt with in excellent reviews, we shall present here only the approaches generally chosen to elucidate carbohydrate structures and shall refer wherever we can to appropriate publications for reference. We have selected some specific examples to illustrate the results of a particular technique or approach. The choice of the references used or omission of other references is not meant to be an indication of merit. Further, this section because of space limitations, cannot be a complete treatise of the topic.

PURIFICATION OF THE CARBOHYDRATE STRUCTURE

An absolute requirement for the successful analysis of carbohydrate is that the structure first be purified to physicochemical homogeneity. The criteria for homogeneity are the same as those demanded for intact isolated glycoconjugates. Major difficulties in purification of carbohydrates arise from their naturally occurring structural heterogeneity, which in some cases can be of remarkable complexity [Fournet et al, 1978; Yoshima et al, 1981a]. Although analytic procedures, such as high-resolution proton nuclear magnetic resonance (NMR) spectroscopy [Endo et al, 1982], allow an assessment of microheterogeneity in a given sample, most other methods of analysis depend on the purity of the sample for unambiguous interpretation of the data.

In addition to purity, the quantity of sample material necessary for a complete structural analysis is of major importance. In cases where either structural analogies are known, comparative material is available, or highly informative techniques such as mass spectroscopy, methylation or NMR spectroscopy are accessible, only microgram quantities of pure compounds are needed. However, when one deals with unknown structures or lacks sophisticated and extensive instrumentation, a complete analysis requires considerably more material.

Glycolipids

In mammals, two major types of glycolipids can be found, glycosphingolipids and glycoglycerolipids. The difference between these two categories lies in the nature of the noncarbohydrate moiety. The hydrophobic part of glycosphingolipids (ceramide) is composed of one molecule sphinganine, 4-sphingenine (=sphingosine) or their derivatives, to which one molecule of fatty acid is bound through amide linkages. The carbohydrate is attached by a glycosidic linkage to the primary hydroxyl group of the sphingolipid base. In glycoglycerolipids, the lipid moiety consists of a 1,2-diacyl-*sn*-glycerol ("diglyceride"). The carbohydrates are bound via a glycosidic linkage to the C-3 position of the glycerol [Burton, 1976; Sweeley and Siddiqui, 1977]. Based on the nature of the residues attached to the carbohydrate moieties, the glycosphingolipids are divided into neutral glycosphingolipids and sialic acid containing glycosphingolipids or gangliosides. Furthermore, in both glycolipid categories, there is an additional subgroup which is characterized by the content of sulfate esters on the carbohydrate moiety (Sulfatoglycophingolipids (=sulfatides) and sulfatoglycoglycerolipids) [Kiss, 1969; Ishizuka et al, 1973; Murray et al, 1976; Sweeley and Siddiqui, 1977].

Besides the two principal glycolipid classes found in animals, there is a third class, the glycosides of steroids, which are, however, confined mainly to plants [Elbein and Forsee, 1976]. For that reason, and due to their specialized chemistry [Lepage, 1964], we will not discuss these glycolipids.

Since all glycolipids contain one carbohydrate structure per molecule, it is sufficient in most cases to purity the intact glycolipid to obtain their carbohydrate moieties in homogenous form which are then suitable for structural analysis [Sweeley and Siddiqui, 1977; Ledeen and Yu, 1982]. There is one property of glycolipids which can contribute to some complications: When glycolipid species which have compositional homogeneity of the carbohydrate structure are purified, there might still be heterogeneity in the physicochemical behavior of the preparation. In general, this heterogeneity is due either to difference in lipid moieties, such as fatty acids [Svennerholm and Stallberg-Stenhagen, 1968] or due to differences in glycosidic linkages and anomeric configuration [Esselman et al, 1972].

Methods for purification of glycolipids are described in detail by Kaufer [1969], Esselman et al [1972], Laine et al [1972, 1974], Hakomori and Watanabe [1976], Sweeley and Siddiqui [1976], Veerkamp and Broekhuyse [1976], Ledeen and Yu [1982]. The principal elements consist of (1) extraction of lipid, (2) separation of the different lipid classes, (3) separation of glycosphingolipids into ganglioside and neutral glycosphingolipids, (4) separation of the individual glycolipid species, and under some circumstances, (5) removal and purification of the carbohydrate moiety. Extraction of glycolipids from tissue, cells, subcellular fractions, or preextracted material is generally achieved by homogenization of the material in mixture of chloroform-methanol followed by partition with aqueous KCl [Folch et al, 1957]. To improve the yield of glycolipids, modification of the basic organic solvent extraction have been described [Suzuki, 1965; Hakomori and Siddiqui, 1974; Svennerholm and Fredman, 1980; Ledeen and Yu, 1982]. Extraction of gangliosides with tetrahydrofuran [Tettamanti et al, 1973] was found to be more efficient than with chloroform-methanol and had at the same time the advantage that fewer glycoproteins were extracted [Brunngraber et al, 1976]. The same extraction technique was also used for neutral glycosphingolipids [Slomiany and Slomiany, 1975; Slomiany et al, 1976].

To avoid losses of specific glycolipids, especially glycolipids with fewer sugar residues or gangliosides, during partition between the chloroform-methanol phase and the methanol-water phase, chromatography procedures were developed allowing isolation of whole glycolipid classes. Isolation of the broadest spectrum of glycosphingolipid can be achieved by the method of Saito and Hakomori [1971]. The total lipid fraction is subjected to a peraceylation reaction and then chromatographed on Florisil [Radin, 1969]. The derivatized glycosphingolipids can be recovered from the column free of other lipids and be converted to the native forms by deacetylation with sodium methoxide. This method is, however, not applicable for isolation of glycolipids with alkaline sensitive residues.

A good recovery with relatively high purity of all ganglioside species, including the less polar ones, can be achieved by the procedure described by Ledeen and Yu [1982] which consists of the following steps: chromatography of the total lipid extract over anion exchange resins (DEAE-Sephadex), treatment of the fractions containing acidic lipids with sodium hydroxide to hydrolyze acidic phospholipids, and removal of contaminants (fatty acids, sulfatides, peptide proteins) by chromatography over a column of silicic acid. This technique requires caution during the base treatment step because alkaline labile residues such as O-acyl-groups could be lost.

To separate the glycolipid fraction into individual species, a variety of chromatography procedures have been employed. A combination of several techniques generally results in the biochemical homogeneity required for structural analysis. The following techniques are routinely used: column chromatography on DEAE-cellulose [Rouser et al, 1969; Hakomori and Watanabe, 1976], DEAE-Sephadex or Sepharose [Ando and Yu, 1977, 1979; Itoh et al, 1981; Fredman et al, 1980b]; chromatography on Anasil [Penick et al, 1966; Siddiqui and Hakomori, 1970]; silicic acid like silica gel G, H, or HR [Svennerholm, 1972; Yohe et al, 1980]; Biosil [Hakomori and Watanabe, 1976]; or Iatrobeads [Momoi et al, 1976; Sweeley and Siddiqui, 1977]. High-pressure liquid chromatography has been successful in separation with high resolution of complex glycolipid mixtures [Bremer et al, 1979; Watanabe and Arao, 1981; Fukuda and Hakomori, 1982; Kannagi et al, 1982].

A most commonly used technique in separation of glycolipids for both analytical and preparative purposes is chromatography on thin-layer silica gel plates [Penick et al, 1966; Skipski and Barclay, 1969; Esselman et al, 1972; Verkamp and Broekhuyse, 1976; Ledeen and Yu, 1982]. Homogeneity of the purified material is normally assessed by thin-layer chromatography using several solvent systems [Sweeley and Siddiqui, 1977] or by thin-layer chromatography before and after peracetylation [Stellner et al, 1973b; Hakomori and Watanabe, 1976]. The separated glycolipids are visualized by colorimetric staining such as with resorcinol [Svennerholm, 1957], orcinol [Esselman et al, 1972; DeVries and Norton, 1974], or anthrone [Roughan and Batt, 1968].

If only the carbohydrate structures of the glycolipids are of interest, these can be purified by a different approach—ie, after liberation from the lipid moiety. Three principal techniques are employed to release the intact oligosaccharides. Cleavage of the glycosidic linkage to the ceramide can be achieved first by oxidation of the double bond in the sphingosine base with ozone followed by a β-elimination reaction catalyzed by alkaline carbonate [Ozonolysis: Wiegandt and Bashang, 1965; Wiegandt and Bucking, 1970; Ledeen and Yu, 1982]. A similar cleavage reaction is obtained when glycolipids are treated with a combination of osmium tetroxide and periodate [Hakomori, 1966; Hakomori and Siddiqui, 1972; MacDonald et al, 1980]. Despite improvements of the original methods, yields of released carbohydrates are not quantitative. A more efficient and specific cleavage of the glycosidic bond to the springosine moiety is obtained when glycolipids are treated with a mixture of trifluoroacetic acid/ trifluoroacetic acid anhydride [Lundsten et al, 1979; Lindblad et al, 1980]. The trifluoroactolysis reaction does not lead to degradation of the carbohydrate structure except that N-acetyl neuraminic acid residues are rapidly destroyed. Only N-acetylneuraminic acids linked to the C-3 of pyranosidic sugars are not cleaved although they are being modified [Svensson, 1980]. Oligosaccharide structures recovered from glycolipids by either technique can be purified like those derived from glycoproteins (see below).

Glycoproteins

Liberation of carbohydrate structures. The carbohydrate moieties in glycoproteins are either attached via a glycosidic linkage to the hydroxyl groups of serine, threonine, hydroxylysine, or hydroxyproline residues, or via an amide linkage to asparagine residue [Zinn et al, 1977]. Since most glycoproteins contain more than one oligosaccharide side group per molecule [Spiro, 1973], the approach to obtain the carbohydrates for structural studies is different from that for glycolipids. The actual strategy employed will in part affect the value of the information of the whole glycoprotein. The following schemes should yield the most information: The glycoprotein is purified to physico-chemical homogeneity; after establishing a proteolytic peptide map, the glycopeptides, whose position within the protein is known, are isolated; from these glycopeptides the carbohydrate moieties are released, subjected to further purification and finally structurally analyzed. Although this procedure is most demanding, and therefore executed usually in only a few cases, the results obtained have the potential to show (1) the carbohydrate structures at a specific polypeptide site, (2) heterogeneity of the structures at a given site, and (3) through comparison with data of other proteins, the causal influence of polypeptide moiety on the biosynthesis of the carbohydrate structures [Anderson and Grimes, 1982; Hsieh et al, 1983a,b].

In most cases where the carbohydrate structures of a purified glycoprotein have been studied, no fragmentation and separation of glycopeptides are carried out prior to the release of the oligosaccharides [eg, Fournet et al, 1978; Takasaki et al, 1979; Yoshima et al, 1980, 1981a]. By this approach, one obtains a catalog of carbohydrate structures for a given glycoprotein but, due to the "homogenization" of oligosaccharides, no assignments of structures to specific sites on the polypeptide chain can be made. Even more restricted information is gained by analyzing the structures of carbohydrates which have been released from non-fractionated components such as cell membranes [Entlicher and Kocourek, 1976; Farrar and Harrison, 1978; Funakoshi and Yamashina, 1982].

Regardless of the biochemical purity of the material which serves as a source for the carbohydrate structure to be analyzed, sugar moieties have to be liberated from this material and purified to homogeneity. The actual strategy or sequence of treatments that has to be adopted in order to release specific groups of intact carbohydrates is dependent upon the structural composition of the glycoproteins—the presence of O-linked and N-linked glycans, or high mannose type and complex type of N-glycans [Kornfeld and Kornfeld, 1976]. In general, the following schemes have been used [eg, Chandrasekaran et al, 1981; Nilsson et al, 1982]: O-linked carbohydrates are removed first by alkaline treatment and separated from the protein. N-linked carbohydrates are then released either enzymatically or chemically mostly after complete destruction of the polypeptide chain. If all carbohydrates are released in a nondiscriminating fashion, eg, by trifluoroacetolysis [Lunbald et al, 1980], a subsequent isolation step has to be included which separates the normally small molecular-sized O-glycan from the larger N-glycan units [Finne and Krusius, 1979]. On the other hand, a most specific release of defined carbohydrate structures is achieved by endoglycosidase; but this enzymatic treatment is at present restricted to N-glycan units. In such cases the isolation of the carbohydrates is simple due to the limited number of structures present in the released material.

What are the methods employed to remove carbohydrate moieties from proteins? One of the most commonly used methods to release carbohydrates linked via N-acetylgalactosamine to serine or threonine residues is by an alkaline β-elimination reaction [Neuberger et al, 1972]. A variety of conditions have been described for this reaction [Downs and Pigman, 1976a]. The glycoproteins are treated with diluted KOH or NaOH (0.05 to 0.5 N) solutions containing 1 M $NaBH_4$ at temperatures between O° to 45°C and for times between a few hours and days. The high concentrations of borohydride results in reduction of the reducing end, which is necessary in order to prevent secondary β-elimination reactions leading to "peeling" of the carbohydrate structures [Carlson, 1968; Mayo and Carlson, 1970]. The degree of β-elimination is assessed either by immediate acid hydrolysis of the carbohydrates followed by quantitating the amount of formed alditol residues [Carlson, 1972; Spiro, 1972] or by measurements of the unsaturated amino acids formed from the β-eliminated hydroxy amino acids [Spiro, 1972; Downs and Pigman, 1976a]. In general, satisfactory removal of O-linked carbohydrates was achieved by using a mixture of 0.05 M KOH and 1 M $NaBH_4$ at 45°C for 15 h. More resistant to elimination are carbohydrates attached to serine or threonine residues with unsubstituted amino or carboxy termini [Zinn et al, 1977]. Such conditions are prevalent when the starting materials are glycopeptides derived from complete protease digestions of glycoproteins. An increase in the hydroxide concentration improved efficiency of the elimination reaction, but also observed were an increased number of side reactions [Downs and Pigman, 1976a].

Under the conditions routinely used for the β-elimination reaction, oligosaccharides linked to asparagine and to hydroxylysine are not significantly affected [Spiro, 1967, 1976a; Marshall and Neuberger, 1976]. In order to achieve a release of N-linked carbohydrates, treatment of the proteins with 1 M NaOH and 4 M $NaBH_4$ at 80°C for 24 h is required [Zinn et al, 1978].

Mild β-elimination reaction does not change the overall carbohydrate structure except that alkaline labile sialic acid residues are destroyed [Schauer, 1982]. During the elimination reaction, the polypeptide chain is subjected to fragmentation. Fragmentation can be reduced if demethylsulfoxide is included in the reaction mixture [Downs et al, 1973].

Since detection of carbohydrates during purification steps is facilitated when they are radioactively labeled, a β-elimination in the presence of NaB^3H_4 allows the introduction of tritium into otherwise nonlabeled oligosaccharide units [Aminoff et al, 1980; Blanchard et al, 1983].

Release of N-linked carbohydrates is achieved by hydrazinolysis [Yosizawa et al, 1966]. Glycopeptides of glycoproteins are suspended in anhydrous hydrazine and reacted for 8 to 12 h at 100°C [Takasaki et al, 1982]. Hydrazinolysis results in cleavage of the linkage between 2-acetamido-2-deoxyglucose and asparagine residues with formation of glycosyl amine or hydrazone. The reaction does not affect glycosidic linkages except that some neuraminiosyl linkages are cleaved. Hydrozinolysis includes also N-deacetylation of amino sugars [Fukuda et al, 1976; Saeed and Williams, 1980]. Complete deacetylation of certain 2 - acetamido - 2 - deoxyglucopyranosides, especially those glycosylated at the C-3 position, requires more vigorous hydrozinolysis [Bayard and Roux, 1975]. However, prolonged hydrozinolysis leads to various isomerization and degradation of the carbohydrate structure [Takasaki et al, 1982].

After hydrazinolysis, excess hydrazine is removed by sublimation [Schroeder, 1972]. To restore the reducing end of the released oligosaccharide moiety, the hydrazone products are treated with acetic acid [Bayard and Roux, 1975]. However, it was found that the conversion to reducing sugar is not always complete. Heterogeneity at the reducing end of oligosaccharides might influence subsequent methylation analysis, nitrous acid deamination, or periodate oxidation. Once the reducing termini are established, a treatment of the oligosaccharides with NaB^3H_4 allows an introduction of a radioactive label [Mizuochi et al, 1979, 1980]. Since hydrazinolysis caused deacetylation, the oligosaccharides have to be subjected to a re-N-acetylation reaction in order to restore the initial structure of the amino sugars [Roseman and Daffner, 1956; Reading et al, 1978]. If [^{14}C]-acetic anhydride is used for re-N-acetylation, a radioactive labeling of the carbohydrate moiety is achieved [Bayard and

Kerckaert, 1980, 1981]. Deacetylated carbohydrate moieties derived from hydrozinolysis can also serve as substrate for further specific cleavage reactions. A treatment of the deacetylated carbohydrates with nitrous acid [Cifonelli, 1968; Horton and Philips, 1973] leads to deamination of amino sugar residues to 2,5-anhydrohexoses, concomitant with specific acid-catalyzed cleavage of anhydrohexosyl linkages [Bayard and Fournet, 1976; Strecker et al, 1981].

Asparagine-linked carbohydrates can be most specifically removed from glycoproteins or glycopeptides by glycosidases [Kobata, 1979]. Enzymatic methods also offer, besides substrate specificity, the advantage that the polypeptide moiety is not affected, as is normally the case with chemical methods. Two kinds of glycosidases are available: One group consists of endo-β-N-acetylglucosaminases (endo H,D,C_I,C_{II},F) which hydrolyze the di-N-acetylchitobiosyl moiety of asparagine-oligosaccharides. The enzymes, therefore, release an intact carbohydrate unit minus the most proximal N-acetylglucosamine, which remains still attached to the asparagine residue. The substrate specificity of these enzymes is quite restricted and is mostly dependent upon the carbohydrate structures. The most well characterized and widely used enzyme is endo-β-N-acetylglucosaminidase H (endo H) of *Streptomyces griseus* [Tarentino and Maley, 1974; Tarentino et al, 1978; Tai et al, 1975]. Endo H cleaves preferentially nonmodified, high mannose types of N-glycans [Trimble and Maley, 1977; Trimble et al, 1978; Maley and Trimble, 1981]. This makes the enzyme a most useful tool to purify either early biosynthetic forms of asparagine-linked carbohydrate structures [Robbins et al, 1977] or forms not further modified during maturation of glycoproteins [Natowicz et al, 1982; Hsieh et al, 1983]. Furthermore, the selectivity of this enzyme allows the collection of distinct N-glycans from a given glycoprotein: those sensitive to the enzyme (high mannose type) and those resistant (modified and complex type) [Howard et al, 1982]. A similar but not identical substrate specificity as for endo H was found for two of the endo-β-N-acetylglucosaminidases C_I and C_{II} of *Clostridium perfringens* [Ito et al, 1975; Kobata, 1978; Tabas et al, 1978]. Somewhat more modified carbohydrate structures can be released by endo-β-N-acetylglucosaminidase D (endo D) from *Diplococcus pneumoniae* [Muramatsu, 1971, 1978]. This enzyme has received only limited application, mainly due to its high substrate specificity [Tai et al, 1975]. Endo D does not cleave the high mannose type of N-glycan but does cleave the complex types provided the peripheral sugars have been removed by exoglycosidases [Koide and Muramatsu, 1974].

Much different from the above endoglycosidases is the substrate specificity of endo-β-N-acetylglucosaminidase F (endo F) from *Flavobacterium meningosepticum* [Elder and Alexander, 1982]. Endo F also hydrolyzes the chitobiose core structure adjacent to asparagine residues, but it does not discriminate between the high mannose and the complex type of N-glycans. It was noted, however, that conformation of the proteins had a determining influence on the rate of hydrolysis. Most of the carbohydrates could be removed only after the proteins were reduced and all sulfhydryl groups were alkylated.

Another type of glycosidase which allows the release of intact N-glycans is found in the almond emulsion peptide: N-glycosidase [Takahashi and Nishibe, 1978]. This enzyme cleaves the N-acetylglucosaminyl-asparagine bond between high mannose or complex type carbohydrates and peptides [Plummer and Tarentino, 1981] or to proteins [Tarentino and Plummer, 1982; Gabel and Kornfeld, 1982]. The release of oligosaccharides is enhanced if the protein is unfolded.

Besides the endoglycosidases, which release whole carbohydrate structures, there are a few other endoglycosidases to be mentioned which allow the recovery of characteristic fragments of carbohydrates. Endo-β-galactosidase of *Escherchia freundii* [Fukuda and Matsumura, 1976] and of *Diplococcus pneumoniae* [Kobata and Takasaki, 1978], both differing in their substrate specificity, were successfully applied in the characterization of complex carbohydrate structures in glycoproteins (N- and O-linked) and in glycosphingolipids [Takasaki and Kobata, 1976; Fukuda et al, 1979; Fukuda and Hakomori, 1982]. For structural studies, endo-α-N-acetylgalactosaminidase from *Diploccous pneumoniae* [Endo and Kobata, 1976] is of very limited value because of its high glycan specificity. The enzyme liberates only the disaccharide galβ1-3GalNAc linked to serine or threonine [Kobata and Takasaki, 1978].

So far we have presented methods which rather specifically release certain carbohydrate moieties from glycoproteins. There are, however, techniques in use which result in the liberation of all carbohydrates at the same time. Since under certain circumstances one obtains a quite complex mixture of oligosaccharides, the subsequent purification steps have to be designed accordingly. A method commonly used to release intact carbohydrate moieties is by extensive digestion of the glycoprotein with a protease of broad substrate specificity. This digestion produces carbohydrates with usually only a few attached amino acids, free amino acids, and small oligopeptides [Spiro, 1970, 1973]. The preferred protease for this procedure is pronase from *Streptomyces griseus*, although other proteases of broad specificity such as papain, subtilisin, and proteinase K or mixtures of specific endo- and exopeptidases also have been used [Spiro, 1972, 1976b; Spiro and Bhoyroo, 1974; Finne and Krusius, 1982]. Treatment of glycoprotein preparations with high concentrations of pronase (1–10% of the samples weight) at 60°C for 2–4 days will result in almost quantitative "solubilization" of the protein bound carbohydrates [Finne and Krusius, 1982]. The oligosaccharides present in the digest then can either be subjected to fractionation or be further treated in order to remove all attached amino acids.

To facilitate detection of glycopeptides during chromatographic steps which are required for their purification, the glycopeptides can first be radioactively labeled at the free amino group by N-acetylation with [^{3}H]-acetic anhydride [Wheat, 1966; Finne and Krusius, 1982]. One has to be cautious, however, to remove all nonglycosylated, free amino acids and oligopeptides from the preparation after radioactive labeling.

A simultaneous release of all types of protein-bound oligosaccharides can be achieved not only by proteolysis but also by trifluoroacetolysis [Svensson, 1980]. When glycoproteins are treated with a mixture of trifluoroacetic acid and trifluoracetic anhydride at 100°C for 48 h, N-glycosidically linked carbohydrates are cleaved by transamidation, while O-glycosidically linked ones are cleaved by acid catalysis [Nilsson et al, 1979; Lundblad et al, 1980]. Because all hydroxyl groups of the carbohydrates are substituted with the highly electronegative trifluoroacetyl groups, the glycosidic bonds within the oligosaccharide structures are protected from solvolysis [Hoffman et al, 1972; Nilsson and Svensson, 1978a]. 2-Acetamido-2-deoxy-sugar residues are converted by transamidation into their corresponding 2-deoxy-2-trifluoroacetamido derivatives [Nilsson and Svensson, 1978b, 1979; Nilsson and Zopf, 1982]. N-acetylneuraminidic acids are rapidly degraded by trifluoroacetolysis. Only N-acetyneuramidinic acid linked to the C-3 position of a pyranosidic sugar residue is not cleaved, although it is modified. When N-acetylneuraminic acid is linked to the C-6 position, it is removed by transamidation and then degraded [Svensson, 1980].

The ratio of trifluoroacetic acid to trifluoracetic anhydride is crucial for preservation of the fucose and the reducing hexosamine residues [Nilsson and Svensson, 1978a; Nilsson and Zopf, 1982]. A ratio of 1:100 was found satisfactory for many glycoproteins [Nilsson et al, 1982]. O-Trifluoroacetyl groups can be removed from the carbohydrate structure after trifluoroacetolysis by treatment with a mixture of methanol and pyridine, while O- and N-trifluoroacetyl groups require a treatment with strong base (0.1 N NaOH) [Nilsson and Svensson, 1979]. A restoration of hydroxyl groups allows further modification such as reduction and permethylation of the carbohydrates. Permethylated oligosaccharide alditols containing N-trifluoroacetyl hexosamine groups have increased volatility enabling a direct analysis of larger-sized carbohydrates (up to seven monosaccharide units) by gas chromatography-mass spectroscopy [Nilsson and Zopf, 1982].

In cases where the glycoproteins contains both O- and N-glycan units, a β-elimination reaction prior to trifluoracetyolysis is advantageous in order to purify these two kinds of carbohydrates separately [Nilsson et al, 1982].

Purification of carbohydrates. Most of the above-described methods produce a complex mixture of carbohydrates released from glycoproteins or even glycolipids. To fractionate this mixture and to purify each species, a combination of chromatographic procedures is required. The techniques employed are in principle the same as those used for purification of other macromolecules, such as proteins, peptides or lipids [volume II, this series]. However, some modifications of the techniques were necessary in order to accommodate particular physicochemical properties of the carbohydrates. Numerous studies involving purification of carbohydrates have resulted during the last few years in a refinement of the techniques to a degree that isolation of most sugar structures, if derived from relatively pure protein or lipid preparation, can be achieved now by applying relatively few chromatographic steps. Some of these chromatographic procedures are not only important for preparative purposes, but represent also an essential tool for analytical measurements (for instance, gel filtration in sequencing of carbohydrates by exoglycosidase digestion). As already mentioned, sensitive detection of carbohydrates during isolation and subsequent analysis is provided by incorporation of a radioactive tracer into the sugar structure. This is achieved either by obtaining the radioactive glycoconjugates from metabolically labeled material such as cells or tissue [Yurchenco et al, 1978] or by chemical modification of the mature carbohydrate structure, eg, by reduction of aldehydes with NaB^3H_4, acetylation of amino groups with [^{3}H]acetic anhydride, or other methods. A stable but nonradioactive marker can be introduced into carbohydrates by fluorescently tagging their reducing end [Hase et al., 1978, 1979a–c]. Oligosaccharides derived from partial hydrolysis or hydrazinolysis are reacted with 2-aminopyridine. The modified molecules can be easily purified by electrophoresis because of the added positive charge.

Separation techniques for the liberated carbohydrate structures rely on the following physicochemical properties of carbohydrates or glycopeptides: size, charge, lectin-binding ability, hydrophobicity, and interaction with organic solvents. Consequently, naturally occurring heterogeneity in certain structures, especially in the content of charged sialic acids on N-linked complex carbohydrates [Schauer, 1982] poses some problems for isolation and characterization of the individual components (see example of α_1-acid glycoprotein [Yoshima et al, 1981a], Sendai virus glycoproteins [Yoshima et al, 1981b], or ovomucoid [Parente et al, 1982b; Yamashita et al, 1982b]).

By far the most frequently employed method in fractionation of glycopeptides and carbohydrates is separation according to size. Although glycoproteins do not always separate according to their true molecular size, either when using gel filtration or electrophoresis [Alhadeff, 1978; Leach et al, 1980a,b], carbohydrates were found to follow rather well the general principles of size separation [Churms, 1970]. The fractionation techniques are almost exclusively gel filtration in columns [Whistler and Anisuzzaman, 1980]. For optimal separation, BioGel P4 or P6 are used as a column packing [Tabas et al, 1978; Natowicz and Benziger, 1980; Gershman and Robbins, 1981; Yamashita et al, 1982b]. Although Sephadex G-25 is also used [Nilsson et al, 1979], BioGel P4 (−400 mesh) in columns of 1 to 2 mm in length allows separation of oligosaccharides of up to 20 carbohydrate units. To minimize carbohydrate-BioGel interactions, separation is performed at 55°C

[Yamashita et al, 1982b]. The nature of the residues being separated and their substituents and linkage arrangements affect the chromatographic migration. This is of importance when standardization with oligomers of glucose is done [Natowicz and Baenziger, 1980].

Separation of neutral carbohydrates by gel filtration is remarkably reproducible. Therefore, fully automatic chromatography systems have been developed for routine analysis [Kennedy and Fox, 1980b]. In cases where the eluate fractions cannot be checked for the presence of carbohydrate by radioactivity, the detection is achieved either by differential refractometry or by their absorption at 190 nm or by a colorimetric test with orcinol [Winzler, 1955]. The sensitivity of the colorimetric measurements is increased by prior hydrolysis of an aliquot of the eluate [Vratny et al, 1980]. Since gel filtration requires long separation times, high-performance liquid chromatography (HPLC) was developed for carbohydrates. Major improvements were achieved in shorter separation times and higher resolution [Ng et al, 1980]. Separation by HPLC follows the same basic principle as regular liquid chromatography; does not require modification of the components, unless separated by reverse phase HPLC [McNeil et al, 1982; Wells et al, 1982]; and losses are virtually nonexistent [McGinnis and Faug, 1980; Parente et al, 1982a]. The resolution is so good that alkaline-induced epimerization occurring during reduction of reducing ends results in the appearance of double peaks [Mellis and Baenziger, 1981].

Ion exchange chromatography is most important in the resolution of charged (=sialic acid, uranic acid, phosphate, sulfate-containing) glycopeptides or carbohydrates as well as to separate the charged from neutral components. The best separation is achieved on anion exchangers such as diethyl aminoethylcellulose or Sephadex [Smith et al, 1978; Samuelson, 1978; McGinnis and Faug, 1980; Endo et al, 1982; Finne and Krusius, 1982]. Neutral carbohydrate structures can be also separated by ion exchange chromatography [Samuelson, 1978]. In the presence of aqueous ethanol, neutral carbohydrates show adsorption to both cation and anion exchange resins. Furthermore, negatively charged groups are introduced by simply incubating the carbohydrates with sodium tetraborate under alkaline conditions resulting in complex formation of borate hydroxy groups with sugar residues [Foster, 1957]. The formation of the borate complex is dependent on accessibility of hydroxyl groups: *cis* configurations of adjacent or nonadjacent hydroxyls, if conformation of oligosaccharide structure allow, are preferred over *trans*. The complex is stable at alkaline pH and, therefore, analyzable by anion exchange chromatography. This procedure yields, as does gel filtration, highly reproducible results, which have aided the development of automatic ion exchange analytic procedures [Kennedy and Fox, 1980a].

Separation based on charge is possible not only by ion exchange chromatography, but also by electrophoresis. For instance natural occurring sialic acid-containing oligosaccharides or released by hydrazinolysis of glycoproteins [Kobata, 1972; Mizuochi et al, 1979; Yoshima et al, 1981a] or acetolysis products [Verma et al, 1977] can be separated by high-voltage paper electrophoresis. Visualization of the separated carbohydrates is achieved either by autoradiography or colorimetric staining [Anet and Reynolds, 1956; Dische, 1962]. Neutral glycopeptides can similarly separated on paper after complex formation with borate [Narasimhan et al, 1980]. An extraordinarily high resolution of borate complexed glycopeptides is achieved by polyacrylamide gel electrophoresis [Weitzman et al, 1979]. This technique also is applicable to analysis of exoglycosidase products during sugar sequence determination.

A traditionally often applied separation technique is partition chromatography (paper and thin-layer chromatography) [Hough, 1954; Zweig and Whitaker, 1971]. The chromatographic behavior of the carbohydrates is a function of the partition coefficient of the molecules in the water organic phases [Isherwood and Jermyn, 1951]. Separation of larger-sized oligosaccharides (more than ten sugar residues) requires long periods of chromatography [Smith et al, 1978; Mizuochi et al, 1980]. A combination of paper chromatography and chromatography on charcoal-celite has been proposed by Strecker and Montreuil [1979]. Compared to paper chromatography, thin-layer chromatography has better resolution and shorter chromatographic times for complex oligosaccharides and glycopeptides [Holmes and O'Brien, 1979; Bayard and Kerckaert, 1980]. The thin-layer material most often is either silica gel [Holmes and O'Brien, 1979] or cellulose [Momoi and Wiegandt, 1980].

Affinity chromatography using immobilized plant lectins allows the separation of glycopeptides and carbohydrate based on a completely different set of molecular properties than those discussed so far. Knowing the ligand specificity of the lectin to be used, predictions about structural features of the oligosaccharide can be made. Concanavalin A, which has a specificity for α-mannosides and α-glucosides [Ogata et al, 1975; Krusius et al, 1976; Baenziger and Fiete, 1979], has received wide application. It was found that O-glycans and tri-antennary N-glycans do not bind to the lectin; biantennary N-glycans bind weakly, and high mannose type of N-glycans bind tightly. Therefore, from a mixture of all possible N- and O-glycans, O-glycans can readily be separated from the comparably sized high mannose type of N-glycan. Concanavalin A chromatography has been successfully used to demonstrate heterogeneity in specific oligosaccharide units [Kornfeld et al, 1981; Anderson and Grimes, 1982]. Because the oligosaccharide structures obtained from concanavalin A chromatography are still heterogenous, a subfractionation using other lectins is possible. For instance, wheat germ agglutinin [Bhavanandan and Katlic, 1979; Krusius and Finne, 1978]. *Lens lentil* lectins [Tsuji et al, 1981], or *Ricinus communis* lectins [Li and Kornfeld, 1978; Baenziger and Fiete, 1979] have been used for this purpose.

A purification of a specific subset of carbohydrate structures is accomplished in the case of polysialylated oligosaccharides. Recently, it was discovered that some membrane glycoproteins contain, like glycosaminoglycans, carbohydrate moieties with polymers of either (N-glycolyl neuraminic acid 2–8)$_n$ [Inoue and Iwasaki, 1980] or (N-acetylneuraminic acids 2–8)$_n$ [Finne, 1982]. The purification of the polysialylated glycopeptides was facilitated by their polyanionic character [Bhavanandan et al, 1977]. These peptides, like glycosaminoglycans, are precipitated by cetyl pyridinium chloride [Scott, 1960; Margolis and Margolis, 1973]. The polysialylated glycopeptides are separated from the glycosaminoglycans by NaCl extraction and from interfering O-glycans by β-elimination. Final purification steps are gel filtration and DEAE-Sephadex chromatography [Finne, 1982].

Application of the methods described in this section can yield homogenous preparations of carbohydrates regardless of the starting material. Criteria of biochemical purity is at this stage the presence of *one* component when tested by several different separation methods and conditions. A further demonstration of homogeneity will be furnished by compositional, methylation and/or NMR analysis.

DETERMINATION OF THE IDENTITY AND MOLECULAR RATIO OF THE MONOSACCHARIDE RESIDUES COMPOSING THE CARBOHYDRATE STRUCTURES

To assess the type and amounts of constituents in a given carbohydrate structure, the oligosaccharide has to be resolved first into its monosaccharides. This step of the procedure has to be done under the best possible conditions capable of preserving the structural integrity of the sugar residues. Once released from the oligosaccharide, the monosaccharides have to be separated from each other and quantitated.

Two methods are currently used for the conversion of oligosaccharides into monosaccharides: acid hydrolysis and methanolysis. The identification of the sugar residues is achieved by ion exchange chromatography, paper, thin-layer, or gas-liquid chromatography (GLC) or a combination of GLC with mass spectrometry.

Glycosidic linkages are susceptible to acid-catalyzed hydrolysis [BeMiller, 1967]. The sensitivity of the linkages is dependent upon the sugar residues. Taking this into consideration complete, partial, or specific hydrolysis of the carbohydrate structure is possible [Spiro, 1966]. Furthermore, to achieve a certain degree of cleavage frequency, hydrolysis conditions have to be optimized, in many instances, because of the differences in oligosaccharide composition [Anderson and Grimes, 1982]. A complete hydrolysis of carbohydrates with optimal recovery of neutral monosaccharides requires an incubation in 1–2N HCl or H_2SO_4 at 100°C for 4–6 h [Spiro, 1972; Conchie, 1976]. Neutral sugars are deionized by chromatography over cation and anion exchangers and separated in borate buffer on an anion exchange column. The recovery of amino sugars is optimal after hydrolysis in 4N HCl at 100°C for 6 h. Without re-N-acetylation, these amino sugars can be identified on an amino acid analyzer. A mixture of glucosamine, galactosamine, glucosaminitol and galactosaminitol—the latter two are generated during β-elimination reactions—are separated on an amino acid analyzer using citrate-borate buffer [Downs and Pigman, 1976b; Fournet et al, 1978].

An alternative route to analyze acid hydrolyzates is to reduce the monosaccharides with $NaBH_4$ to alditols followed by acetylation under water-free conditions [Sawardekar et al, 1965]. The derivatization renders the molecules more volatile, allowing their separation by GLC [Bishop, 1962; Clamp et al, 1972; Laine et al, 1972; Sloneker, 1972]. This analytic procedure yields single peaks for each alditol derivative and enables the analysis of picomolar amounts of material. By the combination of GLC with mass spectrometry (MS), the separated carbohydrates can be directly identified based on their fragmentation pattern [Kochetkov and Chizhov, 1966, 1972; Hanessian, 1972; Radford and DeJongh, 1972, 1980; McFadden, 1973].

An efficient release of N-acetylhexosamine residues requires a hydrolysis of the glycosidic linkage before loss of acetamido groups since the glycosidic linkages of deacetylated amino sugars are very resistant to cleavage. An increased acid concentration, however, results in an elevated breakdown of neutral sugar residues. An improved recovery of amino sugars and at the same time preservation of neutral sugars was noted when carbohydrates were subjected first to acetolysis then to hydrolysis [Lehnhardt and Winzler, 1968; Yang and Hakomori, 1971; Steller et al, 1973a; Gehrke et al, 1979].

The conditions normally used to hydrolyze carbohydrates causes destruction of fucose and sialic acid residues. To obtain fucose residues, the hydrolysis has to be modified by lowering the acid concentration to 0.1–0.3 N [Schrager and Oates, 1968; Spiro and Spiro, 1965], and shortening the hydrolysis time to 1 h [Gyorky and Houck, 1965], or both [Takasaki and Kobata, 1978]. Similarly, sialic acid residues are hydrolyzed under very mild conditions using low concentrations (0.1–0.1 N) of acid and short incubation times (15 min to 1 h) [Neuberger and Ratcliffe, 1972; Spiro and Bhoyroo, 1974; Codington et al, 1976; Schauer, 1978, 1982]. A specific removal of sialic acids was also reported when hydrolysis was performed in 5.6 mM formic acid at 80°C [Ishizuka and Wiegandt, 1972] or 1 M formic acid at 100°C [Svennerholm et al, 1973]. Another way to remove sialic acid residues is enzymatically using sialidase from *Vibrio cholerae, Chlostridium perfringens*, or *Arthrobacter ureafaciens* [Wenger and Wardell, 1973; Schmid et al, 1977; Saito et al, 1979]. Released fucose and sialic acid residues can be measured by colorimetric methods [Dische, 1962; Downs and Pigman, 1976c] or by GLC [Yu and Ledeen, 1970].

A simple radioactive method has been described that allows the analysis of the reducing sugar residues in hydrolyzates [Takasaki and Kobata, 1978]. After hydrolysis, the liberated monosaccharides, including the added marker 2-deoxyribose, are quantitatively reduced with 5 M excess of NaB^3H_4 at alkaline pH [Howell et al, 1971; Mclean et al, 1973]. After re-N-acetylation of the amino sugar derivatives, the radiolabeled sugar alcohols are separated by paper electrophoresis and chromatography. The radioactivities in the individual components reflect the relative amounts present in the hydrolysate. Identification was done by colorimetric visualization and comparison with comigrating alcohols [Dubois et al, 1956].

The second major method used to cleave carbohydrates into monosaccharides is by methanolysis [Bhatti et al, 1970; Chambers and Clamp, 1971; Zanetta et al, 1972; Laine et al, 1972]. Carbohydrates, glycopeptides, or glycolipids are treated with 0.5–2N methanolic HCl at 80°C for 24 h. Sugar residues are released as methyl glycosides. Carboxyl groups of sialic acids and uronic acids are blocked by methyl esters. Methanolysis causes a rapid release of sialic acid residues, but, contrary to acid hydrolysis, there is no destruction of either sialic acid and fucose residues [Chamber and Clamp, 1971]. There is no appreciable extent of deacylation if the reaction is carried out under absolute water-free conditions [Zanetta et al, 1972]. In glycolipids the glucosyl but not galactosyl-sphingosine linkage is more resistant to cleavage than normal glycosidic linkages. Ceramides are degraded to sphingosine and fatty acid methyl esters. In proteins and peptides methanolysis does not affect peptide bonds and cleaves slowly and not quantitatively the N-acetylglucosaminyl-asparagine linkage [Fournet et al, 1981]. When higher acid concentrations (>2N HCl) are used, neuraminic acids are degraded, and the recovery of xylose is reduced [Chambers and Clamp, 1971]. Since methanolysis results in some deacetylation of amino sugars, re-N-acetylation of amino sugars is required.

Separation and quantitation of the methyl glycosides is conveniently achieved by GLC and GLC-MS. To increase their volatility, the glycosides are directly converted to trimethyl silyl ethers [Sweeley et al, 1963]. This modification has the advantage of being reversible [Sweeley et al, 1972], but poses some difficulties with amino sugars [Wong et al, 1980]. Other derivatization methods for GLC analysis include trifluoreacetylation [Ando and Yamakawa, 1971; Zanetta et al, 1972] or acetylation [Fournet et al, 1978]. In the latter two cases, there is no re-N-acetylation of amino sugar methyl esters required. GLC analysis of trimethylsilyl derivatives [Gheorghiu and Oette, 1970; Chambers and Clamp, 1971; Sweeley and Tao, 1972; Laine et al, 1972; Wong et al, 1980] or acetyl and trioflu-oroacetyl derivatives [Clamp et al, 1972; McFadden, 1973] show that each glycoside species is resolved into characteristic multiple peaks. This chromatographic heterogeneity is due to the presence of an equilibrium mixture of anomeric forms of each sugar. Although the existence of multiple peaks is convenient for identification of components, it contributes to a highly complex separation pattern when heteroglycans are studied, again rendering quantitation more difficult.

Information about the composition of a carbohydrate structure will also be obtained when methods are used in conjunction with analysis of sequence and glycosidic substitution (methylation, mass spectrometry, and NMR).

SEQUENCE OF THE CARBOHYDRATE RESIDUES

To establish the sequence of monosaccharides constituting a carbohydrate structure, three approaches are available: (1) the carbohydrate structure is partially fragmented and the composition of the fragments is determined; (2) the monosaccharides are removed sequentially from the carbohydrate structures and at each step the released residue is identified; and (3) fragments of or entire carbohydrate structures are subjected to mass spectrometry. Furthermore, methods used for the determination of linkage points and anomeric configuration will also yield substantial data on sequences [see the following section].

Partial fragmentation of carbohydrate structure can be achieved by taking advantage of difference in the susceptibility of glycosidic bonds to cleavage by acid hydrolysis [Bouveng and Lindberg, 1960] or acetolysis [Bayard and Montreuil, 1972]. The various glycosidic linkages of the sugars in glycolipids and glycoproteins do not differ in their rates of acid-catalyzed hydrolysis as much as do those in polysaccharides. Nevertheless, after treatment of nonderivatized permethylated carbohydrates with aqueous formic acid or low concentrations of HCl or H_2SO_4, fragmentation was observed [Bayard and Montreuil, 1972; Lindberg et al, 1975; Reading et al, 1978]. The resulting oligosaccharides are reduced with $NaBD_4$, permethylated and separated by GLC-MS [Lonngren and Svensson, 1974]. More information about linkages involved can be obtained when the procedure of Lindberg and Lonngren [1978] is followed. Permethylated carbohydrates are subjected to partial acid hydrolysis. The newly exposed reducing ends are reduced with $NaBD_4$ and the hydroxyl groups alkylated with trideuterio methyl or ethyl iodide. The products are either analyzed by GLC-MS [Lindberg et al, 1975] or hydrolized and separated by GLC-MS. Trideuteriomethyl groups in the alditol derivatives mark the position to which the acid-labile residues were linked.

A sequencing method for complex carbohydrates which almost entirely relies on partial acid hydrolysis has been proposed by Valent et al [1980]. Purified carbohydrates are reduced with $NaBD_4$ and fully methylated. Fragments are obtained by either partial acid hydrolysis, treatment with 2 M trifluoroacetic acid at 90°C for 40 min, or formolysis (90% formic acid at 70°C). After reduction with $NaBD_4$ and ethylation of all exposed hydroxyl groups, the derivatives are fractionated by reverse-phase HPLC. Each component is then fully hydrolyzed, reduced, acetylated and analyzed by GLC-MS. Based on composition, and linkage points exposed in the first and second hydrolyses, the sequence of the oligosaccharide fragments can be delineated. By comparing overlapping fragments, the sequence for the entire structure can be deduced.

A fragmentation pattern different from partial acid hydrolysis is obtained by partial acetolysis [Guthrie and McCarthy, 1967; Kuhn and Wiegandt, 1963]. Carbohydrates are treated with a mixture of acetic anhydride, acetic acid, and sulfuric acid. The susceptibility of the glycosidic linkages to cleavage under such conditions is different from that during acid hydrolysis. For instance, 1-6 linkages are relatively more stable than others tc acid hydrolysis but are preferentially cleavec during acetolysis. The reverse is true for neuraminic acid linkages [Lindberg et al, 1975]. To study specifically high mannose structures, acetolysis condition has been developed under which preferentially mannose αl-6 mannose linkages were cleaved, while others (Man αl-2 Man, Man αl-3 Man, Man βl-4 GlcNAC) remained intact [Yamashita et al, 1980b]. Oligosaccharides which are produced by partial acetolysis are suitable for separation and identification by GLC-MS [Fournet et al, 1980].

To promote cleavage at specific sites in order to achieve controlled fragmentation, acid-labile or resistant glycosidic linkages can be introduced by chemical modification of the sugar residues [Lindberg et al, 1975]. Glycosidic linkages more resistant to acid hydrolysis are those of uronic acids and 2-amino-2-deoxy-hexose obtained during hydrazinolysis and galacturonic acid residues generated by treatment with galactose oxidase followed by oxidation of the produced aldehyde at C-6 with hypoiodite. To obtain linkages which are more sensitive to acid hydrolysis, such as those of deoxyglycosides, oligosaccharides are first de-N-acetylated by hydrazinolysis. Then, the amino sugars are deaminated with nitrous acid [Horton et al, 1972], resulting in depolymerization yielding fragments terminating at their reducing end by 2,5-anhydrohexose [Bayard and Fournet, 1976; Strecker et al, 1981; Mononen, 1981].

One of the most sensitive and specific procedures to sequence oligosaccharide structures is offered by glycosidases. The usefulness of specific exoglycosidases as a tool for structural analysis, ie, by sequential release of monosaccharides from the reducing end of the oligosaccharides, was recognized years ago [Spiro, 1967, 1972]. But only when a sufficient repertoire of enzymes was available did their application for determination of entire structures became feasible [Li and Li, 1976a, 1977; Li et al, 1980; Flowers and Sharon, 1979; Kobata, 1979]. At the present a large spectrum of specific enzymes of high purity is available. These have contributed significantly to the increasing number of successful analyses involving different types of carbohydrates from glycolipids [eg, Chien et al, 1979; Ohashi, 1980], from glycoproteins [Bahl et al, 1980; Tsuji et al, 1980, 1981; Mizuochi et al, 1979, 1980; Yamamoto et al, 1981; Yoshima et al, 1981a; Yamashita et al, 1981, 1983], and from urinary oligosaccharides [Ohkura et al, 1981].

The use of exoglycosidases for sequence determinations offers two major advantages: (1) the method is very simple and does not require any sophisticated instrumentation; and (2) the enzyme activity is highly specific for the monosaccharide and anomeric configuration of the glycosidic linkage (glycan specificity) and in some instances even for the kind of glycosidic substitution (aglycan specificity) [Kobata, 1979]. If radiolabeled oligosaccharides are analyzed, trace amounts of carbohydrate are sufficient to obtain sequence data. The principle of the methodology consists in treatment of the oligosaccharide with one specific enzyme, the choice of which depends primarily on the information derived from comparable, already known structures. Either only the released monosaccharides or the released monosaccharide and the remaining oligosaccharides are analyzed by paper or thin-layer chromatography or GLC [Chien et al, 1979]. The action of the enzyme, essential for the removal of monosaccharides, is readily apparent in the reduction in size of the oligosaccharide when assayed by gel filtration [Yamashita et al, 1982a,b]. This property is most commonly used to assess the result of a particular digestion. To carry out sequential enzyme treatment on the same structure, it is advantageous to have the oligosaccharide labeled at the reducing end.

Although exoglycosidase digestions yield information about possible sequence arrangements of monosaccharides, there are still possibilities for ambiguities [Kawasaki and Ashwell, 1976; Lowe and Nilsson, 1983]. Therefore, it is required in many cases to confirm data from enzyme analysis by independent techniques such as Smith degradation, methylation analysis, NMR, or mass spectrometry [Mizuochi et al, 1979; Tsuji et al, 1981; Yamashita et al, 1982a; Parente et al, 1982b].

Mass spectrometry has been established as a powerful tool to determine carbohydrate sequences. Besides sequence arrangements of monosaccharides, mass spectrometry in conjunction with appropriate substitution of the sugar residues provides information about branching of the oligosaccharide chain and positions of substitutions [Lonngren and Svensson, 1974; Karlson et al, 1974]. The reduced carbohydrates, intact glycosphingolipids, or glycopeptides are derivatized by trimethylsilylation, acetylation, trifluoroacetylation, or methylation and introduced directly or after separation on GLC or HPLC into the mass spectrograph [Dawson and Sweeley, 1971; Egge, 1978; Karlsson, 1980; Fournet et al, 1981; McNeil et al, 1982; Nilsson and Zopf, 1982].

Fragmentation patterns generated by electron impact ionization [Karlsson, 1980; Samuelsson, 1980; Fredman et al, 1980b], field desorption [Morris and Thompson, 1976; Macher and Klock, 1980; Linscheid et al, 1981], or chemical ionization [Ariga et al, 1980] are interpreted for sequence and linkage based on data from established fragmentation patterns [Hanessian, 1972; Lonngren and Svensson, 1974; Fournet et al, 1981; Nilsson and Zopf, 1982]. Detection of molecular ions is possible only for smaller sized carbohydrates [Karlsson, 1974; Breimer et al, 1981]. The potential of mass spectrometry for analysis of carbohydrate sequences as illustrated by its wide range of applications (Glycolipids [Hanfland and Egge, 1975, 1976; Egge and Hanfland, 1981; Breimer et al, 1981, 1982; Lee et al, 1981; Ångstrom et al, 1982]) and glycans from glycoproteins [Morris et al, 1978; Matsuura et al, 1981; Fournet et al, 1981; Iwasaki and Inoue, 1981]. Since mass spectrometry alone does not allow definite identification of isomeric forms of hexose and hexasamine residues, these data have to be obtained by different methods, eg, ^{1}H-NMR [Ångstrom et al, 1982] or ^{13}C-NMR [Sweeley et al, 1980].

ARRANGEMENT OF THE GLYCOSIDIC LINKAGES

Some of the sequencing methods described above, like exoglycosidic digestion and mass spectroscopy, yield at the same time data on linkage points. One of the most important methods for elucidation of carbohydrate structures, and in particular for determination of the glycosidic substitution sites, is methylation analysis. The basic principle and application for studies of polysaccharides are reviewed by Bouveng and Lindberg [1960], and a detailed description of the technique, its major difficulties, and data interpretation is updated by Lindberg [1972], Lindberg et al [1975], Lindberg and Lonngren [1980] and Rauvala et al [1981].

Methylation analysis requires submilligram quantities of carbohydrates and is based on the following principle: All hydroxyl groups of an intact oligosaccharide structure are blocked by methylation. Then, the permethylated structure is hydrolyzed. The newly exposed hydroxyl groups, representing the original points of substitution, are tagged by groups which are different from methyl ethers. Subsequent separation and identification of the partially methylated sugar residues will indicate all possible linkage arrangements in the carbohydrate structure. To obtain more detailed information about sequence arrangement, the procedure has been modified first by performing partial degradations and then specific labeling of the uncovered reducing ends and hydroxyl groups prior to total hydrolysis [Lindberg and Lonngren, 1978].

Permethylation of glycolipids, glycopeptides, or reduced carbohydrates is achieved in a mixture of dimethyl sulfoxide and either sodium hydride [Hakomori, 1964] or potassium butoxide [Finne et al, 1980]. Both mixtures generate the methylation reagent methylsulfinyl carbonion. During the methylation reaction, O-acyl groups are replaced by O-methyl groups. Amino sugars are not de-N-acetylated but modified to N-methylacetamido sugars [Lindberg, 1972]. To prevent side reactions with uronic and sialic acids [Bjorndal et al, 1970], and to enhance volatility of the oligosaccharides for GLC-MS analysis, these residues are reduced with $LIAlH_4$ or $LiAlD_4$ prior to methylation [Lindberg, 1972; Lindberg and Lonngren, 1978]. Following the methylation reaction, the modified carbohydrate structure can be directly analyzed by mass spectrometry [Breimer et al, 1981; Rauvala et al, 1981] or be hydrolyzed. Since some methylated samples are not water soluble, an optimal depolymerization is obtained by formolysis followed by acid hydrolysis [Stellner et al, 1973]. There is no significant demethylation during acid hydrolysis. The hydrolyzate can at this stage be analyzed by thin-layer chromatography (TLC) [Li et al, 1978], or the sugar residues can be reduced to alditols with $NaBH_4$ or $NaBD_4$. $NaBD_4$ is preferred because it introduces a asymmetry into the alditols. This is of importance in the case of 3-0- and 4-0-methyl hexose where $NaBH_4$ reduction would yield alditols with identical substitution patterns. All available hydroxyl groups in the alditols are derivatized either by acetylation, trifluoroacetylation, or trideuteriomethylation [Lindberg, 1972; Endo et al, 1979; Nilsson and Zopf, 1982]. The alditol derivatives are separated and quantitated by GLC [Jones, 1972; Dutton, 1974] and GLC-MS [Bjorndal et al, 1970; Fournet et al, 1981]. Each alditol derivative containing 1, 2, 3, or 4 methoxyl groups has a characteristic retention time on GLC. Furthermore, GLC allows identification of isomers. Difficulties in analysis of amino deoxy- and acetamido-deoxy sugars, sialic and uronic acids are specifically addressed by Finne and Rauvala [1977], Rauvala and Karkkainen [1977], Lindberg and Lonngren [1978], Nilsson and Zopf [1982]. To release and quantitate sialic acid residues, methanolysis instead of acid hydrolysis is recommended [Finne et al, 1977; Schauer, 1978]. Methylation analysis can also combined with specific degradation to establish certain linkages with specific sequences. One application is the removal of terminal sugar residues, such as sialic acids of fucose, by exoglycosidases. The residues attachment sites are then determined by methylation analysis of the carbohydrate structure before and after enzyme treatment [Rauvala et al, 1981]. Similar assessment of these residue-linkage points can be made by release of the sugars by partial acid hydrolysis.

Another approach is to use partial degradation of the oligosaccharide where at each step of the degradation the involved linkage sites are specifically tagged [Lindberg et al, 1975; Lindberg and Lonngren, 1978; Rauvala et al, 1981]. Permethylated carbohydrates are fragmented by partial acid hydrolysis, partial acetolysis or de-N-acetylation followed by acid hydrolysis. The uncovered reducing ends are reduced with $NaBD_4$ and all available hydroxyl groups ethylated or trideuteriomethylated. The products are then completely hydrolyzed and the sugars reduced with $NaBD_4$ modified via trifluoracetylation. The substitution pattern of each residue is readily identified by GLC-MS. The presence of various markers will indicate the relative location of linkage of the residue.

Residue specific cleavage is achieved by Smith degradation and provides an additional tool to determine positions of monosaccharides. With slightly acid conditions periodate oxidizes two or more vicinyl hydroxyl groups attached to adjacent carbon atoms to aldehydes [Dyer, 1956; Bobbit, 1965]. Oxidation of cyclic *cis* glycols are preferred over cyclic *trans* glycols. In the case of three adjacent hydroxy groups, such as terminal sugars and nonterminal sugars in 1-6 linkages, oxidation leads to two aldehyde and one formic acid. Nonterminal sugars wth 1-2 or 1-4 linkages are cleaved without formic acid, and nonterminal sugars with 1-3 linkages or with branching at C-2 and C-4 are not affected [Hay et al, 1965]. Therefore, formic acid production, the surviving carbohydrate structures, and the degradation products provide information about the site of linkage and the types of sugars involved [Abdel-Akher et al, 1952; Smith, 1956]. After oxidation and subsequent reduction of the aldehydes with $NaBH_4$, the modified carbohydrate can be subjected to mild acid hydrolysis (=Smith degradation). Under such conditions, the noncyclic acetals derived from the oxidized sugars are cleaved but glycosidic linkages remain intact [Goldstein et al, 1965]. The products generated by Smith degradation can be identified by GLC [Yamaguchi et al, 1970; Dutton, 1973, 1974]. The structure of the aglycon indicates the identity of the sugar residue from which it is derived [Lindberg et al, 1975]. More structural information is obtained when the polyalcohol is first methylated, and then treated under mild acid conditions. The consequently exposed hydroxyl groups are tagged by trideuteriomethylation or ethylation. The analysis of the derivatized monosaccharides will indicate the type of residue and the site of substitution [Lindberg and Lonngren, 1978]. However, one has to be cautious in interpretation of oxidation product analysis in regard to the position of residues. It is of importance to have independent confirmation of the results by other methods [Gathmann and Aminoff, 1981].

The ability to expose by Smith degradation or partial fragmentation a unique or a limited number of hydroxyl groups within an otherwise fully methylated carbohydrate structure, lead to the design of a sequential degradation method [Svensson, 1978; Svensson and Kenne, 1980]. The free hydroxyl group is oxidized with chlorine-dimethylsulfoxide to a keto or adlehyde. Subsequent treatment with sodium ethoxide in ethanol results in the removal (β-elimination) of the methoxyl or glycosyl residue at the β-position to the oxidized group. The resulting α-β-unsaturated, oxidized sugar is readily degraded by mild acid hydrolysis resulting in the release of the ring constituents. This reaction frees a hydroxyl group on the next sugar residue, which is now amenable to a new round of oxidation-β-elimination–mild acid hydrolysis. At each step, an aliquot of the sample is analyzed for the identity of the removed sugar residue and its site of attachment.

Glycosidic substitutions can also be assessed by the specificity of exoglycosydases for a glycan structure [Kobata, 1979; Li et al, 1980]. An example is presented by Paulson et al [1982]. In O- and N-glycans, N-acetylneuraminic acid is linked to C-3 or C-6 of galactose, or C-6 of N-acetyl galactosamine. Sialidases from *Vibrio cholerae, Clostridium perfringens,* and *Streptococcus pneumoniae* are capable of hydrolyzing all the possible bonds. On the sialidase purified from Newcastle disease virus specifically cleaves N-acetylneuraminic acid α2-3 galactose linkages. Other examples are α1,2 galactose linkage-specified fucosidase [Kochibe, 1973], αl,3/ 1,4 N-acetylglucosamine linkage-specific fucosidase [Ogata et al, 1977], α1,4 N-acetylglucosamine linkage-specific galactosidase [Paulson et al, 1978] and α1,2 linkage-specific mannosidase [Yamashita et al, 1980a].

Besides exoglycosidases, the extraordinarily high acceptor specificity of glycosyltransferase can be exploited for sequence analysis [see detailed review by Beyer et al, 1981]. In these enzymes analyses, one simply determines whether an oligosaccharide structure can serve as substrate for a specific transferase. When a transfer is observed, a chromatographic test, methylation analysis, or exoglycosidase digestion will indicate whether the formed structure is identical to the native oligosaccharide. For instance, Van de Eijnden et al [1977, 1980] prepared as substrate radioactive asialo α_1-acid glycoprotein. Treatment with rat liver of bovine colostrum derived sialyl transferase resulted only in the attachment of N-acetylneuraminic acid to the position C-6 of the terminal galactose residue. A similar application was used to identify the formation of a N-acetylneuraminic acid 2-8 N-acetylneuraminic acid linkage during transformation of the ganglioside G_{DIa} to G_{TIa} and G_{TIb} to G_{QIb} [Murakami-Murofushi et al, 1981; Yohe and Yu, 1980, 1981; Yohe et al, 1982].

One of the most sensitive and informative methods to study carbohydrate structure is at the present proton NMR-spectroscopy. This method is also rapid and non-destructive. Proton NMR-spectroscopy developed within the last few years from an auxiliary technique [Kotowycz and Lemieux, 1973; Kamerling et al, 1975] to a major, almost self-sufficient method [Vliegenthart et al, 1981; Dabrowski et al, 1982a]. Spectra of native glycolipids are recorded in deuterated pyridine [Ando et al, 1976] or dimethylsulfoxide [Dabrowski et al, 1980] and oligosaccharides or glycopeptides in D_2O [Vliegenthart et al, 1981]. Interpretations of the spectra is based on chemical shifts of the resonance frequency due to the chemical environment of the nucleus and on spin-spin coupling produced by magnetic interaction of chemically linked nuclei. The coupling constant J is therefore a sensitive parameter for the type of bonds involved and for their spatial orientation within the molecule. Integration of the resonance signal is an indicator of the number of nuclei. The chemical shifts and coupling constants of anomeric proteins with resonance well separated in the spectra were defined as structural reporter groups containing the essential information about primary sequence [Dorland et al, 1977; Strecker et al, 1977; Vliegenthart et al, 1981; Dabrowski et al, 1982a]. Many carbohydrate sequences were established with the aid of 360-MHz ^{1}H-NMR spectroscopy, normally in conjunction with methylation analysis [Fournet et al, 1978; Dabrowski et al, 1980; Carver and Grey, 1981]. The recently introduced 500-MHz high-resolution ^{1}H-NMR spectroscopy provided an increased sensitivity (signal to noise ratio) and higher spectral resolution due to the stronger magnetic field [Van Halbeek et al, 1980c; Vliegenthart et al, 1981]. A better resolution of coupling constants multiples was also achieved by two-dimensional J-resolved ^{1}H-NMR [Bruch and Bruch, 1982]. ^{1}H-NMR spectroscopy was used to study glycosphingolipids [Falk et al, 1979a,b; Hanfland et al, 1981; Hori et al, 1981; Dabrowski et al, 1980, 1982b], glycopeptides [Dorland et al, 1978; Bradbury and Collins, 1979; Cohen and Ballou, 1980; Van Halbeek et al, 1980a; Debray et al, 1981; Parente et al, 1982b], or oligosaccharides [Bradbury and Grant, 1979; Van Halbeek et al, 1980b; Blanchard et al, 1983]. Based on proton NMR data, heterogeneity in the carbohydrate structure in glycopeptides can be assessed [Atkinson et al, 1981; Hakimi et al, 1981]. The sensitivity of the method even allows the detection of heterogeneity in a sequence at a level as low as 5% of the total [Endo et al, 1982]. A further application of proton NMR spectroscopy is to deduce the carbohydrate sequences solely from chemical shift data using the nuclear Overhauser effect [Noggle and Schirmer, 1971]. The method appears to be especially suited to identifying the sequences at branching points in glycolipids [Dabrowski et al, 1982b], oligosaccharides [Lemieux et al, 1980], or glycopeptides [Brisson and Carver, 1983].

ANOMERIC CONFIGURATION OF THE GLYCOSIDIC BONDS

The absolute specificity of exoglycosidases for anomeric linkages is widely used as an indicator for the linkage type [eg, Laine et al, 1974; Uda et al, 1977; Kobata, 1979; Li et al, 1980; Natowicz et al, 1982]. The linkage configuration can also be tested by its susceptibility to chromium trioxide oxidation [Laine and Renkonen, 1975]. It was found that CrO_3 oxidizes preferentially β-anomeric glycosidic-linked sugar residues. This method was applied in analysis of novel gangliosides and the results verified by enzyme treatments [Ohashi, 1980].

As already described above, NMR spectroscopy provides, in addition to sequence data, information about linkage configuration. The preferred method to study anomeric configuration of linkages and overall conformation of carbohydrate structures is ^{13}C-NMR spectroscopy [Coxon, 1972; Taravel and Vottero, 1975; Jennings and Smith, 1978; Seymour, 1979; Becchi et al, 1981; Gorin, 1981]. In principle, ^{13}C-NMR spectroscopy could also provide information about primary sequence [Barker et al, 1982]. With a Fourier transform method, spectra of natural abundance ^{13}C atoms in carbohydrates can be recorded. The following parameters are obtained: chemical shift of ^{13}C nuclei, as a function of the chemical environment; spin-spin coupling constants between ^{13}C and other nuclei with spin, indicating the angle between the coupled nuclei; nuclear Overhauser enhancement and nuclear relaxation times, both indicating the internuclear distance and mobility. Interpretation of spectra are based on correlation with data established for monosaccharides and model polysaccharides [Sugiyama, 1978] or by using isotopically enriched carbohydrates [Barker and Walker, 1980].

With the help of ^{13}C-NMR spectroscopy, configuration of glycosidic linkages could be determined in glycolipids [Sillerud et al, 1978, 1982; Koerner et al, 1981; Sweeley et al, 1980], glycopeptides [Shibata and Nakanishi, 1980; Bush et al, 1980], and oligosaccharides [Shashkov et al, 1979; Collins et al, 1981].

NONCARBOHYDRATE RESIDUES

A few examples should illustrate the diversity of methodological approaches to identify noncarbohydrate residues within a oligosaccharide structure. The presence of O-lactyl groups such as in sialic acids can be recognized by GLC-MS analysis [Schauer et al, 1976]. When the site of substitution (eg, acetyl groups) is not recognized by GLC analysis, the following chemical test has been employed [deBelder and Norrman, 1968]: First all hydroxyl groups are converted into acetals or, under certain steriochemical conditions, to acyclic acetals [Wolform et al, 1968]. The modified carbohydrate is then methylated, whereby the O-acetyl groups are replaced by methoxy groups. After release by acid hydrolysis, the partially methylated sugar residue will indicate the site of attachment of the original acetyl group.

Having available suitable instrumentation, assignment of O-acetyl esters in N-glycolylneuraminic acid was readily achieved by a combination of direct probe analysis of the carbohydrate structure in a mass spectrograph and 400-MHz ^{1}H-NMR spectrograph [Gasa et al, 1983].

GLC-MS analysis of the constituents of oyster glycolipids of alditol acetate derivatives indicated the presence of 3-0-methyl-N-acetylgalactosamine [Matsubara and Hayashi, 1978]. Parital methanolysis of the glycolipids oligosaccharide and characterization of the fragments as trimethylsilyl derivatives by GLC-MS allowed an assignment of the methylated sugar residue to the terminal position.

During characterization of the carbohydrate structure of bovine lutropin, modified N-acetylhexosamines were observed [Parsons and Pierce, 1980]. The modified residues were localized by Smith degradation at the distal end of the carbohydrate structure. The residues were resistant to periodate oxidation and to release by hexosaminidases. Colorimetric test for phosphate [Leloir and Cardini, 1957] was negative, but the test for sulfate [Spencer, 1960] yielded two residues per mole oligosaccharides. Mild acid hydrolysis and analysis of products indicated that to each of the terminal N-acetylglucosamine and N-acetylgalactosamine residues one sulfate was linked. The site of attachment has still to be determined.

In N-glycosidic-linked high mannose oligosaccharide of yeast carboxypeptidase, a phosphate residue was suspected [Hashimoto et al, 1981]. By ^{31}P-NMR spectroscopy of endo-β-N-acetylglucosaminidase-released oligosaccharide, a characteristic phosphodiester was determined [Van Wazer et al, 1956]. Hydrolysis of the carbohydrate structure in 1 M trifluoroacetic acid at 120°C for 2 h yielded mannose-6-phosphate when tested by an enzymatic assay [Hashimoto et al, 1980]. These findings suggested the presence of a mannose glycosidically esterified to phosphate (mild acid labile), which in turn is linked to C-6 of mannose (mild acid stable). Identification was achieved by mild acid hydrolysis followed by exhaustive digestion of α-mannosidase (mannose-6-phosphate is not cleaved) and analysis of products by gel filtration, and partial acetolysis.

A comparable route was employed for identification of the phosphate present in the high mannose type of N-linked oligosaccharides of β-glucuronidase of the mouse [Tabas and Kornfeld, 1980; Varki and Kornfeld, 1980] and man [Natowicz et al, 1982]. In these cases, endo H-sensitive carbohydrate contained one or two mannose-6-phosphates/oligosaccharide. The majority of the phosphate was resistant to alkaline phosphatase but became sensitive to the enzyme after mild acid hydrolysis, indicating the presence of phosphodiester linkages. By combination of mild acid hydrolysis, digestion with α-1, -2, linkage-specific mannosidase and with alkaline phosphatase, and by analysis of size and composition of the products, the sites of phosphate attachments to the carbohydrate moiety was determined. In the case of the phosphodiester groups α-glycosidically linked N-acetylglucosamine was found.

CONCLUDING STATEMENT

From the previous discussion, it should be obvious that within the past decade, the methodology for the analysis of carbohydrates in complex structures such as glycolipids and glycoproteins has become sophisticated, reliable, and most important, capable of supplying the absolute composition, sequence, and configuration of the residues in the glycan structure. The methodology is at present more complex than is the methodology for the determination of the primary amino acid sequence of polypeptides or the base sequences of nucleic acids. However, despite the complexity, complete determination of carbohydrate structure has in many cases paralleled the determination amino acid sequences in glycoproteins. Hopefully, in the future, the methodology will be developed further in the direction of requiring less starting material, involving simpler chemistry, utilizing more the enzymatic approach than the chemical, and increasing the validity of instrumentation screening for more rapid structural and configurational analyses.

REFERENCES

Abdel-Akher M, Hamilton JK, Montgomery R, Smith F (1952): A new procedure for the determination of the fine structure of polysaccharides. J Am Chem Soc 74:4970–4971.

Alhadeff JA (1978): Gel filtration of sialoglycoproteins. Biochem J 173:315–319.

Aminoff D, Gathmann WP, McLean CM, Yadomae T (1980): Quantitation of oligosaccharides released by the β-elimination reaction. Anal Biochem 101:44–53.

Anderson DR, Grimes WJ (1982): Heterogeneity of asparagine-linked oligosaccharides of five glycosylation sites on immunoglobulin M heavy chain from mineral oil plasmacytoma 104 E. J Biol Chem 257:14858–14864.

Ando S, Yamakawa T (1971): Application of trifluoroacetyl derivatives to sugar and lipid chemistry. I. Gas chromatographic analysis of common constituents of glycolipids. J Biochem 35–340.

Ando S, Kon K, Isobe M, Nagai Y, Yamakawa T (1976): Existence of glucosaminyl lactosyl ceramide (amino-CTH-I) in human erythrocyte membranes as a possible precursor of blood group-active glycolipids. J Biochem 79:625–632.

Ando S, Yu RK (1977): Isolation and characterization of a novel trisialoganglioside G_{T1a} from human brain. J Biol Chem 252:6247–6250.

Ando S, Yu RK (1979): Isolation and characterization of two isomers of brain tetrasialogangliosides. J Biol Chem 254:12224–12229.

Anet EFLJ, Reynolds TM (1956): Isolation of mucic acid from fruits. Nature 174:930.

Angstrom J, Breimer ME, Falk KE, Hansson GC, Karlsson KA, Leffler H (1982): Chemical characterization of penta-, hepta-, and nonaglycosyl ceramides of rat small intestine having a globoside-like terminus. J Biol Chem 257:682–688.

Ariga T, Murata T, Oshima M, Maezawa M, Miyatake T (1980): Characterization of glycosphingolipids by direct inlet chemical ionization mass spectrometry. J Lipid Res 21:879–887.

Atkinson PH, Grey A, Carver JP, Hakimi J, Ceccarini C (1981): Demonstration of heterogeneity of chick ovalbumin glycopeptides using 360-MHZ proton magnetic resonance spectroscopy. Biochemistry 20:3979–3986.

Baenziger JU, Fiete D (1979): Structural determinants of concanavalin A specificity for oligosaccharides. J Biol Chem 254:2400–2407.

Bahl OP, Reddy MS, Bedi GS (1980): A novel carbohydrate structure in bovine and ovine luteinizing hormones. Biochem Biophys Res Commun 96:1192–1199.

Barker R, Nunez HA, Rosevear P, Serianni AS (1982): ^{13}C-NMR analysis of complex carbohydrates. Methods Enzymol 83:58–69.

Barker R, Walker TE (1980): ^{13}C-NMR spectroscopy of isotopically enriched carbohydrates. Methods Carbohydr Chem 83:151–165.

Basu S, Basu M (1982): Expression of glycosphingolipid glycosyltransferases in development and transformation. In Horowitz MI (ed): "The Glycoconjugates." New York: Academic Press, Vol 3, pp 265–285.

Bayard B, Fournet B (1976): Hydrazinolysis and nitrous acid deamination of the carbohydrate moiety of α_1-acid glycoprotein. Carbohydr Res 46:75–86.

Bayard B, Kerckaert JP (1980): Evidence for uniformity of the carbohydrate chains in individual glycoprotein molecular variants. Biochem Biophys Res Commun 95:777–784.

Bayard B, Kerckaert JP (1981): Uniformity of carbohydrate chains within molecular variants of rat α_1-fetoprotein with distinct affinity for concanavalin A. Eur J Biochem 113:405–414.

Bayard B, Montreuil J (1972): Etudes sur les glycoproteins. XLVIII description d'un procedure fractionnement des acetolysats. Carbohydr Res 24:427–443.

Bayard B, Roux D (1975): Hydrozinolysis and nitrous deamination of glycoproteins evidence for a common core moiety in carbohydrate moiety. FEBS Lett 55:206–214.

Becchi M, Bruneteau M, Pontagnier H, Michel G (1981): Structure confirmation of the oligosaccharide chain of chrysantellin A by carbon-13 NMR. Planta Med 42:265–267.

BeMiller JN (1967): Acid-catalyzed hydrolysis of glycosides. Adv Carbohydr Chem 22:25–108.

Bennett M, Schmid K (1980): Immuno suppression by human plasma alpha-1 acid glycoprotein. Importance of the carbohydrate moiety. Proc Natl Acad Sci USA 77:6109–6113.

Bernhard BA, Yamada KM, Olden K (1982): Carbohydrate selectively protect a specific domain of fibronectin against proteases. J Biol Chem 257:8549–8554.

Beyer TA, Sadler JE, Rearick JI, Paulson JC, Hill RL (1981): Glycosyltransferases and their use in assessing oligosaccharide structure and structure-function relationships. Adv Enzymol 52:23–175.

Bhatti T, Chambers RF, Clamp JR (1970): The gas chromatographic properties of biologically important N-acetylglycosamic derivatives monosaccharides, disaccharides, trisaccharides, tetrasaccharides and pentasaccharides. Biochim Biophys Acta 222:333–347.

Bhavanandan VP, Katlic A (1979): The interaction of wheat germ agglutinin with sialoglycoproteins. The role of sialic acid. J Biol Chem 254:4000–4008.

Bhavanandan VP, Umemoto J, Banks JR, Davidson EA (1977): Isolation and partical characterization of sialoglycopeptides produced by a murine melanoma. Biochemistry 16:4426–4436.

Bishop CT (1962): Separation of carbohydrate derivatives by gas liquid partition chromatography. Methods Biochem Anal 10:1–42.

Bjorndal H, Hellerqvist CG, Lindberg B, Svensson S (1970): Gas-liquid chromatography and mass spectrometry in methylation analysis of polysaccharides. Angew Chem [Engl] 9:610–619.

Blanchard D, Cartron JP, Fournet B, Montreuil J, van Halbeek H, Vliegenthart JFG (1983): Primary structure of the oligosaccharide determinant of blood group cad specificity. J Biol Chem 258:7691–7695.

Bobbit JM (1965): Periodate oxidation of carbohydrate. Adv Carbohydr Chem 11:1–41.

Bouveng HO, Lindberg B (1960): Methods in structural polysaccharide chemistry. Adv Carbohydr Chem 15:53–89.

Bradbury JH, Collins JG (1979): An approach to the structural analysis of oligosaccharides by NMR spectroscopy. Carbohydr Res 71:15–24.

Breimer ME, Hansson GC, Karlsson KA, Leffler H, Pimlott W, Samuelsson BE (1981): Sequencing of oligosaccharides by mass spectrometry applied on a 12-sugar glycolipid. FEBS Lett 124:299–303.

Breimer ME, Hansson GC, Karlsson KA, Leffler H (1982): Isolation and partial characterization of blood group A and H active glycosphingolipids of rat small intestine. J Biol Chem 257:906–912.

Bremer EG, Gross SK, McCluer RH (1979): Quantitative analysis of monosialogangliosides by high-performance liquid chromatography of their perbenzoyl derivatives. J Lipid Res 20:1028–1035.

Brisson JR, Carver JP (1983): Sequence determination for N-linked oligosaccharides through the use of the nuclear Overhauser effect. J Biol Chem 258:1431–1434.

Brunngraber EG, Tettamanti G, Berra B (1976): Extraction and analysis of materials containing lipid-bound sialic acid. In Witting LA (ed): "Glycolipid Methodology." Champaign, IL: American Oil Chemists' Society, pp 159–186.

Bruch RC, Bruch MD (1982): Two-dimensional J-resolved proton NMR spectroscopy of oligomannosidic glycopeptides. J Biol Chem 257:3409–3413.

Burton RM (1976): The naming of glycolipids. In Witting LA (ed): "Glycolipid Methodology." Champaign, IL: American Oil Chemists' Society, pp 1–11.

Bush CA, Duben A, Ralapati S (1980): Conformation of the glycopeptide linked glycoproteins. Biochemistry 19:501–504.

Byrd JC, Tarentino AL, Maley F, Atkinson PH, Trimble RB (1982): Glycoprotein synthesis in yeast. Identification of $Man_8GlcNAc_2$ as an essential intermediate in oligosaccharide processing. J Biol Chem 257:14657–14666.

Carlson DM (1968): Structures and immunochemical properties of oligosaccharides isolated from pig submaxillary mucins. J Biol Chem 243:616–626.

Carlson DM (1972): Assay for N-acetyl galactosaminitol. Methods Enzymol 28:43–45.

Carver JP, Grey AA (1981): Determination of glycopeptide primary structure by 360-MHz proton magnetic resonance spectroscopy. Biochemistry 20: 6607–6616.

Chambers RE, Clamp JR (1971): An assessment of methanolysis and other factors used in the analysis of carbohydrate-containing materials. Biochem J 125:1009–1018.

Chandrasekaran EV, Mendicino A, Garver FA, Mendicino J (1981): Structures of sialylated O-glycosidically and N-glycosidically linked oligosaccharides in a monoclonal immunoglobulin light chain. J Biol Chem 256:1549–1555.

Chien JL, Li SC, Li YT (1979): Isolation and characterization of a heptaglycosylceramide from bovine erythrocyte membranes. J Lipid Res 20:669:673.

Churms SC (1970): Gel chromatography of carbohydrates. Adv Carbohydr Chem Biochem 25:13–51.

Cifonelli JA (1968): Reaction of heparitin sulfate with nitrous acid. Carbohydr Res 8:233–242.

Clamp JR, Bhatti T, Chambers RE (1972): The determination of carbohydrate in biological material by gas-liquid chromatography. Methods Biochem Anal 10:229–344.

Codington JF, Linsley KB, Silber C (1976): Removal of sialic acids from glycoproteins by chemical methods and determination of sialic acids. Methods Carbohydr Chem 7:226–232.

Cohen RE, Ballou CE (1980): Linkage and sequence analysis of mannose-rich glycoprotein core oligosaccharides by proton NMR spectroscopy. Biochemistry 19:4345–4358.

Collins JG, Bradbury JH, Trifonoff E, Messer M (1981): Structures of four new oligosaccharides from marsupial milk, determined mainly by ^{13}C-NMR spectroscopy. Carbohydr Res 92:37–41.

Conchie J (1976): Acid-catalyzed hydrolysis and methanolysis of glycoproteins. Methods Carbohydr Chem 7:195–199.

Coxon B (1972): Conformational analysis via nuclear magnetic resonance spectroscopy. Methods Carbohydr Chem 6:513–539.

Critchley DR (1979): Glycolipids as membrane receptors important in growth regulation. In Hynes RO (ed): "Surface of Normal and Malignant Cells." Chichester: John Wiley and Sons, pp 63–101.

Cummings RD, Trowbridge IS, Kornfeld S (1982): A mouse lymphoma cell line resistant to the leukoagglutinating lectin from *Phaseolus vulgaris* is deficient in UDP-GlcNac: α-D-mannoside β-1,6-N-acetylglucosaminyltransferase. J Biol Chem 257:13421–13427.

Dabrowski J, Hanfland P, Egge H (1980): Structural analysis of glycosphingo lipids by high resolution proton NMR spectroscopy. Biochemistry 19:5652–5658.

Dabrowski J, Hanfland R, Egge H (1982a): Analysis of glycosphingolipids by high-resolution proton nuclear magnetic resonance spectroscopy. Methods Enzymol 83:69–86.

Dabrowski J, Hanfland P, Egge H, Dabrowski U (1982b): Immunochemistry of the Lewis blood system: Proton nuclear magnetic resonance study of plasmatic Lewis blood group-active glycosphingolipids and related substances. Arch Biochem Biophys 210:405–411.

Dawson G (1978): Disaccharide units from complex carbohydrates of animals. Methods Enzymol 50:272–284.

Dawson G, Sweeley CC (1971): Mass spectrometry of neutral monosialo glycosphingolipids and disialo glycosphingolipids. J Lipid Res 12:56–64.

deBelder AN, Norrman B (1968): The distribution of substituents in partially acetylated dextran. Carbohydr Res 8:1–6.

Debray H, Fournet B, Montreuil J, Dorland L, Vliegenthart JFC (1981): Structure determination by 360 megahertz proton NMR spectroscopy and methylation analysis of a biantennary glycan of the N-acetyl lactosaminic type isolated from rat liver plasma membrane. Eur J Biochem 115:559–564.

DeVries GH, Norton WT (1974): The lipid composition of axons from bovine brain. J Neurochem 22:259–264.

Dische Z (1962): General color reactions of carbohydrates. Methods Carbohydr Chem 1:477–514.

Dorland L, Haverkamp J, Schut BL, Vliegenthart JFG, Spik G, Strecker G, Fournet B, Montreuil J (1977): The structure of the asialo-carbohydrate units of human serotransferrin as proven by 360 MHz proton magnetic resonance spectroscopy. FEBS Lett 77:15–20.

Dorland L, Haverkamp J, Vliegenthart JFG, Fournet B, Strecker G, Spik G, Montreuil J, Schmid K, Binette JP (1978): Determination by 360 megahertz proton NMR spectroscopy of the type of branching in complex asparagine linked glycan chains of glycoproteins. FEBS Lett 89:149–152.

Downs F, Pigman W (1976a): Determination of O-glycosidic linkages to L-serine and L-threonine residues of glycoproteins. Methods Carbohydr Chem 7:200–204.

Downs F, Pigman W (1976b): Determination of hexosamines and hexosaminitols using the amino acid analyzer. Methods Carbohydr Chem 7:244–248.

Downs F, Pigman W (1976c): Qualitative and quantitative determination of sialic acids. Methods Carbohydr Chem 7:233–243.

Downs F, Herp A, Moschera J, Pigman W (1973): β-elimination and reduction reactions and some applications of dimethylsulfoxide on submaxillary glycoproteins. Biochim Biophys Acta 328:182–192.

Dubois M, Gilles KA, Hamilton KJ, Rebers RA, Smith F (1956): Colorimetric method for determination of sugars and related substances. Anal Chem 28:350–356.

Dutton GGS (1973): Applications of gas-liquid chromatography to carbohydrates. Part I. Adv Carbohydr Chem Biochem 28:11–160.

Dutton GGS (1974): Application of gas-liquid chromatography to carbohydrates. Part II. Adv Carbohydr Chem Biochem 30:9–110.

Dyer JR (1956): Use of periodate oxidations in biochemical analysis. Methods Biochem Anal 3:111–152.

Egge H (1978): The application of mass spectrometry in the structural elucidation of glycosphingolipids. Chem Phys Lipids 21:349–360.

Egge H, Hanfland R (1981): Immunochemistry of the Lewis blood group system. Mass spectrometric analysis of permethylated LeA-, LeB-, and H-type 1 (LeDH) blood group active and related glycosphingolipids from human plasma. Arch Biochem Biophys 210:396–404.

Elbein AD, Forsee WT (1976): Biosynthesis of steryl glycosides and acylated steryl glycosides in plants. In Witting LA (ed): "Glycolipid Methodology." Champaign, IL: American Oil Chemists' Society, pp 345–368.

Elder JH, Alexander S (1982): Endo-β-N-acetylglucosaminidase F: Endoglycosidase from *Flavobacterium menigosepticum* that cleaves both high-mannose and complex glycoproteins. Proc Natl Acad Sci USA 79:4540–4544.

Endo Y, Kobata A (1976): Partial purification and characterization of an endo-α-N-acetylgalactosaminidase from the culture medium of *Diplococcus penumoniae*. J Biochem 80:1–8.

Endo M, Suziki K, Schmid K, Fournet B, Karamanos Y, Montreuil J, Dorland L, Van Halbeek B, Vliegenthart JFG (1982): The structures and microheterogeneity of the carbohydrate chains of human plasma ceruloplasmin. A study employing 500-MHz

^{1}H-NMR spectroscopy. J Biol Chem 257:8755–8760.

Endo Y, Yamashita K, Tachibana Y, Tojo S, Kobata A (1979): Structure of the asparagine-linked sugar chains of human chorionic gonadotropin. J Biochem 85:669–679.

Entlicher G, Kocourek J (1976): Isolation of glycopeptides from cell membrane glycoproteins. Methods Carbohydr Chem 7:191–194.

Esselman WJ, Laine RA, Sweeley CC (1972): Isolation and characterization of glycosphingolipids. Methods Enzymol 28:140–156.

Falk KE, Karlsson KA, Samuelsson BE (1979a): Proton nuclear magnetic resonance analysis of anomeric structure of glycosphingolipids blood group ABH-active substances. Arch Biochem Biophys 192:177–190.

Falk KE, Karlsson KA, Samuelsson BE (1979b): Proton nuclear magnetic resonance analysis of anomeric structures of glycosphingolipids Lewis-active and Lewis-like substances. Arch Biochem Biophys 192:191–202.

Farrar GH, Harrison R (1978): Isolation and structural characterization of alkali-labile oligosaccharides from bovine milk fat globule membrane. Biochem J 171:549–557.

Finne J (1982): Occurrence of unique polysialosyl carbohydrate units in glycoproteins of developing brain. J Biol Chem 257:11966–11970.

Finne J, Krusius T (1979): Structural features of the carbohydrate unit of plasma glycoproteins. Eur J Biochem 102:583–588.

Finne J, Krusius T (1982): Preparation and fractionation of glycopeptides. Methods Enzymol 83:269–277.

Finne J, Rauvala H (1977): Determination by methylation analysis of the substitution pattern of 2-amino-2-deoxyhexitols obtained from O-glycosylic carbohydrate units of glycoproteins. Carbohydr Res 58:57–64.

Finne J, Krusius T, Rauvala H (1980): Use of potassium tert-butoxide in the methylation of carbohydrates. Carbohydr Res 80:336–339.

Finne J, Krusius T, Rauvala H, Hemmink K (1977): The disialosyl group of glycoproteins. Occurrence in different tissues and cellular membranes. Eur J Biochem 77:319–323.

Flowers HM, Sharon N (1979): Glycosidases- properties and application to the study of complex carbohydrates and cell surfaces. Adv Enzymol 48:29–95.

Folch J, Lees MB, Sloane-Stanley GH (1957): A simple method for the isolation and purification of total lipids from animal tissues. J Biol Chem 226:497–509.

Foster AB (1957): Zone electrophoresis of carbohydrates. Adv Carbohydr Chem 12:81–115.

Fournet B, Dhalluin JM, Strecker G, Montreuil J, Bosso C, Defaye J (1980): Gas-liquid chromatography and mass spectrometry of oligosaccharides obtained by partial acetolysis of glycans of glycoproteins. Anal Biochem 108:35–56.

Fournet B, Montreuil J, Strecker G, Dorland L, Haverkamp J, Vliegenthart JFG, Binette JP, Schmid K (1978): Determination of the primary structures of 16 asialo-carbohydrate units derived from human plasma α_1-acid glycoprotein by 360 MHz ^{1}H-NMR spectroscopy and permethylation analysis. Biochemistry 17:5206–5214.

Fournet B, Strecker G, Leroy Y, Montreuil J (1981): Gas-liquid chromatography and mass spectrometry of methylated and acetylated methyl glycosides. Application to the structural analysis of glycoprotein glycans. Anal Biochem 116:489–502.

Fredman P, Mansson JE, Svennerholm L, Karlsson KA, Pascher I, Samuelsson BE (1980a): The structure of the tetrasialoganglioside from human brain. FEBS Lett 110:80–84.

Fredman P, Nilsson O, Tayot JL, Svennerhom L (1980b): Separation of gangliosides on a new type of anion-exchange resin. Biochim Biophys Acta 618:42–52.

Freed JH, Nathenson SG (1977): Similarity of the carbohydrate structures of H-2 and IA glycoproteins. J Immunol 199:477–482.

Fukuda MN, Hakomori S (1982): Structure of branched blood group A-active glycosphingolipids in human erythrocytes and polymorphism of A- and H-glycolipids in A_1 and A_2 subgroups. J Biol Chem 257:446–455.

Fukuda MN, Matsumura G (1976): Endo-β-galactosidase of *Escherichia freundii*. Purification and endoglycosidic action on keratan sulfates oligosaccharides and blood group active glycoprotein. J Biol Chem 251:6218–6225.

Fukuda MN, Fukuda M, Hakomori S (1979): Cell surface modification by endo-β-galactosidase change of blood group activities and release of oligosaccharides from glycoproteins and glycosphingolipids of human erythrocytes. J Biol Chem 254:5458–5465.

Fukuda M, Kondo T, Osawa T (1976): Studies on the hydrazinolysis of glycoproteins. Core structures of oligosaccharides obtained from porcine thyroglobulin and pineapple stem bromelain. J Biochem 80:1223–1232.

Funakoshi I, Yamashina I (1982): Structure of O-glycosidically linked sugar units from plasma membranes of an ascites hepatoma AH 66. J Biol Chem 257:3782–3787.

Gabel CA, Kornfeld S (1982): Lysosomal enzyme phosphorylation in mouse lymphoma cell lines with altered asparagine-linked oligosaccharides. J Biol Chem 257:10605–10612.

Gasa S, Makita A, Kinoshita Y (1983): Further study of the chemical structure of the equine erythrocyte hematoside containing O-acetyl ester. J Biol Chem

258:876–881.

Gathmann WD, Aminoff D (1981): Periodate oxidation studies in the elucidation of the structures of sialic-acid containing oligosaccharides. Biochem Biophys Res Commun 100:1453–1458.

Gehrke CW, Waalkes TP, Borek E, Swartz WF, Cole TF, Kuo KC, Abeloff M, Ettinger DS, Rosenhein N, Young RC (1979): Quantitative gas-liquid chromatography of neutral sugars in human serum glycoproteins. Fucose, mannose, and galactose as predictors in ovarian and small cell lung carcinoma. J Chromatogr 162:507–528.

Geoghegan KF, Osuga DT, Ahmed AI, Yeh Y, Feeney RE (1980): Antifreeze glycoproteins from polar fish. Structural requirements for function of glycopeptide 8. J Biol Chem 255:663–667.

Gershman H, Robbins PW (1981): Transitory effect of glucose starvation on the synthesis of dolichol-linked oligosaccharides in mammalian cells. J Biol Chem 256:7774–7780.

Gheorghiu T, Oette K (1970): Die gaschromatische Analyse von Zuckern und Zuckerderivaten aus nicht dialysierbaren Substanzen biologischer Flussigkeiten. J Chromatogr 48:430–445.

Glaudemans CPJ (1975): The interaction of homogenous murine myeloma immunoglobulins with polysaccharide antigens. Adv Carbohydr Chem Biochem 31:313–346.

Goldstein IJ, Hay GW, Lewis BA, Smith F (1965): Controlled degradation of polysaccharides by periodate oxidation, reduction, and hydrolysis. Methods Carbohydr Chem 5:361–370.

Gorin PAJ (1981): Carbon-13 nuclear magnetic resonance spectroscopy of polysaccharides. Adv Carbohydr Chem Biochem 38:13–104.

Guthrie RD, McCarthy JF (1967): Acetolysis. Adv Carbohydr Chem 22:11–23.

Gyorky G, Houck JC (1965): The determination of terminal protein-bound fucose. Can J Biochem 43:1807–1812.

Hakimi J, Carver J, Atkinson PH (1981): Application of high-field proton magnetic resonance spectroscopy in the structural determination of membrane-derived Sindbis virus glycopeptides. Biochemistry 20:7314–7319.

Hakomori S (1964): A rapid permethylation of glycolipid and polysaccharide catalyzed by methylsulfinyl carbanion in demethyl sulfoxide. J Biochem 55:205–208.

Hakomori S (1966): Release of carbohydrates from sphingoglycolipid by osmium-catalyzed periodate oxidation followed by treatment with mild alkali. J Lipid Res 7:789–792.

Hakomori S (1975): Structures and organization of cell surface glycolipids. Dependency on cell growth and malignant transformation. Biochim Biophys Acta 417:55–89.

Hakomori S, Gahmberg CG, Laine RA, Kiehn D (1974): Growth behavior of transformed cells as related to the surface structures of glycolipids and glycoproteins. Miami Winter Symp 8:69–102.

Hakomori S, Siddiqui B (1972): Release of oligosaccharides from glycolipids. Methods Enzymol 28:156–159.

Hakomori S, Siddiqui B (1974): Isolation and characterization of glycosphingolipids from animal cells and their membranes. Methods Enzymol 32:345–367.

Hakomori S, Watanabe K (1976): Blood group glycolipids of human erythrocytes. In Witting LA (ed): "Glycolipid Methodology." Champaign, IL: American Oil Chemists' Society, pp 13–48.

Hanessian S (1972): Mass spectrometry in the determination of structure of certain natural products containing sugars. Methods Biochem Anal 19:105–228.

Hanfland P, Egge H (1975): Mass spectrometric analysis of permethylated glycosphingolipids. Part 1. Sequence analysis of 2 blood group B-active glycosphingolipids from human B erythrocyte membranes. Chem Phys Lipids 15:243–247.

Hanfland P, Egge H (1976): Mass spectrometric analysis of permethylated glycosphingolipids. Part 2. Comparative studies on different blood group active and related erythrocyte membrane glycosphingolipids. Chem Phys Lipids 16:201–214.

Hanfland P, Egge H, Dabrowski U, Kuhn S, Roelcke D, Dabrowski J (1981): Isolation and characterization of an I-active ceramide decasaccharide from rabbit erythrocyte membranes. Biochemistry 20:5310–5319.

Hase S, Hara S, Matsushima Y (1979a): Tagging of sugars with a fluorescent compound 2-aminopyridine. J Biochem 85:217–220.

Hase S, Ikenaka T, Matsushima Y (1978): Structure analysis of oligosaccharides by tagging of the reducing end sugars with a fluorescent compound. Biochem Biophys Res Commun 85:257–263.

Hase S, Ikenaka T, Matsushima Y (1979b): Analyses of oligosaccharides by tagging the reducing end with a fluorescent compound. I. Application to glycoproteins. J Biochem 85:989–894.

Hase S, Ikenaka T, Matsushima Y (1979c): Analyses of oligosaccharides by tagging the reducing end with a fluorescent compound. II. Linkage point analyses. J Biochem 85:995–1002.

Hashimoto C, Cohen RE, Ballou CE (1980): Characterization of phosphorylated oligomannosides from *Hansenula wingii* mannoprotein. Biochemistry 19:5932–5938.

Hashimoto C, Cohen RE, Zhang WJ, Ballou CE (1981): Carbohydrate chains on yeast saccharomyces-cerevisiae carboxypeptidase Y are phosphorylated. Proc Natl Acad Sci USA 78:2244–2248.

Hay, GW, Lewis BA, Smith F (1965): Periodate oxidation of polysaccharides: General Procedures. Meth-

ods Carboh Chem 5:357–361.
Herp A, Wu AM, Moschera J (1979): Current concepts of the structure and nature of mammalian salivary mucous glycoproteins. Mol Cell Biochem 23:27–44.
Hoffman J, Lindberg B, Svensson S (1972): Determination of the anomeric configuration of sugar residues in acetylated oligo- and polysaccharides by oxidation with chromium trioxide in acetic acid. Acta Chem Scan 26:661–666.
Holmes EW, O'Brien JS (1979): Separation of glycoprotein derived oligosaccharides by thin layer chromatography. Anal Biochem 93:167–170.
Hori T, Sugita M, Ando S, Kuwahara M, Kumauchi K, Sugie E, Itasaka O (1981): Characterization of a novel glycosphingolipid, ceramide nonasaccharide, isolated from spermatozoa of the fresh water bivalve, *Hydriopsis schlegelli.* J Biol Chem 256:10979–10985.
Horton D, Philips, KD (1973): The nitrous acid deamination of glycosides and acetates of 2-amino-2-deoxy-D-glucose. Carbohydr Res 30:367–374.
Horton D, Philips KD, Defaye J (1972): The nitrous acid deamination of 2-amino-2-deoxy-D-mannose hydrochloride to D-glucose. Carbohydr Res 21:417–419.
Hough L (1954): Analysis of mixtures of sugars by paper and cellulose column chromatography. Methods Biochem Anal 1:205–242.
Howard DR, Natowicz M, Baenzinger JV (1982): Structural studies of the endoglycosidase H-resistant oligosaccharides present in human β-glucuronidase. J Biol Chem 257:10861–10868.
Howell HM, Conrad HE, Voss EW (1971): Radiochromatographic carbohydrate analysis and its application to gamma globulin systems. Fed Proc 30:594.
Hsieh P, Rosner MR, Robbins PW (1983a): Host-dependent variation of asparagine linked oligosaccharides at individual glycosylation sites of Sindbis virus glycoproteins. J Biol Chem 258:2548–2554.
Hsieh P, Rosner MR, Robbins PW (1983b): Selective cleavage by endo-β-N-acetylglucosaminidase H at individual glycosylation sites of Sindbis virion envelope glycoproteins. J Biol Chem 258:2555–2561.
Hunt LA (1981): Sindbis virus glycoproteins acquire unusual neutral oligosaccharides in both normal and lectin-resistant Chinese hamster ovary cell lines. Virology 113:534–543.
Hunt LA, Lamph W, Wright SE (1981): Transformation dependent alterations in the oligosaccharides of Prague C Rous sarcoma virus glycoproteins. J Virol 37:207–215.
Inoue S, Iwasaki M (1980): Characterization of a new type of glycoprotein saccharide containing polysialosyl sequence. Biochem Biophys Res Commun 33:162–165.
Isherwood RA, Jermyn MA (1951): Relationship between the structure of the simple sugars and their behavior on the paper chromatogram. Biochem J 48:515–524.
Ishizuka I, Suzuki M, Yamakawa T (1973): Isolation and characterization of a novel sulfaglycolipid, seminolipid from boar testis and spermatozoa. J Biochem 73:77–87.
Ishizuka I, Wiegandt H (1972): An isomer of trisialoganglioside and the structure of tetra- and pentasialoganglioside from fish brains. Biochim Biophys Acta 260:279–289.
Ito S, Muramatsu T, Kobata A (1975): Endo-β-N-acetylglucosaminidases acting on carbohydrate moieties of glycoproteins. Purification and properties of the two enzymes with different specificities from clostridium perfringens. Arch Biochem Biophys 171:78–86.
Itoh T, Li YT, Li SC, Yu RK (1981): Isolation and characterization of a novel monosialosylpentahexosyl ceramide from Tay-Sachs brain. J Biol Chem 256:165–169.
Iwasaki M, Inoue S (1981): Structures of O-glycosidically linked carbohydrate units of herring Clupea pallasii egg sialoglycoprotein. J Biochem 89:1067–1074.
Jacob R (1977): Mouse teratocarcinoma and embryonic antigens. Immunol Rev 33:3–32.
Jennings HJ, Smith ICP (1980): Determination of polysaccaride structures with ^{13}CNMR. Methods Carbohydr Chem 8:97–105.
Jones HG (1972): Gas-liquid chromatography of methylated sugars. Methods Carbohydr Chem 6:25–41.
Kalyan NK, Bahl OP (1981): Effect of deglycosylation on the subunit interactions and receptor binding activity of human chorionic gonadotropin. Biochem Biophys Res Commun 102:1246–1253.
Kalyan NK, Bahl OP (1983): Role of Carbohydrate in human chorionic gonadotropin. Effect of deglycosylation on the subunit interaction and on its in vitro and in vivo biological properties. J Biol Chem 258:67–74.
Kalyan NK, Lippes H, Bahl OP (1982): Role of carbohydrate in human chorionic gonadotropin. Effect of periodate oxidation and reduction in its in vitro biological properties. J Biol Chem 257:12624–12631.
Kamerling JP, Gerwig GJ, Vliegenthart JFG, Clamp JR (1975): Characterization by gas-liquid chromatography-mass spectrometry and proton-magnetic-resonance spectroscopy of pertrimethylsilyl methylglycosides obtained in the methanolysis of glycoproteins and glycopeptides. Biochem J 151:491–495.
Kannagi R, Nudelman E, Levery SB, Hakamori S (1982): A series of human erythrocyte glycosphingolipids reacting to the monoclonal antibody directed to a developmentally regulated antigen, SSEA-1. J Biol Chem 257:14865–14874.
Kapadia M, Feizi T, Evans MJ (1981): Changes in the expression and polarization of blood group I and i

antigens in post-implantation embryos and teratocarcinomas of mouse associated with cell differentiation. Exp Cell Res 131:185–195.

Karlsson KA (1974): Carbohydrate composition and sequence analysis of a derivative of brain disialoganglioside by mass spectrometry, with molecular weight ions at M/E 2245. Potential use in the specific microanalysis of cell surface components. Biochemistry 13:3643–3647.

Karlsson KA (1980): Structural fingerprinting of gangliosides and other glycoconjugates by mass spectrometry. Adv Exp Med Biol 125:47–61.

Karlsson KA, Pascher I, Pimlott W, Samuelsson BE (1974): Use of mass spectrometry for the carbohydrate composition and sequence analysis of glycosphingolipids. Biomed Mass Spectrom 1:49–56.

Kaufer JN (1969): Preparation of gangliosides. Methods Enzymol 14:660–664.

Kawasaki T, Ashwell G (1976): Carbohydrate structure of glycopeptides isolated from an hepatic membrane-binding protein specific for asialoglycoproteins. J Biol Chem 251:5292–5299.

Kennedy JF, Fox JF (1980a): Fully automatic ion-exchange chromatographic analysis of neutral monosaccharides and oligosaccharides. Methods Carbohydr Chem 8:3–12.

Kennedy JF, Fox JF (1980b): Fully automatic gel permeation chromatographic analysis of neutral oligosaccharides. Methods Carbohydr Chem 8:13–19.

Kiss J (1969): Glycosphingolipids (sugar-sphingosine conjugates). Adv Carbohydr Chem Biochem 24:381–433.

Kobata A (1972): Isolation of oligosaccharides from human milk. Methods Enzymol 28:262–271.

Kobata A (1977): Milk glycoproteins and oligosaccharides. In Horowitz MI, Pigman W (eds): "The Glyconjugates." New York: Academic Press, Vol 1, pp 423–440.

Kobata A (1978): Endo-β-N-acetylglucosaminidase C_I and C_{II} from Clostridium perfringens. Methods Enzymol 50:567–574.

Kobata A (1979): Use of endo- and exoglycosidases for structural studies of glycoconjugates. Anal Biochem 100:1–14.

Kobata A, Takasaki S (1978): Endo-β-galactosidase and endo-α-N-acetylgalactosaminidase from *Diplococcus pneumoniae*. Methods Enzymol 50:560–567.

Kobata A, Katsuko Y, Tachibana Y (1978): Oligosaccharides from human milk. Methods Enzymol 50:216–220.

Kochetkov NK, Chizhov OS (1966): Mass spectrometry of carbohydrate derivatives. Adv Carbohydr Chem 21:39–93.

Kochetkov NK, Chishov OS (1972): Mass spectrometry of carbohydrates. Methods Carbohydr Chem 6:540–554.

Kochibe N (1973): The broad aglycon specificity of α-L-fucosidase from marine gastropods. J Biochem 75:509–517.

Koerner TAW, Cary LW, Li SC, Li YT (1981): Carbon-13 NMR spectroscopy of Forssman hapten. Biochem J 195:529–534.

Koide N, Muramatsu T (1974): Endo-β-N-acetylglucosaminidase acting on carbohydrate moieties of glycoproteins. Purification and properties of the enzyme from *Diplococcus pneumoniae*. J Biol Chem 249: 4897–4904.

Kornfeld S (1982): Oligosaccharide processing during glycoprotein biosynthesis. In Horowitz MI (ed): "The Glycoconjugates." New York: Academic Press, Vol 3, pp 3–23.

Kornfeld R, Kornfeld S (1976): Comparative aspects of glycoprotein structures. Annu Rev Biochem 45:217–237.

Kornfeld K, Reitman ML, Kornfeld R (1981): The carbohydrate-binding specificity of pea and lentil lectins. Fucose is an important determinant. J Biol Chem 256:6633–6640.

Kotowycz G, Lemieux RU (1973): Nuclear magnetic resonance in carbohydrate chemistry. Chem Rev 73:669–698.

Krusius T, Finne J (1978): Characterization of a novel sugar sequence from rat brain glycoproteins containing fucose and sialic acid. Eur J Biochem 84:395–403.

Krusius T, Finne J, Rauvala H (1976): The structural basis of the different affinities of two types of acidic N-glycosidic glycopeptides for concanavalin A-Sepharose. FEBS Lett 71:117–120.

Kuhn R, Wiegandt H (1963): Consitution of ganglio-N-tetrose and the ganglioside G_1. Chem Ber 96:866–880.

Laine RA, Renkonen O (1975): Analysis of anomeric configurations in glyceroglycolipids and glycosphingolipids by chromium trioxide oxidation. J Lipid Res 16:102–106.

Laine RA, Esselman WJ, Sweeley CC (1972): Gas-liquid chromatography of carbohydrates. Methods Enzymol 28:159–178.

Laine RA, Stellner K, Hakomori S (1974): Isolation and characterization of membrane glycosphingolipids. Methods Membr Bio 205–230.

Larraga V, Edidin M (1979): Immunochemical characterization of surface antigens of Ter C, a teratocarcinoma-derived cell line. Proc Natl Acad Sci USA 76:2912–2916.

Leach BS, Collawn JF, Fish W (1980a): Behavior of glycopolypeptides with empirical molecular weight estimation methods. I. In sodium dodecyl sulfate. Biochemistry 19:5734–5741.

Leach BS, Collawn JF, Fish W (1980b): Behavior of glycopolypeptides with empirical molecular weight estimation methods. II. In random coil producing solvents. Biochemistry 19:5741–5747.

Ledeen RW, Yu RK (1982): Gangliosides: structure, isolation and analysis. Methods Enzymol 83:139-191.

Ledley FD, Mullin BR, Lee G. Aloj SM, Fishman PH, Hunt LT, Dayhoff MO, Kohn LD (1976): Sequence similarity between cholera toxin and glycoprotein hormones: Implications for structure activity relationship and mechanism of action. Biochem Biophys Res Commun 69:852–859.

Lee WMF, Klock JC, Macher BA (1981): Isolation and structural characterization of human lymphocyte neutral glycosphingolipids. Biochemistry 20:3810-3814.

Lehnhardt WR, Winzler RJ (1968): Determination of neutral sugars in glycoproteins by gas-liquid chromatography. J Chromatogr 34:471–479.

Leloir LF, Cardini CE (1957): Characterization of phosphorus compounds by solubility. Methods Enzymol 3:840–850.

Lemieux RU, Bock K, Delbaere LTJ, Koto S, Rao US (1980): The conformation of oligosaccharides related to the ABH and Lewis human blood group determinants. Can J Chem 58:631–653.

Lepage M (1964): Isolation and characterization of an esterified form of steryl glycoside. J Lipid Res 5:587-592.

Li E, Kornfeld S (1978): Structure of the altered oligosaccharide present in glycoproteins from a clone of chinese hamster ovary cells deficient in N-acetylglucosaminyl transferase activity. J Biol Chem 253:6426–6431.

Li YT, Li SC (1976): Enzymic methods for the structure analysis of the carbohydrate chain of glycoproteins. Methods Carbohydr Chem 7:221–225.

Li YT, Li SC (1977): Use of enzymes in elucidation of glycoprotein and glycosphingolipid structure. In Horowitz MI, Pigman W (eds): "The Glycoconjugates." New York: Academic Press, Vol 1, pp 51-67.

Li E, Tabas I, Kornfeld S (1978): The initial stages of processing of protein-bound oligosaccharides in vitro. J Biol Chem 254:4560–4567.

Li YT, Koerner TAW, Li SC (1980): General principle of using structural analysis of glycoconjugates by using glycosidases and carbon-13 NMR spectroscopy. Int Congr Pure Appl Chem Proc 27:199–204.

Lindberg B (1972): Methylation analysis of polysaccharides. Methods Enzymol 28:178–195.

Lindberg B, Lonngren J (1978): Methylation analysis of complex carbohydrates: General procedure and application for sequence analysis. Methods Enzymol 50:3–33.

Lindberg B, Lonngren J, Svensson S (1975): Specific degradation of polysaccharides. Adv Carbohydr Chem Biochem 31:185–240.

Linscheid M, D'Angona J, Burlingame AL, Dell A, Ballou CE (1981): Field desorption mass spectrometry of oligosaccharides. Proc Natl Acad Sci USA 78:1471–1475.

Lonngren J, Svensson S (1974): Mass spectrometry in structural analysis of natural carbohydrates. Adv Carbohydr Chem Biochem 29:141–160.

Lowe M, Nilsson B (1983): The structure of the complex type oligosaccharide from rabbit hepatic binding protein. A re-examination. J Biol Chem 258:1885–1887.

Lundblad A (1977): Urinary glycoproteins, glycopeptides and oligosaccharides. In Horowitz MI, Pigman W (eds): "The Glycoconjugates." New York: Academic Press, Vol 1, pp 441–458.

Lundblad A (1978): Oligosaccharides from human urine. Methods Enzymol 50:226–235.

Lundblad A, Svensson S, Low B, Messeter L, Cedergren B (1980): Release of oligosaccharides from human erythrocyte membranes of different blood group P phenotypes by trifluoroacetolysis. Eur J Biochem 104:323–330.

Lundsten J, Svensson S, Svennerhold L (1979): Increased urinary excretion of a glucose-containing tetrasaccharide in patients with Duchenne muscular dystrophy. FEBS Lett 97:249–252.

MacDonald DL, Pratt LM, Hakomori S (1980): Notes on improved procedures for the chemical modification and degradation of glycosphingolipids. J Lipid Res 21:642–645.

Macher BA, Klock JC (1980): Structural studies of neutral glycosphingolipids of human neutrophils by electron impact/desorption mass spectrometry. ACS Symp Ser 128:127–134.

Macher BA, Sweeley CC (1978): Glycosphingolipids: Structure, biological source, and properties. Methods Enzymol 50:236–272.

Maley F, Trimble RB (1981): Revision of the structure for an endo-β-N-acetyl-glucosaminidase H substrate using a novel modification of the Smith degradation. J Biol Chem 256:1088–1090.

Margolis RK, Margolis RU (1973): Extractability of glycoproteins and mucopolysaccharides of brain. J Neurochem 20:1285–1288.

Marshall RD, Neuberger A (1976): Determination of the 4-N-2-acetamido-2-deoxy-β-D-glucopyranosyl-1-asparagine linkage in glycoproteins. Methods Carbohydr Chem 7:212–220.

Matsubara T, Hayashi A (1978): Structural studies on glycolipid of shellfish. II. Occurrence of 3-O-methylgalactosamine in oyster glycolipid. J Biochem 83:1195–1197.

Matsuura F, Laine RA, Jones MZ (1981): Oligosaccharides accumulated in the kidney of a goat with β-mannosidosis: Mass spectrometry of intact permethylated derivatives. Arch Biochem Biophys 211:485-493.

Mayo JM, Carlson DM (1970): Effect of alkali of sodium borohydride of alkaline pH on N-acetylchondrosine: Reduction versus cleavage. Carbohydr Res 15:300–303.

McFadden W (1973): Techniques of combined gas chromatography/mass spectrometry: application in organic analysis. New York: John Wiley and Sons.

McGinnis GD, Faug P (1980): High-performance liquid chromatography. Methods Carbohydr Chem 8:33–43.

McLean C, Werner DA, Aminoff D (1973): Quantitative determination of reducing sugars, oligosaccharides and glycoproteins with tritiated borohydride. Anal Biochem 55:72–84.

McNeil M, Darvill AG, Aman P, Franzen LE, Albersheim P (1982): Structural Analysis of complex carbohydrates using high-performance liquid chromatography, gas chromatography and mass spectrometry. Methods Enzymol 83:3–45.

Mellis SJ, Baenziger JU (1981): Separation of neutral oligosaccharides by high-performance liquid chromatography. Anal Biochem 114:276–280.

Mizuochi T, Yamashita K, Fujikawa K, Kiseiel W, Kobata A (1979): The carbohydrate of bovine prothrombin. Occurrence of Gal $\beta \rightarrow 3$ Gal NAc grouping in asparagine-linked sugar chains. J Biol Chem 254:6419–6425.

Mizuochi T, Yamashita K, Fujikawa K, Titani K, Kobata A (1980): The structures of the carbohydrate moieties of bovine blood coagulation factor X. J Biol Chem 255:3526–3531.

Momoi T, Ando S, Nagai Y (1976): High resolution preparative column chromatographic system for gangliosides using Deae-sephadex and a new porus silica latrobead. Biochem Biophys Acta 441:488–497.

Momoi T, Wiegandt H (1980): Separation and microdetection of oligosaccharides of glycosphingolipids by high-performance cellulose thin-layer chromatography-autoradiofluorography. Physiol Chem 361-1201–1210.

Mononen I (1981): Quantitative analysis, by gas-liquid chromatography and mass fragmentography, of monosaccharides after methanolysis and deamination. Carbohydr Res 88:39–50.

Montreuil J (1975): Recent data on the structure of the carbohydrate moiety of glycoproteins metabolic and biological implications. Pure Appl Chem 42:431–477.

Morris HR, Thompson MR (1976): Studies on carbohydrates, peptides and glycopeptides by electron impact, field desorption and chemical ionization mass spectrometry. Adv Mass Spectrom Biochem Med 1:239–245.

Morris HR, Thompson MR, Osuga DT, Ahmed AI, Chan SM, Vandenheede JR, Feeney RE (1978): Anti-freeze glycoproteins from the blood of an antarctic fish the structure of the proline containing glycopeptides. J Biol Chem 253:5155–5162.

Mort AJ (1977): A new approach to the structural determination of glycoproteins and polysaccharides anhydrous HF solvolysis. J Supramol Struct Suppl 1:28.

Murakami-Murofushi K, Tadano K, Koyama I, Ishizuka I (1981): A trisialosyl ganglioside GT_3 of hog kidney structure and biosynthesis in vitro. J Biochem (Tokyo) 90:1817–1820.

Muramatsu T (1971): Demonstration of an endo-glycosidase acting on a glycoprotein. J Biol Chem 256:5335–5337.

Muramatsu T (1978): Endo-β-N-acetylglucosaminidase D from *Diplococcus pneumoniae*. Methods Enzymol 50:555–559.

Muramatsu T, Gachelin G, Pamonneville M, Delabre C, Jacob F (1979): Cell surface carbohydrates of embryonal carcinoma cells: Polysaccharidic side chains of F9 antigens and of receptors to two lectins, FBP and PNA. Cell 18:183–191.

Murray RK, Levine M, Kornblatt MJ (1976): Sulfatides: principal glycolipids of the testes and spermatozoa of chordates. In Witting LA (ed): "Glycolipid Methodology." Champaign, IL: American Oil Chemists' Society, pp 305–328.

Narasimhan S, Harpaz N, Longmore G, Carver JP, Grey AA, Schachter H (1980): Control of glycoprotein synthesis. The purification by preparative high voltage paper electrophoresis in borate of glycopeptides containing high mannose and complex oligosaccharide chains linked to asparagine. J Biol Chem 255:4876–4884.

Natowicz M, Baenziger JU (1980): Rapid method for chromatographic analysis of oligosaccharides on biogel P-4. Anal Biochem 105:159–164.

Natowicz MN, Baenziger JU, Sly WS (1982): Structural studies of the phosphosylated high mannose-type oligosaccharides on human β-glucuronidase. J Biol Chem 257:4412–4420.

Neuberger A, Gottschalk A, Marshall RD, Spiro RG (1972): Carbohydrate-peptide linkages in glycoproteins and methods for their elucidation. In Gottschalk A (ed): "Glycoproteins." Amsterdam: Elsevier, Vol 2, pp 450–490.

Neuberger A, Ratcliffe WA (1972): The acid and enzymic hydrolysis of O-acetylated sialic acid residues from rabbit Tamm-Horsfall glycoprotein. J Biochem 129:683–693.

Ng Y, Kin NMK, Wlafe LS (1980): High-performance liquid chromatographic analysis of oligosaccharides and glycopeptides accumulating in lysosomal storage disorders. Anal Biochem 102:213–219.

Nilsson B, Svensson S (1978a): Studies of the stability of reducing sugar towards trifluoroacetalysis: A method for specific elimination of 2-acetamido-2-deoxyhexase residues at reducing ends of oligosaccharides. Carbohydr Res 65:169–177.

Nilsson G, Svensson S (1978b): A new method for N-deacetylation of 2-acetamido-2-deoxy sugars. Carbohydr Res 62:377–380.

Nilsson B, Svensson S (1979): Studies of the reactivity of methylglycosides, oligosaccharides and polysaccharides towards trifluoroacetolysis. Carbohydr Res 69:292–296.

Nilsson B, Zopf D (1982): Gas chromatography and mass spectrometry of hexosamine containing oligosaccharide alditols as their permethylated, N-trifluoroacetyl derivatives. Methods Enzymol 83:46–58.

Nilsson B, DeLuca S, Lohmander S, Hascall VC (1982): Structures of N-linked and O-linked oligosaccharides on proteo glycan monomers isolated from swarm rat chondrosarcoma. J Biol Chem 257:10920–10927.

Nilsson B, Norden NE, Svensson S (1979): Structural studies on the carbohydrate portion of fetuin. J Biol Chem 254:4545–4553.

Noggle JH, Schirmer RE (1971): "The Nuclear Overhauser Effect." New York: Academic Press.

Ogata S, Muramatsu T, Kobata A (1975): Fractionation of glycopeptides by affinity column chromatography on concanavalin A-sepharose. J Biochem (Tokyo) 78:687–696.

Ogata AM, Muramatsu T, Kobata A (1977): α-L-fucosidases from almond emulsin: Characterization of the two enzymes with different specificities. Arch Biochem Biophys 181:353–358.

Ohashi M (1980): A new type of ganglioside. The structures of three novel gangliosides from the fat body of the frog. J Biochem (Tokyo) 88:583–589.

Ohkura KT, Yamashita K, Kobata A (1981): Urinary oligosaccharides of G_{M2}-gangliosidosis. Structures of oligosaccharides excreted in the urine of type 1 but not in the urine of type 2 patients. J Biol Chem 256:8485–8490.

Parente P, Strecker G, Leroy Y, Montreuil J, Fournet B (1982a): Analytical and semi-preparative high-performance liquid chromatography of oligosaccharides obtained by hydrazinolysis of hen ovomucoid. J Chromatogr 249:199–204.

Parente JP, Wieruszeski JM, Strecker G, Montreuil J, Fournet B, van Halbeek H, Dorland L, Vliegenthart JFG (1982b): A novel type of carbohydrate structure present in hen ovomucoid. J Biol Chem 257:13173–13176.

Parsons TF, Pierce JG (1980): Oligosaccharide moieties of glycoprotein hormones bovine lutropin resists enzymatic deglycosylation because of terminal O-sulfated N-acetylhexosamines. Proc Natl Acad Sci USA 77:7089–7093.

Paulson JC, Pricels JP, Glasgow LR, Hill RL (1978): Sialyl- and fucosyltransferase in the biosynthesis of asparaginyl-linked oligosaccharides in glycoproteins mutually exclusive glycosylation by β-galactoside α2→6 sialytransferase and N-acetylglucosamine α1→3 fucosyltransferase. J Biol Chem 253:5617–5624.

Paulson JC, Weinstein J, Dorland L, van Halbeek H, Vliegenthart JFG (1982): Newcastle disease virus contains a linkage-specific glycoprotein sialidase. Application to the localization of sialic acid residues in N-linked oligosaccharides of α_1-acid glycoprotein. J Biol Chem 257:12734–12738.

Pazur JH, Forsberg LS (1978): Determination of the sugar sequences and the glycosidic bond arrangements of immunogenic heteroglycans. Carbohydr Res 60:167–178.

Pazur JH, Miller KB, Dreher KL, Forsberg LS (1976): Antiglycosyl antibodies: Preparation and characterization of rabbit anti-galactose and anti-lactose antibodies. Biochem Biophys Res Commun 70:545–552.

Penick RJ, Meisler MH, McCluer RH (1966): Thin-layer chromatographic studies of human brain gangliosides. Biochem Biophys Acta 116:279–287.

Pierce JG, Parsons TF (1981): Glycoprotein hormones: Structure and function. Annu Rev Biochem 50:465–495.

Plummer TH, Tarentino AL (1981): Facile cleavage of complex oligosaccharides from glycopeptides by almond emulsin peptide: N-glycosidase. J Biol Chem 256:10243–10246.

Radford T, De Jongh DC (1972): Carbohydrates. In Walter GR (ed): "Biochemical Applications of Mass Spectrometry." New York: John Wiley and Sons, pp 313–350.

Radford T, De Jongh DC (1980): Carbohydrates. In Walter GR, Demer OC (eds): "Biochemical Applications of Mass Spectrometry," First supplementary volume. New York: John Wiley and Sons, pp 255–310.

Radin NS (1969): Florisil chromatography. Methods Enzymol 14:268–272.

Rauvala H, Finne J (1979): Structural similarity of the terminal carbohydrate sequences of glycoproteins and glycolipids. FEBS Lett 97:1–8.

Rauvala H, Finne J, Krusius T, Karkkainen J, Jarneflet J (1981): Methylation techniques in the structural analysis of glycoproteins and glycolipids. Adv Carbohydr Chem Biochem 38:389–416.

Rauvala H, Karkkainen J (1977): Methylation analysis of neuraminic acids by gas chromatography-mass spectrometry. Carbohydr Res 56:1–9.

Reading CL, Penhoet E, Ballou C (1978): Carbohydrate structure of vesicular stomatitis virus glycoprotein. J Biol Chem 253:5600–5612.

Reitman ML, Trowbridge IS, Kornfeld S (1982): A lectin-resistant mouse lymphoma cell line is deficient in glucosidase II, a glycoprotein-processing-enzyme. J Biol Chem 257:10357–10363.

Robbins PW, Hubbard SC, Turco SJ, Wirth D (1977): Proposal for a common oligosaccharide intermediate in the synthesis of membrane glycoproteins. Cell 12:893–900.

Roseman S, Daffner I (1956): Colorimetric method for determination of glucosamine and galactosamine. Anal Chem 28:1743–1746.

Roughan PG, Batt RD (1968): Quantitative analysis of sulfolipid (sulfoquinovosyl diglyceride) and galactolipids (monogalactosyl and digalactosyl diglycerides) in plant tissues. Anal Biochem 22:74–88.

Rouser G, Kritchevsky G, Yamamoto A, Simon G, Galli C, Bauman AL (1969): Diethylaminoethyl and triethylaminoethyl cellulose column chromatographic procedures for phospholipids, glycolipids and pigments. Methods Enzymol 14:272–317.

Saeed MS, Williams JM (1980): Model studies pertaining to the hydrazinolysis of glycopeptides and glycoproteins. Hydrazinolysis of the 1-N-acetyl and 1-N-L-β aspartyl derivatives of 2-acetamido-2-deoxy-β-D-glucopyranosylamine. Carbohydr Res 84:83–94.

Saito T, Hakomori S (1971): Quantitative isolation of total glycosphingolipids from animal cells. J Lipid Res 12:257–259.

Saito M, Sugano K, Nagai Y (1979): Action of arthrobacter ureafaciens sialidase on sialoglycolipid substrates. Mode of action and highly specific recognition of the oligosaccharide moiety of gangliside G_{M1}. J Biol Chem 254:7845–7854.

Samuelson O (1978): Chromatography of oligosaccharides and related compounds on ion-exchange resins. Adv Chromatogr 16:113–149.

Samuelsson BE (1980): Mass spectra and NMR spectra of gangliosides containing fucose. Adv Exp Med Biol 125:63–70.

Sawardeker JS, Sloneker JH, Jeanes A (1965): Quantitative determination of monosaccharides as their alditol acetates by gas liquid chromatography. Anal Chem 37:1602–1604.

Schauer R (1978): Characterization of sialic acids. Methods Enzymol 50:64–89.

Schauer R (1982): Chemistry, metabolism and biological functions of sialic acids. Adv Carbohydr Chem Biochem 40:131–234.

Schauer R, Haverkamp J, Wember M, Vliegenthart JFG, Kamerling JP (1976): N-acetyl-9-O-L-lactyl neuraminic acid. A new neuraminic acid from bovine submandibular gland. Eur J Biochem 62:237–242.

Schmid K, Nimberg RB, Kimura A, Yamaguchi H, Binette JP (1977): The carbohydrate units of human plasma α_1-acid glycoprotein. Biochim Biophys Acta 492:291–302.

Schrager J, Oates MDG (1968): The carbohydrate components of hydrolysates of gastric secretion and extracts from mucous blanck of the gastric body mucosa and antrum. Biochem J 106:523–529.

Schroeder W (1972): Hydrazinolysis. Methods Enzymol 25:138–143.

Scott JE (1960): Aliphatic ammonium salts in the assay of acidic polysaccharides from tissue. Methods Biochem Anal 8:145–197.

Seymour FR (1979): Polysaccharides branching and carbon-13 NMR. In Pasika WM (ed): "Carbon-13 NMR in Polymer Science." Washington, DC: Am Chem Soc pp 27–51.

Shashkov AS, Arbatsky N, Derevitskaya VA, Kochetkov NK (1979): Application of carbon-13 NMR spectroscopy in the structural study of complex heterooligosaccharides. Carbohydr Res 72:218–221.

Shibata S, Nakanishi H (1980): The α-D configuration of the glycosyl-N linkage in the nephritogenic glycopeptide isolated from rat glomerular basement membrane. Carbohydr Res 86:316–320.

Siddiqui B, Hakomori S (1970): Change of glycolipid pattern in Morris hepatomas 5123 and 7800. Cancer Res 30:2930–2936.

Sillerud LO, Prestegard JH, Yu RK, Schafer DE, Konigsberg WH (1978): Assignment of the ^{13}C nuclear magnetic resonance spectrum of aqueous ganglioside G_{M1} micelles. Biochemistry 17:2619–2628.

Sillerud LO, Yu RK, Schafer DE (1982): Assignment of the carbon-13 nuclear magnetic resonance spectra of gangliosides G_{M4}, G_{M3}, G_{M2}, G_{M1}, G_{D1a}, G_{D1b}, and G_{T1b}. Biochemistry 21:1260–1271.

Skipski VP, Barclay M (1969): Thin-layer chromatography of lipids. Methods Enzymol 14:530–598.

Slomiany A, Slomiany BL (1975): Bloodgroup A active difucosyl glycolipid from hog gastric mucosa. Biochim Biophys Acta 388:135–145.

Slomiany A, Slomiany BL, Horowitz MI (1976): Blood group A-active glycolipid variants from hog gastric mucosa. In Witting LA (ed): "Glycolipid methodology." Champaign, IL: American Oil Chemists' Society, pp 49–75.

Sloneker JH (1972): Gas-liquid chromatography of alditol acetates. Methods Carbohydr Chem 6:20–24.

Smith F (1956): End group analysis of polysaccharides. Methods Biochem Anal 3:153–212.

Smith DF, Zopf DA, Ginsburg V (1978): Sialyl oligosaccharides from milk. Methods Enzymol 50:221–226.

Solter D, Knowles BB (1979): Developmental stage-specific antigens during mouse embryogenesis. Curr Top Dev Biol 13:139–165.

Spencer B (1960): The ultramicro determination of inorganic sulphate. Biochem J 75:435–440.

Spiro RG (1966): Analysis of sugars found in glycoproteins. Methods Enzymol 8:3–25.

Spiro RG (1967): The structure of the disaccharide unit of the renal glomerular basement membrane. J Biol Chem 242:4813–4823.

Spiro RG (1970): Glycoproteins. Annu Rev Biochem 39:599–638.

Spiro RG (1972): Study of the carbohydrates of glycoproteins. Methods Enzymol 28:3–43.

Spiro RG (1973): Glycoproteins. Adv Protein Chem

27:349–467.

Spiro RG (1976a): Determination of the 5-O-β-D-galactopyranosyl hydroxy-L-lysine linkage in glycoproteins. Methods Carbohydr Chem 7:205–211.

Spiro RG (1976b): Isolation of glycopeptides from glycoproteins by proteolytic digestion. Methods Carbohydr Chem 7:185–190.

Spiro RG, Bhoyroo VD (1974): Structure of the O-glycosidically linked carbohydrate units of fetuin. J Biol Chem 249:5704–5717.

Spiro RG, Spiro MJ (1965): The carbohydrate composition of the thyroglobulins from several species. J Biol Chem 240:997–1001.

Stanley P, Sudo T (1981): Micro heterogeneity among carbohydrate structures at the cell surface may be important in recognition phenomena. Cell 23:763–770.

Stellner K, Saito H, Hakomori S (1973a): Determination of aminosugar linkages in glycolipids by methylation. Aminosugar linkages of ceramide pentasaccharides of rabbit erythrocytes and of Forssman antigen. Arch Biochem Biophys 155:464–472.

Stellner I, Watanabe K, Hakomori S (1973b): Isolation and characterization of glycosphingolipids with blood group H specificity from membranes of human erythrocytes. Biochemistry 12:656–661.

Stern PL, Willison KR, Lennox E, Galfre G, Milstein C, Secher D, Ziegler A, Springer T (1978): Monoclonal antibodies as probes for differentiation and tumor associated antigens: A Forssman specificity on teratocarcinoma stem cells. Cell 14:775–78s.

Strecker G, Montreuil J (1979): Glycoproteines et glycoproteinoses. Biochimie 61:1199–1241.

Strecker G, Herlant-Peers MC, Fournet B, Montreuil J, Dorland L, Haverkamp J, Vliegenthart JFG, Farriaux JP (1977): Structure of seven oligosaccharides excreted in the urine of a patient with Sandhoff's disease (GM_2-gangliosidosis variant O). Eur J Biochem 81:165–171.

Strecker G, Pierce CA, Fournet B, Spik G, Montreuil J (1981): Characterization by gas-liquid chromatography-mass spectrometry of oligosaccharides resulting from the hydrazinolysis-nitrous acid deamination reaction of glycopeptides. Anal Biochem 111:17–26.

Sugiyama H (1978): Carbon-13 NMR study of saccharides. Heterocycles 11:615–629.

Suzuki K (1965): The pattern of mammalian brain gangliosides. II. Evaluation of the extraction procedure, post-mortem changes and the effect of formalin preservation. J Neurochem 12:629–638.

Svennerholm L (1957): Quantitative estimation of sialic acids. II. A colorimetric resorciano-hydrochloric acid method. Biochim Biophys Acta 24:604–611.

Svennerholm L (1972): Gangliosides, isolation. Methods Carbohydr Chem 6:464–474.

Svennerholm L, Fredman P (1980): A procedure for the quantitative isolation of brain gangliosides. Biochim Biophys Acta 617:97–109.

Svennerholm L, Mansson JE, Li YT (1973): Isolation and structural determination of a novel ganglioside a disialosylpentahexosy ceramide from human brain. J Biol Chem 248:740–742.

Svennerholm L, Stallberg-Stenhagen S (1968): Changes in the fatty acid composition of cerebrosides and sulfatides of human nervous tissue with age. J Lipid Res 9:215–225.

Svensson S (1978): Degradation of polysaccharides by osidation and elimination. Methods Enzymol 50:33–38.

Svensson S (1980): Trifluoroacetolysis, a new method for structural studies of glycolipids. Adv Exp Med Biol 125:71–76.

Svensson S, Kenne L (1980): Sequential degradation of methylated polysaccharides. Methods Carbohydr Chem 8:67–71.

Sweeley CC, Bentley R, Makita M, Wells WW (1963): Gas-liquid chromatography of trimethylsilyl derivatives of sugars and related substances. J Am Chem Soc 85:2497–2507.

Sweeley CC, Moskal JR, Nunez H, Matsuura F (1980): Structural analysis of glycoconjugates by mass spectrometry of permethylated derivatives and by carbon-13 NMR spectroscopy. Int Congr Pure Appl Chem Proc 27:233–244.

Sweeley CC, Siddiqui B (1977): Chemistry of mammalian glycolipids. In Horowitz MI, Pigman W (eds): "The Glycoconjugates." New York: Academic Press, Vol 1, pp 459–540.

Sweeley CC, Tao RVP (1972): Gas chromatographic estimation of carbohydrates in glycosphingolipids. Methods Carbohydr Chem 6:8–13.

Sweeley CC, Wells WW, Bentley R (1972): Gas chromatography of carbohydrates. Methods Enzymol 8:95–107.

Szulman AE (1977): The ABH and Lewis antigens of human tissues during prenatal and postnatal life. In: "Human Blood Groups." 5th Int Convocation Immunol, Buffalo, NY. Basel: Karger, pp 426–436.

Tabas I, Kornfeld S (1980): Biosynthetic intermediates of β-glucuronidase contain high mannose oligosaccharides with blocked phosphate residues: J Biol Chem 255:6633–6639.

Tabas I, Schlesinger S, Kornfeld S (1978): Processing of high mannose oligosaccharides to form complex type oligosaccharides on the newly synthesized polypeptides of the vesicular stomatitis virus G protein and the IgG heavy chain. J Biol Chem 253:716–722.

Tai T, Yamashita K, Ogata-Arakawa M, Koide N, Muramatsu T, Iwashita S, Inoue Y, Kobata A (1975): Structural studies of two ovalbumin glycopeptides in relation to the endo-β-N-acetylglucosiminidase specificity. J Biol Chem 250:8569–8575.

Takahashi N, Nishibe H (1978): Some characteristics of a new glycopeptidase acting on aspartylglycosylamine linkages. J Biochem 84:1467–1473.

Takasaki S, Kobata A (1976): Chemical characterization and distribution of ABO blood group active glycoproteins in human erythrocyte membrane. J Biol Chem 251:3610–3615.

Takasaki S, Kobata A (1978): Microdetermination of sugar composition by radio-isotope labeling. Methods Enzymol 50:50–54.

Takasaki S, Mizuochi T, Kobata A (1982): Hydrazinolysis of asparagine-linked sugar chains to produce free oligosaccharides. Methods Enzymol 83:263–268.

Takasaki S, Yamashita K, Suzuki K, Iwanaga S, Kobata A (1979): The sugar chains of cold-insoluble globulin: A protein related to fibronectin. J Biol Chem 254:8548–8553.

Taravel FR, Vottero PJA (1975): Application of carbon-13 NMR to the study of oligosaccharides and polysaccharides anomeric configuration. Tetrahedron Lett 28:2341–2343.

Tarentino AL, Maley F (1974): Purification and properties of an endo-β-N-acetylglucosaminidase from *Streptomyces griseus*. J Biol Chem 249:811–817.

Tarentino AL, Plummer TH (1982): Oligosaccharide accessibility to peptide: N-glycosidase as promoted by protein-unfolding reagents. J Biol Chem 257:10776–10780.

Tarentino AL, Trimble RB, Maley F (1978): Endo-β-N-acetylglucosaminidase from *Streptomyces plicatus*. Methods Enzymol 50:574–580.

Tettamanti G, Bonati F, Marchesini S, Zambotti V (1973): A new procedure for the extraction, purification and fractionation of brain gangliosides. Biochim Biophys Acta 296:160–170.

Trimble RB, Maley F (1977): The use of endo-β-N-acetyl glucosaminidase H in characterizing the structure and function of glycoproteins. Biochem Biophys Res Commun 78:935–944.

Trimble RB, Tarentino AL, Plummer TH, Maley F (1978): Asparaginyl glycopeptides with a low mannose content are hydrolyzed by endo-β-N-acetylglucosaminidase H. J Biol Chem 253:4508–4511.

Tsuji T, Irimura T, Osawa T (1980): The carbohydrate moiety of band 3 glycoprotein of human erythrocyte membranes. Biochem J 187:677–686.

Tsuji T, Irimura T, Osawa T (1981): The carbohydrate moiety of band 3 glycoprotein of human erythrocyte membranes. Structures of lower molecular weight oligosaccharides. J Biol Chem 256:10497–10502.

Uda Y, Li SC, Li YT, McKibbin JM (1977): α-N-acetyl galactosaminidase from limpet, *Patella vulgata*. J Biol Chem 252:5194–5200.

Valent BS, Darvill AG, McNeil M, Robertsen BK, Albersheim P (1980): A general and sensitive chemical method for sequencing the glycosyl residues of complex carbohydrates. Carbohydr Res 79:165–192.

Van den Eijnden DH, Jozrasse DH, Dorland L, Van Halbeek H, Vliegenthart JFG, Schmid K (1980): Specificity in the enzymic transfer of sialic acid to the oligosaccharide branches of bi- and tri-antennary glycopeptides of α1-acid glycoprotein. Biochem Biophys Res Commun 92:839–845.

Van den Eijnden DH, Stoffyn P, Stoffyn A, Schiphorst-Wietske ECM (1977): Specificity of sialytransferase: structure of alpha 1-acid glycoprotein sialylated in vitro. A new methodology for the determination of the structure of the linkage formed by glycosyltransferase action on galactosyl-oligosaccharide protein acceptors. Eur J Biochem 81:1–7.

Van Halbeek H, Dorland L, Veldink GA, Vliegenthart JFG, Michalski JC, Montreuil J, Strecker G, Hull WE (1980a): Structure elucidation of oligomannoside-type asparagine-bound carbohydrate chains of glycoproteins by 500 MHz proton NMR spectroscopy. FEBS Lett 121:65–70.

Van Halbeek H, Dorland L, Veldink A, Vliegenthart JFG, Strecker G, Michalski JC, Montreuil J, Hull WE (1980b): A 500 MHz proton NMR study of urinary oligosaccharides from patients with mannosidosis. FEBS Lett 121:71–77.

Van Halbeek H, Dorland L, Vliegenthart JFG, Schmid K, Montreuil J, Fournet B, Hull WE (1980c): The application of 500 MHz ^{1}H-NMR spectroscopy for the structure elucidation of N-acetyllactosamine type asparagine-bound carbohydrate chains of glycoproteins. FEBS Lett 114:11–16.

Van Wazer JR, Callis CF, Shoolerg JN, Jones RC (1956): Principles of phosphorus chemistry. II. Nuclear magnetic resonance measurements. J Am Chem Soc 78:5715–5726.

Varki A, Kornfeld S (1980): Structural studies of phosphorylated high mannose-type oligosaccharides. J Biol Chem 255:10847–10858.

Veerkamp JH, Broekhuyse RM (1976): Techniques for the analysis of membrane lipids. In Maddy AH (ed): "Biochemical Analysis of Membranes." New York: John Wiley and Sons, pp 252–282.

Verma HK, Raizada MK, Schutzbach JS (1977): Formation of 1,2-mannosyl mannose by an enzyme preparation from rabbit liver. J Biol Chem 252:7235–7242.

Vliegenthart JFG, Van Halbeek H, Dorland L (1981): The applicability of 500 MHz high resolution ^{1}H-NMR spectroscopy for the structure determination of carbohydrates derived from glycoproteins. Pure Appl Chem 53:45–77.

Vratny P, Ouhrabova J, Copikova J (1980): Liquid chromatography of nonreducing oligosaccharides: A new detection principle. J Chromatogr 191:313–317.

Watanabe K, Arao Y (1981): A new solvent system for

the separation of neutral glycosphingolipids. J Lipid Res 22:1020–1024.

Watanabe K, Hakomori S, Childs RA, Feizi T (1979): Characterization of blood group I-active ganglioside. Structural requirements for I and i specificities. J Biol Chem 254:3221–3228.

Weinstein J, Souza-e-Silva V, Paulson JC (1982a): Sialylation of glycoprotein oligosaccharides N-linked to asparagine. Enzymatic characterization of a Galβ1-3(4) GlcNAc α2-3 sialyl-transferase and a Galβ1-4 GlcNAc α2-6 sialyl-transferase from rat liver. J Biol Chem 247:13845–13853.

Weinstein J, Souze-e-Silva V, Paulson JC (1982b): Purification of a Galβ1-4 GlcNac α2-6 sialyltransferase and a Galβl-3(4) GlcNAc α2-3 sialyltransferase to homogeneity from rat liver. J Biol Chem 257:13835–13844.

Weitzman S, Scott V, Keegstra K (1979): Analysis of glycopeptides as borate complexes by polyacrylamide gel electrophoresis. Anal Biochem 97:438–449.

Wells GZ, Kontoyiannidou V, Turco SJ, Lester RL (1982): Resolution of acetylated oligosaccharides by reverse-phase high-pressure liquid chromatography. Methods Enzymol 83:132–137.

Wenger DA, Wardell S (1973): Action of neuraminidase (EC 3.3.1.18) from *Clostridium perfringens* on brain gangliosides in the presence of bile salts. J Neurochem 20:607–612.

Wheat RW (1966): Analysis of hexasamines in bacterial polysaccharides by chromatographic procedures. Methods Enzymol 8:60–78.

Whistler RL, Anisuzzaman AKM (1980): Gel permeation chromatography. Methods Carbohydr Chem 8:45–53.

Wiegandt H, Bashang G (1965): Die Gewinnung des Zuckeranteils der Glykospingolipide durch Ozonolyse and Fragmentierung. Z Naturforsch B20:164–166.

Wiegandt H, Bucking HW (1970): Carbohydrate components of extraneuronal gangliosides from bovine and human spleen and bovine kidney. Eur J Biochem 15:287–292.

Winzler RJ (1955): Determination of serum glycoproteins. Methods Biochem Anal 2:279–313.

Wong CG, Sung SSJ, Sweeley CC (1980): Analysis and structural characterization of amino sugars by gas-liquid chromatography and mass spectrometry. Methods Carbohydr Chem 8:55–65.

Wolform ML, Beattie A, Bhattacharijee SS (1968): Reaction of alkyl vinyl ethers with methyl α-D-glucopyranodide. J Org Chem 33:1067–1070.

Yamaguchi H, Ikenaka T, Matsushima Y (1970): Gas-liquid chromatographic analysis of Smith degradation products from oligosaccharides. J Biochem (Tokyo) 68:253–254.

Yamamoto K, Tsuji T, Irimura T, Osawa T (1981): The structure of carbohydrate unit B of porcine thyro globulin. Biochem J 195(3):701–714.

Yamashita K, Hitos A, Matsuda Y, Tsuji A, Katuruma N, Kobata A (1983): Structural studies of the carbohydrate moieties of rat kidney γ-glutamyl transpeptidase. An extremely heterogeneous pattern enriched with nonreducing terminal N-acetyl glucosamine residues. J Biol Chem 258:1098–1107.

Yamashita K, Ichishima E, Arai M, Kobata A (1980a): An α-mannosidase purified from *Aspergillus saitai* is specific for 1,2 linkages. Biochem Biophys Res Commun 96:1335–1342.

Yamashita K, Kamerling JP, Kobata A (1982a): Structural study of the carbohydrate moiety of hen oromucoid. Occurrence of a series of pentaantemary complex-type asparagine-linked sugar chains. J Biol Chem 257:12809–12814.

Yamashita K, Liang CJ, Funakoshi S, Kobata A (1981): Structural studies of asparagine-linked sugar chains of human ceruloplasmin. Structural characteristics of the triantennary complex type sugar chains of human plasma glycoproteins. J Biol Chem 256:1283–1289.

Yamashita K, Mizuochi T, Kobata A (1982b): Analysis of oligosaccharides by gel filtration. Methods Enzymol 83:105–126.

Yamashita K, Tachibana Y, Mihara K, Okada S, Yabuuchi H, Kobata A (1980b): Urinary oligosaccharide of mannosidosis. J Biol Chem 255:5126–5133.

Yang HJ, Hakomori S (1971): A sphingolipid having a novel type of ceramide and lacto-N-fucopentase III. J Biol Chem 246:1192–1200.

Yohe HC, Macala LJ, Yu RK (1982): In vitro biosynthesis of the tetrasialoganglioside G_{QIb}. J Biol Chem 257:249–252.

Yohe HC, Yu RK (1980): In vitro biosynthesis of an isomer of brain trisialoganglioside G_{TIa}. J Biol Chem 255:608–613.

Yohe HC, Yu RK (1981): The cristal structure of D-ribose-1,4-lactose at 150°. Carbohydr Res 93:1–9.

Yohe HC, Veno K, Chang NC, Glaser GH, Yu RK (1980): Incorporation of N-acetyl-mannosamine into rat brain subcellular gangliosides effect of pentylenetetrazae-induced convulsions on brain gangliosides. J Neurochem 34:560–568.

Yoshima H, Matsumoto A, Mizuochi T, Kawasaki T, Kobata A (1981a): Comparative study of the carbohydrate moieties of rat and human plasma α_1-acid glycoproteins. J Biol Chem 256:8476–8484.

Yoshima H, Nakanishi M, Okada Y, Kobata A (1981b): Carbohydrate structures of HVJ (Sendai virus) glycoproteins. J Biol Chem 256:5355–5361.

Yoshima H, Takasaki S, Kobata A (1980): The asparagine-linked sugar chains of the glycoproteins in calf thymocyte plasma membrane. Structural studies of acidic oligosaccharides. J Biol Chem 255:10793–10784.

Yosizawa A, Sato T, Schmid K (1966):Hydrazinolysis of α_1-acid glycoprotein. Biochim Biophys Acta 121:417–420.

Yu RK, Ledeen RW (1970): Gas-liquid chromatographic assay of lipid-bound sialic acids: Measurement of gangliosides in brain of several species. J Lipid Res 11:506–516.

Yurchenco PD, Cecsariumi C, Atkinson PH (1978): Labeling complex carbohydrates of animal cells with monosaccharides. Methods Enzymol 50:175–204.

Zanetta JP, Breckenridge WC, Vincendon G (1972); Analysis of monosaccharides by gas liquid chromatography of the O-methyl glycosides as trifluoroacetate derivatives. J Chromatogr 69:291–304.

Zinn AB, Marshall JS, Carlson DM (1978): Preparation of glycopeptides and oligosaccharides from thyroxine-binding globulin. J Biol Chem 253:6761–6767.

Zinn AB, Plantner JJ, Carlson DM (1977): Nature of linkages between protein core and oligosaccharides. In Horowitz MI, Pigman W (eds): "The Glycoconjugates." New York: Academic Press, Vol 1, pp 69–85.

Zweig G, Whitaker JR (1971): "Paper Chromatography and Electrophoresis." New York: Academic Press, Vol II, pp 152–171.

Molecular and Chemical Characterization of Membrane Receptors, pages 161–178

6

Determination of the Hydrodynamic Properties of Detergent-Solubilized Proteins

Alan Davis

Psychopharmacology Section, Clarke Institute of Psychiatry, Toronto, Ontario M5T 1R8, Canada

INTRODUCTION

Characterization of membrane receptors is a difficult process, and only in the last decade has this area begun to be fully explored. The initial approach to most receptor characterization has been to study pharmacological profile, anatomical localisation, and correlation with physiological functioning of the endogenous transmitter. However, it is becoming apparent that receptors can exist in families and may have derived from a common gene pool. The structural specificity of closely related receptors (for example, β_1- and β_2-adrenoceptors) can often, therefore, be extremely close. When defined on a molecular basis they do, however, appear as distinctly different molecular entities. Clearly, as more and more receptor types are defined, such molecular definitions will become more and more essential. The aim of this chapter is to explore the methods that have developed over the years to produce standardized methodology to define such molecular parameters.

Only a few techniques are available to study integral, membrane-bound proteins in situ. The technique most recently applied to receptors is that of target size analysis by radiation inactivation [see Guellaen et al, this volume], but this requires at least a Van der Graaf generator and is thus not generally applicable. A more convenient method is to extract the receptor from the membrane by detergent solubilization [see Hjelmeland and Chambrach, this series] and to characterize the protein. However, most of the relevant methodology was derived from studies on water-soluble proteins, and so modifications and assumptions have had to be made in extending this approach to integral membrane proteins. Essentially, allowance has to be made for the contribution of detergent to the detergent-receptor complex formed upon membrane solubilization. Maintaining the receptor in a soluble form, by keeping it associated with detergent, is a second problem. This chapter will deal with such problems, the inherent assumptions involved, and possible sources of error.

The hydrodynamic properties that will be discussed and defined are the Stokes' radius, sedimentation coefficient, frictional ratio, and partial specific volume. From these the molecular weight of the receptor can be deduced. In the final section the validity of this ap-

---

Alan Davis is now Assistant Director of Clinical Research, Boehringer Ingelheim, Burlington, Ontario L7L 5J8, Canada.

proach will be tested by comparing this determination of molecular weight with that determined by gel electrophoresis [see Goding and Roberts et al, this volume] and, where possible, by radiation inactivation.

Definitive description of any protein resides within the primary, secondary and tertiary structures (ie, its aminoacid sequence, subunit composition, and overall conformation). These parameters have yet to be completely determined for a membrane receptor. It is probable that the nicotinic acetylcholine receptor of *Torpedo californica* and other related species will be among the first such receptors completely characterized. This receptor has been purified to homogeneity and its coding genes isolated. From here the complete amino acid sequence for the receptor can be worked out (see, for example, Raftery et al [1980]; Raftery and Changeux [1982], for review).

TERMINOLOGY OF PROTEIN PROPERTIES

Stokes' Radius

The Stokes' radius of a molecule is a theoretical concept. It is defined as the radius of a sphere that would have the same frictional ratio (see below) as the particle under study [Tanford, 1961]. It is a function not only of the protein but also of any associated lipid, detergent, and solvent. Since proteins differ widely in their shape—from globular, water-soluble proteins to elongated, membrane-bound, intrinsic proteins—the Stokes' radius obviously leads to cases of misinterpretation. This is because this parameter is essentially the radius of the macromolecular complex under study *if it were completely globular.* Hence the study of linear, double-stranded DNA molecules leads to difficulties; Stokes' radii determined by viscosity measurements were up to twofold different from those determined by diffusion or sedimentation velocity experiments [Eigner and Doty, 1965]. The normal differences are, however, generally less than 10% [Tanford et al, 1974]. Rodbard [1976] has graphically illustrated this connection between the geometric mean radius and the molecular asymmetry.

Two methods are commonly employed to obtain this parameter:

Measurement of the velocity of diffusion. The Stokes' radius is determined from the following relationship:

$$a = kT/\pi\eta_{20,w}D_{20,w}, \tag{1}$$

where: a = Stokes' radius (m), k = Boltzmann constant, T = temperature (Kelvin), $\eta_{20,w}$ = viscosity of water at 20°C, and $D_{20,w}$ = diffusion coefficient corrected to water at 20°C.

Since the method involves diffusion in an aqueous media, it is obviously inappropriate for application to membrane-bound receptors and so will not be discussed further.

Gel permeation chromatography. This is the most commonly used method since it is relatively simple and quick. Proteins are passed over a column containing a gel (made from, for example, cross-linked dextran), and as elution proceeds, the proteins diffuse into and out of the gel pores. The extent of diffusion depends on the size of the protein, and diffusion results in retardation of a protein relative to the solvent front. Very large molecules are not able to enter the pores and are eluted first. The extent of retardation has been shown to be proportional to the Stokes' radius of the protein (see below). A calibration curve for the column and gel type in use is constructed using proteins of known Stokes' radii. Many investigators have constructed calibration curves using molecular weight rather than Stokes' radii. This is entirely inappropriate for the reasons outlines above; the "molecular weight" will be artificial since, in addition to the protein under study, detergent, lipid, and solvent are all involved (to varying degrees) in determining the elution volume of the micellar complex. Other cautions and drawbacks will be further discussed below.

Sedimentation Coefficient

The sedimentation coefficient (S) value is often used as an approximate guide to the molecular weight of proteins. It is measured by sedimenting the protein under study, by centrifugal force, through a solvent of increasing density. The final location of the molecule is directly related to its molecular mass.

The sedimentation coefficient is thus defined as the sedimentation rate per unit centrifugal force [Schachman, 1959]:

$$S = \frac{\delta r/\delta t}{w^2 r} \quad (2)$$

where r = distance from the centre of rotation to the boundary of the sample and solvent (in cm), w = angular velocity of the rotor (in radians per second), t = time, S = Svedberg units (1×10^{-13} s).

The S value is correct only under the conditions of the centrifugation; that is, at the temperature (T) of the run and the medium density (m). The nature of the solvent obviously influences the sedimentation coefficient by virtue of its density and its viscosity (which in turn is influenced by temperature). This is compensated for by converting all measurements to values relative to water at 20°C, the $S_{20,w}$ value. This is performed using the following equation:

$$S_{20,w} = S_{T,m} \frac{\eta_{T,m}(P_p - P_{20,w})}{\eta_{20,w}(P_p - P_{T,m})}, \quad (3)$$

where T,m = viscosity of the medium at the temperature (°C) of the centrifugation, $\eta_{20,w}$ = viscosity of water at 20°C (0.010002 g/cm), P_p = density of the protein in solution (ie, the reciprocal of the partial specific volume), P_{20} = density of water at 20°C (0.99823 g/ml) [see Martin and Ames, 1961].

The partial specific volume (and therefore density) of most proteins does not vary greatly with temperature, and so the term P_p is considered a constant. The assumption of a partial specific volume of 0.725 ml/g, and therefore a density of 1.379 g/ml, has been calculated to lead to a maximum 3% error in estimating the $S_{20,w}$ of water-soluble proteins [Martin and Ames, 1961]. Of course, this error will be larger for intrinsic proteins that show a wider range of partial specific volumes.

Equation 4 (below) is applicable to solvents of uniform density but can be applied to sucrose density gradients in which the density increases continuously down the gradient. This is dealt with in length by Martin and Ames [1961]. The most convenient and most commonly used alternate to this approach is to construct a standard curve using proteins of known $S_{20,w}$ values, since Martin and Ames found that a linear relationship exists between the distance a protein travels down a sucrose gradient and its $S_{20,w}$ value. All standard proteins used to date are water soluble. An approximate molecular weight can be obtained from the calibration curve:

$$\frac{S_{20,w}\text{ unknown}}{S_{20,w}\text{ known}} = \left[\frac{\text{Mr unknown}}{\text{Mr known}}\right]^{2/3}, \quad (4)$$

where Mr = molecular weight (in daltons).

However, Equation 4 is obviously only applicable if the unknown and the standard are of the same shape. Since the standard proteins are water-soluble, the approximation is only useful for globular proteins. Obviously then, to be applicable to membrane-bound receptors held in solution by detergent, molecular shape must be accounted for. A pencil-like molecule will exhibit less frictional resistance than a spherical one and will thus migrate further. The parameter used to define molecular shape, the frictional ratio, is discussed below.

The combination of methods used to calculate these parameters (sedimentation coefficient, frictional ratio, and molecular weight) are discussed below.

Frictional Ratio

The molecular shape of a protein is embodied by the term "frictional ratio" (f/fo). An equation defining the frictional resistance experienced by a particle moving through a viscous medium was first derived by Stokes [1851]:

$$f = 6\,\pi\,\eta\,r. \quad (5)$$

This only applied to spherical [see Schachman, 1959] particles of radius r moving through a medium of viscosity η. The value obtained for spherical virus particles agreed well with that radius determined by other techniques (eg electron microscopy). It became apparent, however, that this relationship did not hold for the majority of molecules. Equations were derived to yield a frictional ratio, f/fo, (where fo = frictional coefficient of a sphere with a volume equal to that of the particle in question) for which a value of unity implied a sphere. That is, the ratio of the length of each axis was 1.0. (Two sets of equations were derived, one for prolate ellipsoids and one for oblate ellipsoids; most equations are based on the former—see Eq 7). Rodlike particles that would have a f/fo of, for example, 1.433 would have an axial ratio of 8. That is, the length of the prolate (polar) axis would be eight times that of the oblate (equatorial axis) (Table I). The frictional ratio is determined directly by diffusion measurements and indirectly by gel chromatography.

The following relationship between diffusion and the frictional ratio holds:

$$D = RT/N\ [f/fo], \tag{6}$$

where N = Avogadro's number (6.02252×10^{23}).

As for determination of the Stokes' radius, this method is only applicable for water-soluble proteins. For detergent-solubilized proteins, the value is determined indirectly from knowledge of the Stokes' radius and the molecular weight as determined by gel electrophoresis. The frictional ratio is obtained using the following equation [taken from Siegel and Monty, 1966]:

$$f/f_o = a \left[\frac{4\,\pi\,N}{3\ Mr\ [\bar{v} + (\varsigma/P)]} \right]^{1/3}, \tag{7}$$

where ς = solvation factor (g solvent/g protein) and P = density of medium. The solvation factor is normally assumed to be 0.2 solvent/g protein [Edsall, 1953].

TABLE I. Relationship Between the Axial and Frictional Ratios

Axial ratio[a]	Frictional ratio, f/fo (for prolate ellipsoids)
1	1.000
2	1.044
3	1.112
4	1.182
5	1.255
6	1.314
8	1.433
10	1.543
12	1.645
15	1.784
20	1.996

[a]Taken from Schachmann [1959].

The relationship between the sedimentation coefficient, frictional ratio, and molecular weight is given in the following equation:

$$S_{20,w} = \frac{Mr\ (1 - \bar{v}\ p)}{N\ f/f_o}. \tag{8}$$

Partial Specific Volume

The total volume occupied by a protein when in solution is, of course, not representative of its true mass, especially when considering proteins usually residing in a membrane-bound form. Commonly, hydrophilic regions will effectively trap water molecules within the protein. This extent of water hydration (ς) is commonly assumed as 0.2 g (or ml) solvent/g protein. The remaining volume is occupied mainly by the protein and is referred to as the partial specific volume ($\bar{v}$). The usual range is 0.700–0.750, with most (water-soluble) proteins being near 0.730 ml/g. The more hydrophobic the protein, the greater the $\bar{v}$ value and the less the degree of solvation. The difference between the sum of the hydration factor plus the partial specific volume, and unity, is probably related to miscellaneous associated elements such as lipids. The partial specific volume is the most difficult parameter to determine experimentally, especially when the protein is present only in a complex protein mixture. The problems do not arise if the protein is present in a purified form. Then, gross amino acid composition can be used to determine $\bar{v}$. This is because Traube [1899] discovered that apparent molal volumes of organic complexes could be described as the sum of composite atomic volumes. Cohn and Edsall [1943] extended this to amino acids and polypeptides assuming that the volume of the amino acid residue was independent of the tertiary conformation of the protein. The following relationship applied:

$$\bar{v} = \Sigma_i \bar{v}_i \; w_i \, / \, \Sigma_i \, w_i, \qquad (9)$$

where $\bar{v}$ = partial specific volume of the amino-acid residue and w_i = fractional contribution of the polypeptide to the total weight.

The partial specific volumes for the amino acids can be obtained from Fish [1975]. Gibbons [1972] has applied the same technique to sugar residues and glycoproteins. Unfortunately, however, detergent-solubilized proteins are rarely present in a pure form. Even if they were, at least 0.5 mg of pure protein would probably be required [see Edelstein and Schachman, 1967].

For membrane-bound proteins held in solution by detergents, the range of partial specific volumes must be extended. Partial specific volumes of a number of detergents are very close to unity implying very little associated water. This is consistent with their ordered, micellar conformations with hydrophobic regions in the core and hydrophilic regions directed towards the outside. In their monomer form (ie, below their critical micellar concentration), more solvent will be associated with the molecules (reflected in a decreased partial specific volume) due to a loss of this ordered structure (eg, sodium dodecyl sulphate, SDS). The error in assuming a value for $\bar{v}$ will hence be larger the greater the binding of the detergent. Proteins bind SDS to a large extent (see below), and so the detergent will be a major factor in determining the specific volume. Clearly, the assumption of a value of 0.730 ml/g is only applicable to water-soluble proteins or proteins solubilized using a detergent to which the binding is a very small percentage. The problem has been partially resolved by the use of sucrose density gradient ultracentrifugation in two solvents of differing density (H_20 and D_20) in order to calculate the relative contributions of the protein and the detergent.

METHODOLOGY

Gel Permeation Chromatography

Gel permeation chromatography is used to obtain an index of molecular "size" since the position at which a protein elutes is based on a partitioning effect between the solute within the gel and the mobile, exterior, solvent phase. The smaller the molecule, the greater it partitions into the gel and thus the later the elution. Correlations have been found between elution volumes and a number of parameters, principally molecular weight (Mr, daltons), sedimentation coefficient (S, Svedberg units, 10^{-13} s), and Stokes' radii (a, nm). A number of authors have concluded that Stokes' radii can be regarded as the most appropriate parameter, irrespective of the conformational shape [Siegel and Monty, 1966; Fish et al, 1970; Tanford et al, 1974]. Some notable exceptions have, however, been found but mainly among large and asymmetric proteins like myosin and fibrinogen [Warshaw and Ackers, 1971; Nozaki et al, 1976; Le Maire et al, 1980]. It is apparent that most of the controversy arises from the use of inappropriate protein standards in gel calibration. These are usually water-soluble, globular proteins which have very similar frictional ratios and partial specific volumes. As stated above, membrane proteins have a greater range of frictional ratio values. In addition, gel permeation chromatography of intrinsic protein is done in detergent solution.

For example, Siegel and Monty [1966], performing gel filtration on impure enzyme solutions, presented evidence that the Stokes' radius was the critical parameter. Fibrinogen eluted before ferritin, consistent with their order of Stokes' radii, but in the reverse order expected from their molecular weights (3.3 $\times 10^5$ and 1.3 $\times 10^6$ daltons, respectively).

When studying intrinsic proteins it is more readily apparent that the use of molecular weight is inappropriate. The same protein, in differing concentrations and types of detergent, will give different elution volumes. The elution volume depends on the degree of detergent binding, the presence and concentration of detergent in the eluting buffer, the dimensions of the column, and the eluting conditions.

Le Maire et al [1980] also presented data that indicated that some detergents may act to decrease penetration of the protein into the gel. This "repellent" effect was noted by the consistent overestimation of Stokes' radii when proteins were run in the presence of the detergents deoxycholate, dodecyl octaethylene glycol monoether, and dodecyl dimethylamine oxide. Clearly, this effect warrants further investigation.

A very small, fast-running column will probably not result in sufficient dilution of the associated detergent to cause precipitation. Indeed, gel chromatography on mini-Sephadex G-50 columns has been commonly used to assay solubilized receptor proteins [Cuatrecasas, 1972a,b; Caron and Lefkowitz, 1976; Davis et al, 1981, 1982a]. The greater the dilution factor (as with a large column and/or slow eluting conditions), the greater the chance that the protein under study will precipitate. Therefore, one usually has to employ detergent in the eluting buffer. A concentration of detergent below the critical micellar concentration (CMC) is preferable so as to keep the contribution of the detergent to the Stokes' radius to a minimum. If the Stokes' radius, elucidated by gel chromatography, is to be used to determine the molecular weight of the protein in conjunction with sucrose density gradient ultracentrifugation experiments, it is essential that the *same detergent be used in both the column eluting buffer and in the sucrose of the gradient.* If this is not the case, the calculations become invalid since the detergent-protein complex will exist in two different "sizes"—one size on the gel column and one in the sucrose density gradient.

Hence the choice of conditions for gel chromatography are critical. Preferably, the eluting speed and the dimensions of the column will be such to cause most of the original detergent to be diluted out (this may not always be the case since the detergent may elute as a micelle, especially if it has a very low CMC). Protein would be held in solution by the detergent present in the elution buffer. Unfortunately, this may lead to denaturation of the protein. One therefore has to trade recovery of activity against detergent exchange. If the particular protein is a receptor and is being assayed using the binding of a reversible ligand, there is an additional tradeoff. The free ligand will be retarded by the gel in the column and so the concentration at the receptor will decrease as elution progresses. Ligand dissociation will probably occur. Consequently, the lack of good receptor recovery with a reversible ligand can thus reflect not only receptor inactivation by detergent dilution, but also ligand dissociation. Calculations of protein purification must be made with these factors borne in mind.

In summary, the choice of conditions for gel permeation chromatography of detergent-solubilized proteins must be made with care. Preferably an irreversible radioligand can be used to avoid problems of dissociation and denaturation. Then, most of the detergent can be eliminated by dilution with the eluting buffer. To maintain the protein functionally intact detergent should be added to the eluting buffer. This detergent should exhibit a low degree of binding to the protein under question and should be used, at the same concentration, in subsequent determinations of the sedimentation coefficient by sucrose density gradient ultracentrifugation.

Sucrose Density Gradient Ultracentrifugation

The molecular weight of a *pure* protein can be readily obtained by sedimentation equilibrium experiments [Casassa and Eigenberg, 1964; Tanford et al, 1974; Fish, 1975]. However, for intrinsic proteins one invariably deals with impure protein mixtures. In 1961 Martin and Ames described the measurement of sedimentation *velocity* as a means of determining the molecular mass of a single protein within an impure solution, and in 1972 Meunier and co-workers [1972a] derived the methodology required to apply this technique to membrane-bound proteins held in solution by detergents. For sedimentation equilibrium studies, proteins move through density gradients and come to rest at a position determined solely by the molecular mass. In contrast, sedimentation *velocity* experiments are not made at equilibrium but at a time when the protein is still moving through the density gradient. In this situation, information about molecular shape is required to derive the molecular mass.

Proteins of known molecular mass are used to construct a calibration curve, and along with the Stokes' radius, determined by either gel permeation chromatography or from the diffusion coefficient, the molecular mass is determined from the sedimentation coefficient. With pure, soluble proteins this is a theoretically sound method since there are no other interfering proteins and no detergent present. Protein concentration can be measured down the gradient and the calibration curve used to assess the sedimentation coefficient.

Application of this method to detergent-solubilized proteins has three problems:

1. What is the contribution of the detergent (and remaining membrane lipid) to the molecular mass?

2. A specific assay for the protein under study is required since the detergent-solubilized protein is rarely in a pure state.

3. How valid is the application of a calibration curve, constructed invariably from soluble, globular proteins, towards the determination of the sedimentation coefficient of an intrinsic protein (which is invariably nonglobular and, by definition, water insoluble) held in solution by a detergent?

The theoretical approach to allow for the contribution of detergent was initially developed by three main groups. Martin and Ames [1961] first described the use of sucrose density gradients to obtain sedimentation coefficients of water-soluble proteins (in their case, enzymes of the histidine biosynthetic pathway). They used Equation 4 to obtain an approximate molecular weight (which assumes most proteins are essentially spherical) and demonstrated that a linear relationship existed between the distance travelled by the enzyme and the time of centrifugation. Edelstein and Schachman [1967] subsequently described the use of sedimentation equilibrium experiments in H_2O/D_2O solvents to obtain the partial specific volumes of proteins. Finally, Meunier et al [1972a] combined the two approaches (sucrose density gradients in both H_2O and D_2O solvents) to obtain the partial specific volumes and relative detergent contributions to the molecular masses of detergent-solubilized proteins. This latter group found that the sedimentation velocity estimated by sucrose density ultracentrifugation was not compatible with the Stokes' radius as determined by gel permeation chromatography [Meunier et al, 1972b]. It was subsequently discovered that this was due to values of $\bar{v}$ and the frictional ratio that did not corpond to those normally taken for water-soluble proteins.

Essentially, advantage was taken of the differences between the partial specific volume of the protein (on average 0.725 ml/g; see above) and that of the detergent. For example, Triton X-100 has a partial specific vol-

TABLE II. Some Properties of Commonly Used Detergents

Detergent	Partial specific volume (ml g^{-1})	Aggregation number	Micellar Mr (daltons)	Average Mr (daltons)	CMC[a] (mg/ml)	HLb[b]
CHAPS[c]	—	9–10	6,000[d]	650	2.6–3.9	—
Cholate	—	2– 4	900–1,800	408	5.7	18
Deoxycholate	—	4–10	1,700–4,200	392	2.0	16
Digitonin	0.738	60	70,000	1,200	0.01–0.04[e] 0.07–0.09[f]	0.4
Lubrol PX	0.958[g]	—	—	—	—	—
Lubrol WX	—	96	9,200	93	0.04	14.9
SDS	0.815[h]	62	18,000	288	2.4	40
Triton N-100	0.922[g]	—	—	642	0.05	13.4
Triton X-100	0.908[h]	140	90,000	628	0.15–0.2	13.5
Tween 80	0.896[h]	—	—	1,300	0.013	15.0

[a]Critical micellar concentration.
[b]Hydrophilic-lipophilic balance.
[c]3-[(3-cholamidopropyl-1-)dimethylammonio]-1- propane sulphonate [Hjelmeland 1980].
[d]Liscia et al [1982].
[e]Unpurified, expressed as percent.
[f]Purified, expressed as percent [Fleming and Ross, 1980].
[g]Haga et al [1977].
[h]Tanford et al [1974]. Others: see Helenius and Simon [1975] and Furth [1980] for references.

ume of 0.94 ml/g and Lubrol PX 0.98 ml/g (Table II). Obviously, the partial specific volume of the detergent-protein complex will be a function of the relative contribution of each component. By altering the density of the sucrose gradient, by replacing H_2O and D_2O, the $\bar{v}$ of the complex could be deduced. Three immediate considerations arise:

1. The partial specific volume of the detergent must by known.

2. The binding of the detergent to the protein under study is the same in both cases.

3. The detergent partial specific volume must be significantly different from that for the protein.

In the case of 1, this is only the case with a few well-known detergents and is pertinently missing for a number of important detergents (for example, CHAPS). In the case of 2, this is probably a safe assumption. Requirement 3 automatically excludes digitonin, which has proved to be an excellent detergent for the solubilization of receptor molecules including β-adrenoceptors [Caron and Lefkowitz, 1976; Strauss et al, 1979], muscarinic receptors [Aronstam et al, 1978], dopamine receptors [Gorissen et al, 1979; Davis et al, 1981, 1982a; Madras et al, 1982a], opiate receptors [Ruegg et al, 1981], and antidepressant (imipramine) drug receptors [Davis et al, 1982b, 1983]. The partial specific volume of digitonin is very close to that for proteins, and so a change in density would influence the protein and the detergent in the same proportion. Sedimentation coefficients obtained in the presence of digitonin (as performed, for example, for the dopamine receptor, Gorissen and Laduron [1979]; Gorissen et al [1979]) are thus of very limited value. Attempts to obtain meaningful values have been performed on both the α- and β-adrenoceptors [Guellaen et al, 1979; Venter, 1981] and the dopamine receptor [Lilly et al, submitted]. In these experiments, Triton X-100 was added to the gradient to exchange for the digitonin. The extent of exchange, and the amount of digitonin that remains attached, remains to be determined. Detergent binding may be readily deduced using the radiolabelled detergent (eg, Helenius and Simon [1972]; Makino et al [1973], and recently, a radioimmunoassay for digitonin has been described [Brooker et al, 1982]. This is sensitive to as little as 20 mg/ml (15 μM) and can be used in the presence of phospholipids (especially cholesterol with which digitonin forms a tight complex) and other detergents. Determination of residual digitonin and Triton X-100 binding will allow us to make even more accurate molecular weight determinations.

The relationship between the hydrodynamic molecular weight, Stokes' radius and the sedimentation coefficient is

$$Mr = \frac{6 \pi N \eta_{20,w}}{1 - \bar{v} P_{20,w}} a S_{20,w}, \quad (10)$$

where $\eta_{20,w}$ = viscosity of sucrose (relative o water at 20°C at position of protein (g/[cm × s]) and $P_{20,w}$ = density of sucrose (relative to water at 20° C) (g/ml). This equation was derived by Einstein's modification of the Svedberg equation [Siegel and Monty, 1966].

The sedimentation coefficient and Stokes' radius are as defined previously.

From these values, the asymmetry of the molecule can be assessed by calculating the frictional ratio (see Eq. 7 above and Table I).

So far, these calculations apply to soluble, globular proteins and intrinsic proteins that bind minimal detergent. For example, Neer [1974] found that the Lubrol PX solubilized adenylate cyclase enzyme, run in sucrose density gradients containing 0.1% Triton, did not show a difference between sedimentation coefficients obtained from gradients run in both H_2O and D_2O. In contrast, Barchi and Murphy [1981], for example, found a $S_{20,w}$ of 9.1 (for solubilized, saxitonin labelled, sodium-ion channels) in H_2O and a $S_{20,w}$ of 6.1 in D_2O gradient, indicating a markedly differant partial specific volume from soluble, globular proteins (0.73–0.76 ml/g) (assuming that deuterium exchange occurs to the same extent with calibrating proteins). The partial specific volume of the protein-detergent complex is calculated according to the equation derived by Clarke [1975]:

$$S_H = S_D \left[\frac{\eta_D}{\eta_H}\right] \left[\frac{1 - \bar{v} P_H}{1 - \bar{v} P_D},\right] \quad (11)$$

where D = gradients run in D_2O, H = gradients run in H_2O, S = $S_{20,w}$ sedimentation coefficient.

For example, the shift in $S_{20,w}$ described by Barchi and Murphy [1981] implied a $\bar{v}$ of 0.83 ml/g, indicating that only 56% of the total micelle consisted of protein; the rest was attributed to the associated lipids and the detergent. Using this corrected partial specific volume, the sedimentation coefficient is calculated according to Equation 11 above. Criticisms of the validity of this approach have been made. Using the dopamine receptor as an example, a suggested methodology (closely resembling that of Neer [1974] and Haga et al [1977]) and the resulting calculations are given below.

Sucrose density gradients (5 ml, 5–20%) were layered into cellulose nitrate tubes (0.5 × 2 in, Beckmann #305050). The sucrose was dissolved in 5 mM sodium phosphate (pH 7.4 at 4° C) to which Triton X-100 had been added to a final concentration of 0.05%. Gradients were allowed to stand for at least 2 h at 4°C to equilibrate. Samples consisting of 0.25 ml of digitonin-solubilized dopamine receptors [Davis et al, 1981; Madras et al, 1982] and 40 μl of marker enzymes were layered on top of the gradients and centrifuged at 45,000 rpm for approximately 16 h. The rotor (Beckman SW 50 Ti rotor) was allowed to come to halt without the use of a brake. After the run an average of 16 fractions of 0.3-ml volume were collected (by peristaltic pump) from each gradient. Each fraction was assayed for receptor binding activity (using the polyethylene glycol precipitation assay [Chan et al, 1981]) and enzyme activity. The enzymes used as markers—cytochrome c, aldolase, catalase, and malate dehydrogenase—were assayed according to the methods of Clarke [1975].

To allow for viscosity, density, and temperature during sedimentation, a calibration curve was constructed from the marker proteins. The distance a protein travelled down the gradient was calculated from the middle of the sample layer to the middle of the peak enzyme/receptor fraction. A value Of 5.55S was obtained for the ^{3}H-spiperone-labelled dopamine receptor.

The Stokes' radius for the dopamine (D_2) receptor was similarly determined in a 0.05% Triton X-100/phosphate buffer. The chromatographic gel used was Sephacryl S-300 (Pharmacia) and the column size was 2.6 × 100 cm. Recovery of ^{3}H-spiperone-labelled receptor was approximately 10–15%, eluting with a Stokes' radius of 5.8 nm. To illustrate the necessity of obtaining the Stokes' radius and sedimentation coefficient under identical detergent conditions, dopamine receptors were eluted over a Sephacryl S-300 column (0.9 × 60 cm) in 0.1% digitonin. This detergent forms large micelles (approximately 70,000 daltons [Helenius and Simon, 1975]) compared to Triton X-100. A Stokes' radius of 7.1 nm was obtained, thus indicating the large contribution of the digitonin to the apparent molecular size of the receptor protein.

Having obtained the Stokes' radius and sedimentation coefficient, the molecular weight can be calculated using Equation 10 and the following parameters:

N = Avogadro's number, 6.022×10^{23}

$\eta_{20,w}$ = viscosity of water at 23°C, 0.010002 g cm^{-1} s^{-1}

a = Stokes' radius in metres

$S_{20,w}$ = sedimentation coefficient in 10^{-13} s

$\bar{v}$ = partial specific volume of protein, 0.735 ml g^{-1} (assumed value)

$P_{20,w}$ = density of water at 20°C, 0.99823 g ml^{-1}

A value of 136,000 daltons was obtained for the dopamine receptor. Of course, this value, as discussed above, does not, as yet take into account the contribution of the Tirton X-100 (and residual bound digitonin) to the molecular weight. If the solubilized receptor has a partial specific volume comparable to that of the standard proteins used (0.73–0.76 ml g^{-1}), it should migrate at the same rate relative to the standards in either H_2O or D_2O gradients. If different sediment coefficients are obtained, this implies that the partial specific volume is outside the normal range for proteins and the value of 0.735 ml g^{-1} cannot be assumed with any degree of accuracy. In this case, two different molecular weights could be calculated. The influence of deuteration alone is negligible since the weight of deuterium relative to hydrogen is 1.0155. It is also assumed that proteins bind the same amount of detergent in both H_2O and D_2O. Clarke [1975] demonstrated that this assumption led to only a small error. The following equation, also derived by Clarke [1975], is used to calculate the accurate partial specific volume based on the shift in sedimentation coefficient observed between sucrose density gradients run in H_2O and D_2O:

$$\bar{v} = \frac{(S_H\,\eta_H)\,/\,(S_D\,\eta_D) - 1}{P_D\,[(S_H\,\eta_H)\,/\,(S_D\,\eta_D)] - P_H}. \quad (12)$$

This corrected partial specific volume value is then reinserted into Equation 10 to obtain an accurate molecular weight for the protein-detergent complex.

Finally, the relative contributions of protein and detergent are deduced. This is performed by taking an average value for the protein (0.735 ml g^{-1}) and the value for the detergent (most of which have been determined experimentally—for Triton X-100 the value is 0.94 ml g^{-1} [Neer, 1974]). The range of the determined value can be calculated using the extreme $\bar{v}$ values for the protein (0.72–0.76 ml g^{-1}). For example, the NP-40 solubilized sodium channel was found to have an overall molecular weight of 314,000 daltons with a partial specific volume of 0.83 ml g^{-1} [Barchi and Murphy, 1981].

Hence, P + D = 1, and 0.74P + 0.984*D = 0.83 (* = partial specific volume of NP-40), where P = fractional contribution of protein and D = fractional contribution of detergent (and associated lipids). Solving, the protein accounts for approximately 56% of the overall molecular weight and the detergent 44%.

The shape of this complex is calculated using the corrected molecular weight and partial specific volume, and the sedimentation coefficient (using Eq. 7). For detergent-solubilized proteins, which commonly show low solvation, the frequently high f/fo values imply a markedly nonspherical micellar conformation. This is consistent with the concept of receptor proteins extending through the membrane with hydrophilic regions exposed to the extracellular region and lipophilic regions buried deep in the lipid bilayer.

However, in recent years, this concept has been challenged, since it has become apparent that, even though receptor proteins protrude through the membrane (with cytoplasmic and extracellular domains), the hydrophobic region of the protein may be very small. Thus for the nicotonic cholinergic receptor 64% of the total receptor mass protrudes into either the extracellular fluid (50%) or into the cytoplasmic side (14%) [Kistler et al, 1982]. Similarly, limited proteolysis experiments on the muscarinic cholinergic receptor [Venter, 1983] strongly support these findings. Therefore, only a small belt of detergent may be necessary to retain these membrane-bound receptors in solution. The assumption that the partial specific volume of the detergent-protein micelle is equal to an average of the two [eg, Haga et al 1977] may therefore be invalid. This assumption will also have lead to an overestimation of frictional ratio values. Support for these new concepts comes from recent studies on the hydrodynamic properties of the insulin receptor [Pollet et al, 1981]; when hydration of the Triton X-100 receptor complex (normally estimated at 0.2 g/g) was corrected for, the frictional ratio of the complex was found to be 1.2 (axial ratio 4–5), which approaches the range characteristic of water-soluble, globular proteins. The partial specific volume for the insulin receptor was calculated at 0.71 ml g^{-1}, again clearly within the usual range for water-soluble proteins. In contrast, for example, Seigel et al [1981] reported a frictional ratio of 1.4 (axial ratio 6–8).

Problems Associated With the Presence of Detergents

The major problem associated with the presence of detergents is the lack of information on them (see Table II). For example, what types of proteins do they extract? How many detergent molecules are associated with each protein molecule? What is the partial specific volume of the detergent and at what concentration does it form micelles? How hydrophobic is the detergent (on the hydrophile–lipophile balance [HLB] index)? If these questions can be answered, allowance for their presence is made easier. Essentially, however, a specific assay must be available to determine how much detergent is present. Some detergents can be tagged with radioactivity (eg, phenyl ^{3}H-Triton X-100 [Lee et al, 1979a]). Others lend themselves to quantitaanalysis by spectrophotometric methods (eg, absorption of UV light at 208 nm by digitonin [Strauss et al, 1979]). A radioimmunoassay may even be available (as recently reported for digitonin [Brooker et al, 1982]).

The other major problem is the removal of excess detergent. Complete removal will usually result in denaturation and precipitation of the protein since intrinsic proteins are not water soluble. Gel chromatography, in buffer containing the detergent at a concentration below the CMC, is the most common way of removing excess detergent (see review by Furth [1980]) and, as discussed above, provides partial purification as well as determination of the Stokes' radius of the protein-detergent micelle. A popular method of removing Triton X-100 is to adsorb it onto Bio-Beads Sm-2 (Bio-Rad laboratories) as originally suggested by Holloway [1973]. Methods for *complete* removal of detergent (combined with addition of an artificial membrane so as to allow receptor reconstitution to occur) are discussed elsewhere [Klausner et al, this volume]. For the purposes of defining the hydrodynamic properties of intrinsic proteins we are more interested in *exchanging* detergents in order that the protein dominates the properties of the micelle rather than the detergent. Nonionic detergents, because they are milder, are more frequently used for the initial solubilization. However, these are characterized by low CMC values and high aggregation numbers (eg, Triton X-100 forms a micellar molecular weight of 90,000 daltons). In contrast, ionic detergents, which are much harsher in terms of initial solubilization properties, have higher CMC values and lower aggregation numbers (eg, SDS forms a micelle of 18,000 daltons). We have already discussed the most common method of detergent exchange, ie, of sucrose density ultracentrifugation. As seen for catecholamine receptors, digitonin in the original solubilization mixture was replaced with 0.05% Triton X-100 in the sucrose gradient. In summary, future investigations into the hydrodynamic properties of intrinsic proteins must pay more attention to the presence of detergent. This should involve precise measurement of associated detergent, by a specific assay. This will remove one more variable in calculations used in deriving the hydrodynamic molecular weight.

Relationship Between the Hydrodynamic Molecular Weight and Molecular Weight Determined by Gel Electrophoresis

Indirect confirmation that these methods (gel permeation chromatography followed by sucrose density ultracentrifugation) yield a meaningful molecular weight can be obtained by comparing them with molecular weights determined by other means. The most prevalent method used is that of gel electrophoresis; a newer approach is that of radiation inactivation. Photoaffinity labelling is also very useful. Table III compares the molecular weights, as determined by these three methods, for a variety of intrinsic receptor proteins. Under nonreducing conditions, it can be seen that molecular weights are strikingly similar, despite the complete independence of each of the two approaches. With radiation inactivation, some of the results are very close (eg, on the nicotinic receptor) while some are not (eg, the insulin receptor). This method, however, deals only with the moiety that is functionally involved in the assay used. In these examples, therefore, it determines the molecular weight of the part of the receptor that binds the ligand. Hence in the case of the insulin receptor it can be postulated that the 90,000-dalton subunit contains the ligand binding site.

The combination of hydrodynamic and electrophoretic data has the potential to give an indication of the oligomeric molecular weight and that of a protein's constituent subunits (if any). The most complete example to date is that of the nicotinic acetylcholine receptor, purified from either *Torpedo* or *Electrophorus* electric fish. These species possess an enormously rich concentration of these receptors, enabling research to progress rapidly to the point where it will probably be the first receptor completely defined in terms of its molecular parameters. In the 1970s numerous determinations of the hydrodynamic molecular weight, as assessed by sucrose density gradient ultracentrifugation, were reported.

From Table III, it can be seen that this was approximately 315,000 daltons. Sometimes larger species were reported (eg, 380,000 and 1,300,000 daltons [Edelstein et al, 1975]).

Under nonreducing conditions sodium dodecyl sulfate-polyacrylamide gel electrophoresis (SDS-PAGE) gave a molecular weight of 250,000. The causes of the discrepancy between the values still remain to be determined. Conceivably, the contribution of detergent to the hydrodynamic molecular weight has not been completely compensated for. In-

TABLE III. Molecular Weight Characterization of Detergent-Solubilized Receptors

Nicotinic receptor/sodium ion channel (*Torpedo* and *Electrophorus*)		
Hydrodynamic Mr	250–300,000	Ruchel et al [1981]
	316,000	Hartshorne et al [1980]
	314,000	Barchi and Murphy [1981]
	315–470,000	Meunier et al [1972a,b]
	250,000	Reynolds and Karlin [1978]
SDS-gel Mr	250,000	Agnew et al [1980]
	260,000	Biesecker et al [1973]
	250,000 + 39,000	Hartshorne et al [1980]
	250,000	Lindstrom et al [1979]
	270,000	Lo et al [1982]
	50,000	Raferty et al [1980]
	46,000	Carroll et al [1973]
Radiation inactivation	230,000	Levinson and Ellory [1973]
	300,000	Lo et al [1982]
Benzodiazepine/$GABA_A$		
Hydrodynamic Mr	240,000	Asano and Ogasawara [1981]
	355,000	Stephenson et al [1982]
Radiation inactivation	90,000–100,000	Doble and Iversen [1982]
	216,000	Chang et al [1982]
	220,000	Chang and Barnard [1982]
Muscarinic receptor		
Hydrodynamic Mr	86,000	Haga [1980]
SDS-gel Mr	77–86,000	Birdsall et al [1979]
		Venter [1983]
		Avissar et al [1983]
Radiation inactivation	78–82,000	Venter [1983]
Prolactin receptor		
Hydrodynamic Mr	148,000	Jaffe [1982]
SDS-gel Mr	159,000	Carr and Jaffe [1981]
Glucocorticoid receptor		
Hydrodynamic Mr	33–35,000	Climent et al [1977]
		Westphal and Beato [1980]
SDS-gel Mr	40,000	Westphal and Beato [1980]
Insulin receptor		
Hydrodynamic Mr	~300,000	Cuatrecasas [1972a,b]
		Jacobs et al [1977]
		Ginsborg et al [1978]
		Harrison et al [1978]
		Pollet et al [1981]
		Seigel et al [1981]
SDS-gel Mr	125–135,000	Jacobs et al [1977]
Reduced components	90,000	Jacobs et al [1979]
	42–49,000	Yip et al [1980]
Combined subunits	260–275,000	Massagne et al [1980]
		Jacobs and Cuatrecasas [1981]
Radiation inactivation	87,000	Harmon et al [1980]
Glutamate receptor		
Hydrodynamic Mr (H_2O gradients only)	12,750	Michaelis et al [1981]
SDS-gel Mr	13,800	Michaelis [1975]

(Continued on next page)

TABLE III. Molecular Weight Characterization of Detergent-Solubilized Receptors (Continued)

Dopamine receptor			
Hydrodynamic Mr (H_2O gradients only)		136,000	Lilly et al [submitted]
Radiation inactivation		123,000	Lilly et al [1983]
α_1-Adrenoceptor			
Hydrodynamic Mr		96,000	Guellan et al [1979]
SDS-gel Mr		45,000	Guellan et al [1982]
		80–85,000	Fraser et al [1983]
		58,000	Kunos et al [1983]
		45,000	
		40,000	
Radiation inactivation		160,000	Fraser et al [1983]
β_1-Adrenoceptor (turkey erythrocytes)			
Hydrodynamic Mr		65,000	Strauss et al [1979b]
			Venter [1981]
SDS-gel Mr			
Reduced components		70,000	Fraser and Venter [1980]
		31,000	
		22,000	
Non-reduced		70,000	
Reduced components		170,000	Durieu-Trautmann et al [1980]
		33,000	
		30,000	
		42,000	Atlas and Levitzki [1978]
		38,000	
Radiation inactivation		96,000	Nielsen et al [1981]
β_2-Adrenoceptor			
Hydrodynamic Mr		75,000(S49 cells)	Haga et al [1977]
		90,000(lung and liver)	Strauss et al [1979b]
			Venter [1981]
SDS- gel Mr		59,000(lung)	Soiefer and Venter [1980]
		58,000(frog rbc's)	Shorr et al [1981, 1982]
		45,000	Rashidbaigi and Ruoho [1981]
		48,500(lung)	Fraser and Venter [1982]
Radiation inactivation		109,000[a]	Fraser and Venter [1982]
Adenylate cyclase complex[b]			
Hydrodynamic Mr	G/F	130,000	Howlett and Gilman [1980]
	G/F	126,000	Kaslow et al [1980]
	G/F	199,000	Neer et al [1980]
			Neer [1978]
	C.G/F	251,000	Neer et al [1980]
SDS - gel Mr		52,000	Northrup et al [1980]
		45,000	
		35,000	
	C	92,000	Nielsen et al [1981]
	C.G/F	226,000	Nielsen et al [1981]
Radiation inactivation	C.G/F	300,000	Schlegel et al [1979]
	R.C.G/F	316,000	Nielsen et al [1981]
	R	96,000	Nielsen et al [1981]

[a]Receptor proposed as a dimer.
[b]G/F (or N) = guanine nucleotide regulatory unit; C = catalytic unit; R = β_1-adrenoceptor.

terestingly, the discrepancy diminishes the smaller the receptor (eg, as for the glucocorticoid receptor).

The in situ molecular weight of the nicotinic molecular weight, as determined by radiation inactivation, was 230,000 daltons, in good agreement with the electrophoretically determined weight.

Under reducing conditions, the subunit composition of the receptor has now been elucidated. Four subunits have been defined, and the receptor is believed to exist in a pentameric form. This consists of two identical subunits (α, Mr = 40,000) and three other polypeptides (β, δ, γ) of Mr 50,000, 60,000, and 65,000. These form a functional unit of 250,000 daltons (see review by Raftery et al [1980]).

CONCLUSIONS

The methodology described in this chapter is designed to uniquely define various characteristics of the protein portion of a detergent-protein micelle. As discussed previously, the necessity of keeping the protein in solution by having detergent present has led to difficulties in accounting for its presence. Nevertheless, it appears that hydrodynamic properties give values very close to those determined by other means. The fact that they can be applied to impure proteins that are not very stable in solution, or can only be labelled by reversible ligands (like the dopamine D_2 receptor), offers considerable advantages over techniques such as gel electrophoresis.

ACKNOWLEDGMENTS

The author was supported by the C.K. Clarke Psychiatric Research Foundation, The Banting Research Foundation, the Connaught Foundation, and the Hospital for Sick Children. I would like to thank Dr. J.C. Venter for his invaluable assistance during preparation of this manuscript. I thank Sheila McCormick for meticulous preparation of the manuscript.

REFERENCES

Agnew WS, Moore AC, Levinson SR, Raftery MA (1980): Identification of a large molecular weight peptide associated with a tetrodotoxin binding protein from the electroplax of *Electrophorus electricus*. Biochem Biophys Res Commun 92:860–866.

Aronstam RS, Schuessler DC, Eldefrawi ME (1978): Solubilization of muscarinic acetylcholine receptor of bovine brain. Life Sci 23:1377–1382.

Asano T, Ogasawara N (1981): Soluble gamma-amino butyric acid and benzodiazepine receptors from rat cerebral cortex. Life Sci 29:193–200.

Atlas D, Levitzki A (1978): Tentative identification of β-adrenergic subunits. Nature 272:370–371.

Avissar S, Amitai G, Sokolovsky M (1983): Oligomeric structure of muscarinic receptors is shown by photoaffinity labelling: Subunit assembly may explain high- and low-affinity agonist states. Proc Natl Acad Sci USA 80:156–159.

Barchi RL, Murphy LE (1981): Estimate of the molecular weight of the sacrolemmal sodium channel using H_2O - D_2O centrifugation. J Neurochem 36:2097–2100.

Biesecker G (1973): Molecular properties of the cholinergic receptor from *Electrophorus electricus*. Biochemistry 12:4403–4409.

Birdsall NJM, Burgen ASV, Hulme EC (1979): A study of the muscarinic receptor by gel electrophoresis. Br J Pharmacol 66:337–342.

Brooker G, Fleming JW, Ross EM (1982): Radioimmunoassay of digitonin. Biochim Biophys Acta 685:39–43.

Caron MG, Lefkowitz RJ (1976): Solubilization and characterization of the β-adrenergic receptor binding sites of frog erythrocytes. J Biol Chem 251:2374–2384.

Carr FE, Jaffe RC (1981): Solubilization and molecular weight estimation of prolactin receptors from *Rana catesbeiana* tadpole liver and tail fin. Endocrinology 109:943–949.

Carroll RC, Eldefrawi ME, Edelstein SJ (1973): Studies on the structure of the acetylcholine receptor from *Torpedo marmorata*. Biochim Biophys Acta 55:864–872.

Casassa EF, Eisenberg H (1964): Thermodynamic analysis of multi component solutions. Adv Protein Chem 19:287–393.

Chan B, Madras B, Davis A, Seeman P (1981) Assay of soluble dopamine receptors by precipitation. Eur J Pharmacol 74:53–59.

Chang L-R, Barnard EA (1982): The benzodiazepine/GABA receptor complex: Molecular size in brain synaptic membranes and in solution. J Neurochem 39:1507–1518.

Chang L-R, Barnard EA, Lo MMS, Dolly JO (1982): Molecular sizes of benzodiazepine receptors and the interacting GABA receptors in the membrane are identical. FEBS Lett 126:309-312.

Clark S (1975): The size and detergent binding of membrane proteins. J Biol Chem 250:5459-5469.

Climent D, Doenecke D, Beato M (1977): Properties of the partially purified activated glucocorticoid receptor of rat liver. Binding to chromatin subunits. Biochemistry 16:4694-4703.

Cohn EJ, Edsall JT (1943): Proteins, amino-acids and peptides as ions and dipolar ions. New York: Reinhold, (reprinted 1965, New York: Hafner).

Cuatrecasas P (1972a): Isolation of the insulin receptor of liver and fat cell membranes. Proc Natl Acad Sci USA 69:318-322.

Cuatrecasas P (1972b): Properties of the insulin receptor isolated from liver and fat cell membranes. J Biol Chem 247:1980-1991.

Davis A, Madras BK, Seeman P (1981). Solubilization of neuroleptic/dopamine receptors of human brain striatum. Eur J Pharmacol 70:321-329.

Davis A, Madras BK, Seeman P (1982a): Solubilized receptors for (^{3}H) dopamine (D_3 binding sites) from canine brain. Biochem Pharmacol 31:1183-1187.

Davis A, Morris J, Tang SW (1982b): Solubilization of a putative antidepressive receptor (^{3}H imipramine binding site) from human platelets. Soc Neurosci Abstr 9:891.

Davis A, Morris JM, Tang SW (1983): Solubilization and assay of (^{3}H) imipramine binding sites from human platelets. Eur J Pharmacol 86:353-359.

Doble A, Iversen LL (1982): Molecular size of benzodiazepine receptor in rat brain *in situ*: Evidence for a functional dimer. Nature 295:522-523.

Durieu-Trautmann O, Delavier-Klutchko C, Vauqulin G, Strosberg AD (1980): Visualization of the turkey erythrocyte β-adrenergic receptor. J Supramol Struct 13:411-419.

Edelstein SJ, Schachman HK (1967): The simultaneous determination of partial specific volumes and molecular weights with microgram quantities. J Biol Chem 242:306-311.

Edelstein SJ, Beyer WB, Eldefrawi AT, Eldefrawi ME (1975): Molecular weight of the acetylcholine receptors of electric organs and the effect of Triton X-100. J Biol Chem 250:6101-6106.

Edsall JT (1953): In Weurath H, Bailey K (eds): "The Proteins." New York: Academic Press, Vol 1 part B, Chap 7.

Eigner J, Doty P (1965): The native, denatured and renatured states of deoxyribonucleic acid. J Mol Biol 12:549-580.

Fish WW (1975): Determination of the molecular weights of membrane proteins and polypeptides. In Korn ED, (ed),: "Methods in Membrane Biology," New York: Plenum Press, Vol 4.

Fish WW, Reynolds JA, Tanford C (1970): Gel chromatography of proteins in denaturing solvents. J Biol Chem 245:5166-5168.

Fleming JW, Ross EM (1980): Reconstitution of beta-adrenergic receptors into phospholipid vesicles: Restoration of [^{125}I] iodohydroxy-benzylpindolol binding to digitonin-solubilized receptors. J Cyclic Nucleotide Res 6:407-419.

Fraser CM, Venter JC (1980): Monoclonal antibodies to β-adrenergic receptors: Use in purification and molecular characterization of β-receptors. Proc Natl Acad Sci USA 77:7034-7038.

Fraser CM, Venter JC (1982): The size of the mammalian lung β_2-adrenergic receptor as determined by target size analysis and immunoaffinity chromatography. Biochem Biophys Res Commun 109:21-29.

Fraser CM, Greguski R, Eddy B, Venter JC (1983): Autoantibodies and monoclonal antibodies in the purification and molecular characterization of neurotransmitter receptors. J Cell Biochem (in press).

Furth AJ (1980): Removing unbound detergent from hydrophobic proteins. Anal Biochem 109:207-215.

Gibbons RA (1972): Physico-chemical methods for the determination of the purity, molecular size and shape of glycoproteins. In Gottschalk A (ed): "Glycoproteins: Their Composition, Structure and Function." New York: Elsevier, pp 31-140.

Ginsberg BH, Cohen RM, Kahn CR, Roth J (1978): Properties and partial purification of the detergent solubilized insulin receptor. Biochim Biophys Acta 542:88.

Gorissen H, Laduron P (1979): Solubilization of high-affinity dopamine receptors. Nature 279:72-74.

Gorissen H, Aerts G, Laduron P (1979): Characterization of solubilized dopamine receptors from dog striatum. FEBS Lett 100:281-285.

Guellan G, Aggerbeck M, Hanoune J (1979): Characterisation and solubilization of the α-adrenoceptor of rat liver plasma membranes labelled with [^{3}H] phenoxybenzamine. J Biol Chem 254:10761-10768.

Guellan G, Goodhardt M, Barouki R, Hanoune J (1982): Subunit structures of rat liver α_1 adrenergic receptor. Biochem Pharmacol 31:2817-2810.

Haga T (1980): Molecular size of muscarinic acetylcholine receptors of rat brain. FEBS Lett 113:68-72.

Haga T, Haga K, Gilman AG (1977): Hydrodynamic properties of the β-adrenergic receptor and adenylate cyclase from wild type and variant S49 lymphoma cells. J Biol Chem 252:5776-5782.

Harmon JT, Kahn CR, Kempner ES, Schlegel W (1980): Characterization of the insulin receptor in its membrane environment by radiation inactivation. J Biol Chem 255:3412.

Harrison LC, Billington T, East IJ, Nicholas RJ, Clark S (1978): The effect of solubilization on the properties of the insulin receptor of human placental membrane. Endocrinology 102:1485.

Hartshorne RP, Coppersmith J, Catterall WA (1980): Size characteristics of the solubilized saxitonin receptor of the voltage-sensitive sodium channel from rat brain. J Biol Chem 255:10572–10585.

Heleniuis A, Simon K (1972): The binding of detergents to lipophilic and hydrophilic proteins. J Biol Chem 247:3656.

Helenius A, Simons K (1975): Solubilization of membranes by detergents. Biochim Biophys Acta 415:29–79.

Hjelmeland LM (1980): A nondenaturing zwitterionic detergent for membrane biochemistry: Design and synthesis. Proc Natl Acad Sci USA 77:6368–6370.

Holloway PW (1973): A simple procedure for removal of Triton X-100 from protein samples. Anal Biochem 53:304–308.

Howlett AC, Gilman AG (1980): Hydrodynamic properties of the regulatory component of the adenyl cyclase. J Biol Chem 255:2861–2866.

Jacobs S, Cuatrecasas P (1981): Insulin receptor: Structure and function. Endocr Rev 2:251–263.

Jacobs S, Schechter Y, Bissell K, Cuatrecasas P (1977): Purification and properties of insulin receptors from rat liver membranes. Biochem Biophys Res Commun 77:981–988.

Jacobs S, Hazum E, Shechter Y, Cuatrecasas P (1979): Insulin receptor: Covalent labelling and identification of subunits. Proc Natl Acad Sci USA 6:4918–4921.

Jaffe RC (1982): Hydrodynamic characterization of the Triton X-100 solubilized lactogenic hormone receptor of rat liver. Biochemistry 21:2936–2939.

Kaslow HR, Johnson GL, Brothers VM, Bourne HR (1980): A regulatory component of adenylate cyclase from human erythrocyte membranes. J Biol Chem 255:3736–3741.

Kistler J, Stroud RM, Wlymkowsky MW, Lalancette RA, Fairclough RH (1982): Structure and function of an acetylcholine receptor. Biophys J 37:371–383.

Kunos G, Kan WH, Greguski R, Venter JC (1983): Selective affinity labelling and molecular characterization of hepatic α_1-adrenergic receptors with [^{3}H] phenoxybenzamine. J Biol Chem 258:326–332.

Lee SO, Cohen NS, Jacobs AJ, Brodie AP (1979): Isolation, purification and reconstitution of a proline carrier protein from Mycobacterium phlei. Biochemistry 18:2232–2238.

Le Maire M, Rivas E, Moller JV (1980): Use of gel chromatography for determination of size and molecular weight of proteins: Further caution. Anal Biochem 106:12–21.

Levinson SR, Ellory JC (1973): Molecular size of the tetrodotoxin binding site estimated by irradiation inactivation. Nature New Biol 245:122–123.

Lilly L, Fraser CM, Jung CY, Seeman P, Venter JC (1983): Molecular size of canine and human brain D_2 dopamine receptor as determined by radiation inactivation. Mol Pharmacol 24:10–14.

Lindstrom J, Merlie J, Yogeeswaran G (1979): Biochemical properties of acetylcholine receptor subunits from *Torpedo californica*. Biochemistry 18:4465–4470.

Liscia DS, Alhadi T, Vonderhaar BK (1982): Solubilization of active prolactin receptors by a nondenaturing zwitterionic detergent. J Biol Chem 257:9401–9405.

Lo, MMS, Barnard EA, Dolly JO (1982): Size of acetylcholine receptors in the membrane. An improved version of the radiation inactivation method. Biochemistry 21:2210–2217.

Madras BK, Davis A, Seeman P (1982): Comparison of soluble dopamine D_2 receptors form three species. Eur J Pharmacol 78:431–438.

Makino S, Reynolds JA, Tanford C (1973): The binding of deoxycholate and Triton X-100 to proteins. J Biol Chem 248:4926.

Martin RG, Ames BN (1961): A method for determining the sedimentation behaviour of enzymes: Application to protein mixtures. J Biol Chem 236:1372–1379.

Massagne J, Pilch PF, Czech MP (1980): Electrophoretic resolution of three major insulin receptor structures with unique subunit stoichiometries. Proc Natl Acad Sci USA 77:7137–7141.

Meunier JC, Olson RW, Changeaux JP (1972a): Studies on the cholinergic receptor protein from *Electrophorus electricus*. FEBS Lett 24:63–68.

Meunier JC, Olsen RW, Menez A, Fromageot P, Boquet P, Changeaux JP (1972b): Some physical properties of the cholinergic receptor protein from *Electrophous electricus* revealed by a tritiated α-toxin from *Naja Nigricollis* venom. Biochemistry 11:1200–1210.

Michaelis EK (1975): Partial purification and characterisation of a glutamate-binding glycoprotein from rat brain. Biochem Biophys Res Commun 65:1004–1012.

Michaelis EK, Michaelis ML, Chang HH, Grubbs RD, Kuonen DR (1981): Molecular characteristics of glutamate receptors in the mammalian brain. Mol Cell Biochem 38:163–179.

Neer EJ (1974): The size of adenylate cyclase. J Biol Chem 294:6527–6531.

Neer EJ (1978): Multiple forms of adenylate cyclase. Adv Cyclic Nucleotide Res 9:69–83.

Neer EJ, Echeverria D, Knox S (1980): Increase in the size of soluble brain adenylate cyclase with activation by guanosine 5′ -β, γ -imino) triphosphate. J Biol Chem 255:9782–9789.

Nielsen TB, Lad PM, Preston MS, Kempner E, Schlegel W, Rodbell M (1981): Structure of the turkey erythrocyte adenylate cyclase system. Proc Natl Acad Sci USA 78:722–726.

Northrup JK, Sternweis PC, Smiger MD, Schliefer LS, Ross EM, Gilman AG (1980): Purification of the regulatory component of adenylate cyclase. Proc natl Acad Sci USA 77:6516–6520.

Nozaki Y, Schechter NM, Reynolds JA, Tanford C (1976): Use of gel chromatography for the determination of the Stokes radius of proteins in the presence and absence of detergents: A reexamination. Biochemistry 15:3884–3890.

Pollet FJ, Haase BA, Standaert ML (1981): Characterisation of detergent-solubilized membrane proteins. J Biol Chem 256:12118–12126.

Raftery MA, Changeux J-P (1982): The nicotinic acetylcholine receptor (AChR). Neurosci Res Prog Bull 20:277–310.

Raftery MA, Hunkapillar MW, Strader CD, Hood LE (1980): Acetylcholine receptor: Complex of homologous subunits. Science 208:1454–1457.

Rashidbaigi A, Ruoho AE (1981): Iodoazidobenzylpindolol: A photoaffinity probe for the β -adrenergic receptor. Proc Natl Acad Sci USA 78:1609–1613.

Reynolds J, Karlin A (1978): Molecular weight in detergent solution of acetylcholine receptor from *Torpedo californica.* Biochemistry 17:2035–2038.

Rodbard D (1976): Estimation of molecular weight by gel filtration and gel electrophoresis. In Catsimpoulas N (ed): "Methods of Protein Separation," Vol 2. New York: Plenum Press pp 145–179.

Ruchel R, Watters D, Maelicke A (1981): Molecular forms and hydrodynamic properties of acetylcholine receptor from electric tissue. Eur J Biochem 119:215–223.

Ruegg UT, Cuenod S, Hiller JM, Gioannini T, Howells RD, Simon EJ (1981): Characteristics and partial purification of solubilized active opiate receptors from toad brain. Proc Natl Acad Sci USA 78:4635–4638.

Schachman HK (1959): In: "Ultracentrifugation in Biochemistry." New York: Academic Press.

Schlegel W, Kempner ES, Rodbell M (1979): Activation of adenylate cyclase in hepatic membranes involves interactions of the catalytic unit with multimeric complexes of regulatory proteins. J Biol Chem 254:5168–5176.

Seigel TW, Ganguly S, Jacobs S, Rosen OM, Rubin CS (1981): Purification and properties of the human placental insulin receptor. J Biol Chem 256:9266–9273.

Shapiro AL, Vinuela E, Maizel JV (1967): Molecular weight estimation of polypeptide chains by electorphoresis in SDS-polyacrylamide gels. Biochem Biophys Res Commun 28:815–820.

Shorr RGL, Lefkowitz RJ, Caron MG (1981): Purification of the β-adrenergic receptor. J Biol Chem 256:5820–5826.

Schorr RGL, Heald SL, Jeffs PW, Lavin TN, Strohsacker MW, Lefkowitz RJ, Caron MG (1982): The β-adrenergic receptor: Rapid purfication and covalent labelling by photoaffinity cross-linking. Proc Natl Acad Sci USA 79:2778–2782.

Siegel LM, Monty KJ (1966): Determination of molecular weights and frictional ratios of proteins in impure systems by use of gel filtration and density gradient centrifugation. Application to crude preparations of sulfite and hydroxylamine reductases. Biochim Biophys Acta 112:346–362.

Soiefer AI, Venter JC (1980): Mammalian lung β-adrenergic receptor purification utilising an affinity reagent. Fed Proc 39:313.

Stephenson FA, Watkins AE, Olsen RW (1982): Physicochemical characterization of detergent-solubilized γ-aminobutyric acid and benzodiazepine receptor proteins from bovine brain. Eur J Biochem 123:291–298.

Strauss WL, Ghai G, Fraser CM, Venter JC (1979a): Detergent solubilization of mammalian cardiac and hepatic β -adrenergic receptors. Arch Biochem Biophys 196:566–573.

Strauss WL, Ghai G, Fraser CM, Venter JC (1979b): Hydrodynamic properties and sulfhydryl reagent sensitivity of β_1 (cardiac) and β_2 (lung and liver) adrenergic receptors: Molecular evidence for isoreceptors. Fed Proc 38:843.

Tanford C (1961): "Physical Chemistry of Macromolecules." New York: Wiley, Chap 6.

Tanford C, Nozaki Y, Reynolds JA, Makino S (1974): Molecular characterization of proteins in detergent solutions. Biochemistry 13:2369–2376.

Traube J (1899): Ueber den Raum der Atome. In Ahrens FB (ed) "Sammlung Chemischer und Chemishtechnischer Vortrage," Vol 4. Stuttgart: Ferdinard Enke, pp 255–332.

Venter JC (1981): β -Adrenoceptors, adenylate cyclase, and the adrenergic control of cardiac contractility. In Kunos G (ed): "Adrenoceptors and Catecholamine Action," Part A. New York: Wiley-Interscience, pp 213–245.

Venter JC (1983): Muscarinic cholinergic receptor structure 1. Receptor size, membrane orientation and absence of major phylogenetic structural diversity. J Biol Chem (in press).

Warshaw HS, Ackers GK (1971): Molecular sieve studies on interacting protein systems. Anal Biochem 42:405–421.

Westphal HM, Beato M (1980): The activated glucocorticoid receptor. Eur J Biochem 106:395–403.

Williams JG, Gratzer WB (1971): Limitations of the detergent-polyacrylamide gel electrophoresis method for molecular weight determinations of proteins. J Chromatogr 57:121.

Yip CC, Cheung CWT, Moule ML (1980): Photoaffinity labelling of insulin receptor proteins of liver plasma membrane proteins. Biochemistry 19:70.

Molecular and Chemical Characterization of Membrane Receptors, pages 179–191

7

Binding and Equilibrium Studies Using an Air-Driven Ultracentrifuge

Geoffrey J. Howlett

Russell Grimwade School of Biochemistry, University of Melbourne, Parkville 3052, Victoria, Australia

INTRODUCTION

Sedimentation methods have been used extensively to determine the molecular weight of proteins [Svedberg and Petersen, 1940; Schachman, 1959] and to characterize the extent of protein interactions in solution [eg, Teller, 1973; Van Holde, 1975]. As a result, the recent development of a small, commercially available air-driven ultracentrifuge has created considerable interest in its potential biological applications. The high angular velocities available using this instrument permit rapid sedimentation of membranes and cellular organelles. This is the basis of a number of clinical applications [Heidemann and Wittliff, 1979; Moulin et al, 1979; Duck-Chong, 1979]. The present chapter describes a number of the analytical applications of the air-driven ultracentrifuge, particularly those of interest to the characterization of membrane proteins and their interactions in solution.

The analytical applications of the air-driven ultracentrifuge may be divided into two types: The first involves the analysis of binding phenomena such as the interactions of small molecules with proteins. Such studies have been performed using the analytical ultracentrifuge [Steinberg and Schachman, 1966] but are restricted to the use of ligands which can be detected optically. With the air-driven ultracentrifuge, it is possible to analyse the solution after centrifugation and to determine the concentration of unbound ligand, by measuring radioactivity or any other specific property of the ligand [Howlett et al, 1978]. The second type involves the use of sedimentation equilibrium procedures to determine molecular size [Bothwell et al, 1978; Pollet et al, 1979a]. With the air-driven ultracentrifuge it is possible to exploit a range of sensitive and specific assays to define the equilibrium profiles formed during centrifugation. For example, using ^{125}I-labeled proteins, molecular weights can be determined using only nanograms of material [Howlett et al, 1982a]. These analytical procedures are well suited to studies on membrane proteins, where it is often difficult to purify sufficient material for studies using the analytical ultracentrifuge. In contrast to studies using sucrose gradient centrifugation or gel filtration, the results of sedimentation-equilibrium studies are independent of macromolecular shape. A major advantage of the use of the air-driven ultracentrifuge is that it is possible to study heterogeneous mixtures of macromolecules provided a suitable assay is available for the species of interest. In this way it is possible to obtain information on the extent of macromolecular interactions in solution [Howlett et al, 1982a]. The present chapter describes the techniques which have been developed for studies using the air-driven ultracentrifuge and representative examples of the information obtained.

CHARACTERISTICS OF THE AIR-DRIVEN ULTRACENTRIFUGE

Description of the Instrument

The air-driven ultracentrifuge or Airfuge (Beckman Instruments, Inc, Palo Alto, CA) is a tabletop model with an aluminum rotor approximately 4 cm in diameter. The rotor has turbine flutes machined into the bottom so that it can be lifted and driven by ordinary laboratory compressed air. The rotor attains its maximum operating speed of 100,000 rpm in about 30 s. At this speed the centrifugal field at the bottom of the centrifuge tubes is approximately 160,000g. The six sample tubes are oriented at an angle of 18° to the axis of rotation, and each tube has a capacity of approximately 175 μl. For a tube containing 150 μl of solution the radial distances to the center of rotation for the meniscus and base of the solution are approximately 1.1 and 1.45 cm, respectively. Although the temperature of the Airfuge is not controlled, measurements of the rotor chamber during centrifugation indicate that after a period of 2–3 h the temperature is constant and slightly above ambient. This temperature is similar to that of the tube contents measured immediately after centrifugation. The Airfuge is equipped with a timer and a braking device which can be preset and serves to decelerate the rotary smoothly. Deceleration from top speed involves a coasting period of about 2 min, followed by mechanical braking which causes the rotor to stop after an additional 2 min. At the conclusion of centrifugation, the contents can be fractionated, using either an automatic pipette or preferably a device such as the microfractionator (Beckman Instruments), which permits sequential fractionation of the tube contents [Howlett and Markov, 1980]. Concentrations of macromolecular solutes such as bovine serum albumin or dextran of approximately 5 mg/ml are required to provide density stabilization and to prevent stirring during braking of the rotor [Howlett et al, 1978].

Rates of Sedimentation

The high centrifugal fields available using the air-driven ultracentrifuge permit selective depletion of macromolecular solutes from the meniscus. The rates of sedimentation of aspartate transcarbamylase (molecular weight [MW] 310,000) from *Escherichia coli,* its isolated catalytic subunit (MW 100,000), and bovine pancreatic ribonuclease (MW 14,000) are presented in Figure 1. It can be seen that aspartate transcarbamylase is almost completely removed from the meniscus region in about 30 min. For the catalytic subunit ap-

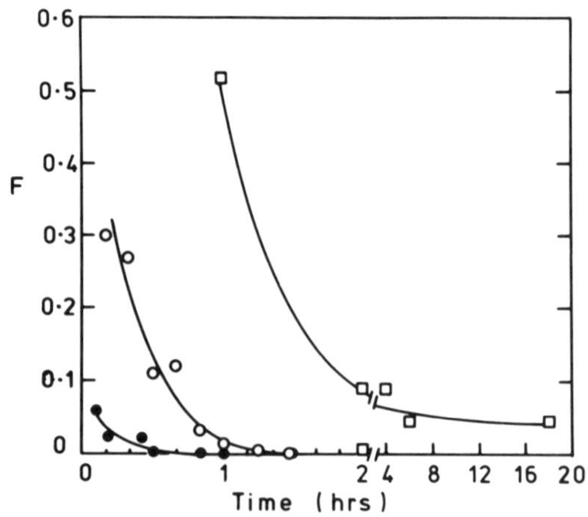

Fig. 1. *The rate of sedimentation of proteins in the air-driven ultracentrifuge: aspartate transcarbamylase of E. coli (●), the corresponding catalytic subunit (○), and bovine pancreatic ribonuclease (□). The ordinate shows the fraction (F) of protein remaining in the upper 20 µl of solution relative to the amount present initially. The times of centrifugation at 100,000 rpm and 20°C are given on the abscissa. The concentrations of protein and Dextran (for density stabilization) were aspartate transcarbamylase, 1 mg/ml; Dextran T40, 2 mg/ml; catalytic subunit, 1 mg/ml; dextran T10, 20 mg/ml; ribonuclease, 20 mg/ml; and Dextran T10, 20 mg/ml.*

proximately 90 min are required, whilst for ribonuclease a plateau concentration of approximately 5% of the initial concentration is obtained after 6 h. The rate of sedimentation of aspartate transcarbamylase is close to that which might be predicted on the basis of the known sedimentation coefficient(s) of 11.7 S [Gerhart and Schachman, 1968]. Using an average value of 1.25 cm for the radius of the boundary to the center of rotation, the movement of the boundary (assuming no convection) in 30 min at 100,000 rpm (ω = 10,470 rad/sec) may be calculated as follows:

$$\frac{dx}{dt} = s\omega^2 x$$

$$\Delta x = 11.7 \times 10^{-13} \times 1.25 \times (10{,}470)^2 \times 30 \times 60 = 0.29 \text{ cm.}$$

The transport of macromolecules to the bottom of the tubes may be even more rapid due to the convective flow of concentrated protein along the outer wall of the tubes [Howlett et al, 1978]. Nevertheless, the above calculations provide a guide to the time required for the sedimentation of other macromolecules. With erythrocyte membranes and vesicles, centrifugation in the Airfuge at 100,000 rpm for 1 min (followed by 4 min deceleration) depletes the meniscus of membrane-bound protein [Hill et al, 1982].

THE BINDING OF SMALL MOLECULES TO PROTEINS

The binding of a low molecular weight ligand (S) to a macromolecular acceptor (A) is a common biological phenomena and an essential feature of hormone-receptor interactions. Such interactions can be analysed using the air-driven ultracentrifuge. For the case where the acceptor has n independent and equivalent binding sites for the ligand, the various equilibria can be described as follows:

$$\begin{aligned} A + S &\rightleftarrows AS \\ AS + S &\rightleftarrows AS_2 \\ AS_{i-1} + S &\rightleftarrows AS_i \\ AS_{n-1} + S &\rightleftarrows AS_n. \end{aligned} \tag{1}$$

where the intrinsic binding constant K describes the binding of S to a single site on A (K = [AS]/[A][S]). During centrifugation the meniscus can be depleted of all the macromolecular species. The concentration of free S may then be determined from aliquots taken from the meniscus region after centrifugation. To determine K and n it is assumed that this concentration is equal to the concentration of ligand initially present in the equilibrium mixture [S]. This assumption implies that effects due to reequilibration of the ligand with the acceptor during centrifugation are negligible. The total molar concentration of ligand ($\bar{c}_S$) and protein ($\bar{c}_A$) in the equilibrium mixture are known and thus it is possible to calculate the amount of ligand bound per mole of acceptor (R) which is related to n and K as follows [Scatchard, 1949]:

$$R = \frac{\bar{c}_S - [S]}{\bar{c}_A} = \frac{nK[S]}{1 + K[S]} \quad . \tag{2}$$

Rearrangement of Equation 2 yields:

$$\frac{R}{[S]} = nK - RK \tag{3}$$

and a plot of bound/free (R/[S]) vs bound (R) yields K and n from the slope and abscissa intercept, respectively.

The procedure is similar in principle to other binding methods and could be extended to studies on cooperative binding or to the binding of ligands to nonequivalent sites on an acceptor [Klotz and Hunston, 1975]. For large macromolecules (molecular weight

$> 10^5$) the method is analogous to sedimentation-velocity experiments using the analytical ultracentrifuge [Steinberg and Schachman, 1966] and offers the advantage that radioactive ligands can be used to determine the concentration of free ligand. A receptor binding assay has recently been described which uses the Airfuge to determine the binding of ouabain to several membrane Na^+K^+ ATPases and the binding of [^{3}H]-etorphine to the opiate receptor from brain membranes [Albers and Krishnan, 1979]. In this case, 30 s were allowed for sedimentation of the complexes, thus minimising effects due to reequilibration of the ligand with the receptor during sedimentation.

Longer times are required to deplete smaller proteins from the meniscus, and the procedures become more similar to those used for sedimentation-equilibrium studies. An example of such an experiment is given in Figure 2. The data were obtained by centrifuging mixtures of [^{14}C]-octanoate and bovine serum albumin at 80,000 rpm for 16 h, followed by the direct determination of the concentration of free octanoate in the supernatant [Howlett and Markov, 1980]. The results are plotted in the form of a Scatchard plot and indicate that there is at least one high-affinity site ($K = 2.1 \times 10^5\ M^{-1}$). An advantage of the use of a sedimentation equilibrium technique is that it permits a theoretical estimation of the effects due to reequilibrium resulting from the differential sedimentation of the ligand-protein complexes relative to pure protein. These effects have been shown to be small [Howlett et al, 1978] for ligands of low molecular weight (< 1000), especially in the case described in Figure 2, where the density of the ligand is close to that of the solvent.

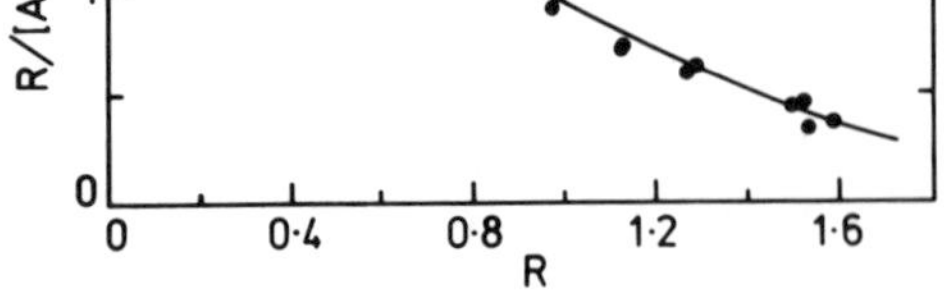

Fig. 2. *The binding of octanoate to bovine serum albumin. Mixtures of octanoate and bovine serum albumin in phosphate-buffered saline, pH 7.4, were centrifuged at 80,000 rpm for 16 h at 8°C and the concentration of unbound ligand in the supernatant measured directly. R, the moles of octanoate bound per mole of bovine serum albumin, was determined as a function of the free ligand concentation (A).*

DETERMINATION OF MOLECULAR WEIGHT

Description of Procedures

During initial studies of ligand binding using the air-driven ultracentrifuge it was apparent that the amount of residual macromolecule remaining in an upper fraction of the Airfuge tube was a function of the molecular weight of the macromolecule. Considerations of sedimentation equilibrium theory indicated that the logarithm of the fraction remaining in an upper fraction of the Airfuge tube (log F) varied linearly with the molecular weight of the macromolecule. Thus a standard curve of log F vs molecular weight obtained for a series of standard proteins could be used for the interpolation of the molecular weight of other proteins [Bothwell et al, 1978]. As indicated in the introduction, these procedures could be applied to studies on impure systems, provided a suitable assay was available for the species of interest. Furthermore, such results are independent of shape and therefore offer an advantage over methods such as gel filtration and sucrose gradient centrifugation. This empirical method of determining molecular weights provides results accurate to 10% [eg, Bothwell and Shooter, 1977; Garwin et al, 1980].

More recent studies have shown that the entire contents of the Airfuge tube can be fractionated using a micromanipulator and 10-μl microcapillaries to obtain the complete sedimentation-equilibrium distribution [Pollet et al, 1979a]. By relating the volumes of the aliquots taken from the Airfuge tube to radial distances, these authors obtained accurate molecular weight values independent of the use of standard proteins. Using a microfractionator these techniques have been used to obtain molecular weights for ^{125}I-labeled serum proteins which agree to within 5% of the literature values [Howlett et al, 1982a].

A typical sedimentation equilibrium experiment using the air-driven ultracentrifuge is performed as follows: A sample volume of 130 μl is added to the Airfuge tube and centrifuged for 22 h. Experiments using a variety of proteins indicate that there are no significant differences between data obtained after centrifugation for 22 h or 30 h. Density stabilization during centrifugation is usually provided by either bovine serum albumin (5 mg/ml) or Dextran T40 (5 mg/ml). The tube contents are fractionated after centrifugation using a microfractionator and the weight of each aliquot determined gravimetrically. The time taken to fractionate each tube is approximately 6 min. The stability of the distributions during this procedure is indicated by experiments using ^{125}I-α_1-acid glycoprotein; no difference was observed for results obtained from four separate tubes fractionated 2, 8, 36, and 56 min after the conclusion of centrifugation [Howlett et al, 1982a].

The experimental data in the form of concentration (c) vs aliquot number are converted to plots of ln c/c_o vs radial distance squared (r^2) where c_o is the initial concentration and r is determined using Equation 1 of Pollet et al [1979a], with the appropriate correction [Pollet et al, 1979b]. The slope of the corresponding ln c/c_o vs r^2 plots is related to the molecular weight, M.

$$\frac{d\ln c/c_o}{dr^2} = \frac{M(1 - \bar{v}\rho)\,\omega^2}{2RT} = \sigma, \qquad (4)$$

where the symbols are as follows: $\bar{v}$ (partial specific volume), ω (angular velocity), ρ (solution density) and T (absolute temperature). A typical result is shown in Figure 3. In Figure 3a the concentration of ^{125}I-labeled α_1-acid glycoprotein is plotted against the radial distance in the Airfuge tube. The slope of the line in Figure 3b yields a value of 40,900 for the molecular weight of α_1-acid glycoprotein, which agrees well with a value of 39,600 obtained using the analytical ultracentrifuge [Howlett et al, 1982a].

Acceptor Molecular Weights Determined From the Distribution of Bound Ligand

The molecular weight of ligand-acceptor complexes can be obtained by analyzing the distribution of bound ligands provided the concentration of free ligand is negligible [Howlett and Markov, 1980]. This method for the determination of the molecular weight of acceptors has been referred to as ligand affinity sedimentation equilibrium and is of interest for the characterization of membrane receptors where one of the chief biological properties is the ability to bind ligands. The method was developed using radioactive fatty acids bound to serum albumin; a typical result is shown in Figure 4. The solid circles refer to the distribution of [^{14}C]-palmitate bound to human serum albumin. Analysis of the ln c/c_o vs r^2 plot yields a value of 67,800 for the molecular weight of the acceptor molecule. This value agrees well with the literature value (66,248) obtained from the amino acid sequence [Behrens et al, 1975]. The open circles in Figure 4 were obtained in a control experiment performed at higher angular velocities. Under these conditions the acceptor is depleted from the meniscus. The results demonstrate that under the conditions used the concentration of free radioactive ligand is close to zero. Results using [^{14}C]-palmitate and human serum indicated that the major palmitic acid-binding species had a molecular weight of 67,300 and was presumably serum albumin. Studies with serum albumin and octanoate indicated that the ligand-affinity sedimentation equilibrium procedure is suitable for studies on the molecular weight of acceptors even in cases when the ligand affinity is as low as 10^5 M^{-1}.

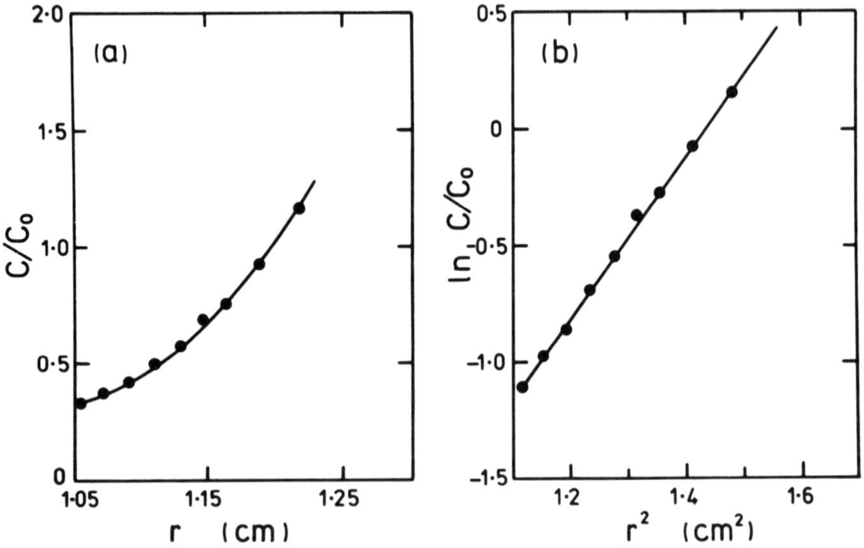

Fig. 3. *The sedimentation equilibrium behavior of ^{125}I-labeled rat serum α₁-acid glycoprotein. The distribution was formed by centrifugation in an air-driven ultracentrifuge for 22 h at 34,900 rpm at 10°C. The concentration distribution was obtained by sequential fractionation of the tube contents at the conclusion of centrifugation. The solution contained 5 mg/ml Dextran T40 for density stabilization. r, the radial distance to the center of rotation; c, the concentration of ^{125}I-labeled α₁-acid glycoprotein at r; and c_o, the initial concentration before centrifugation.*

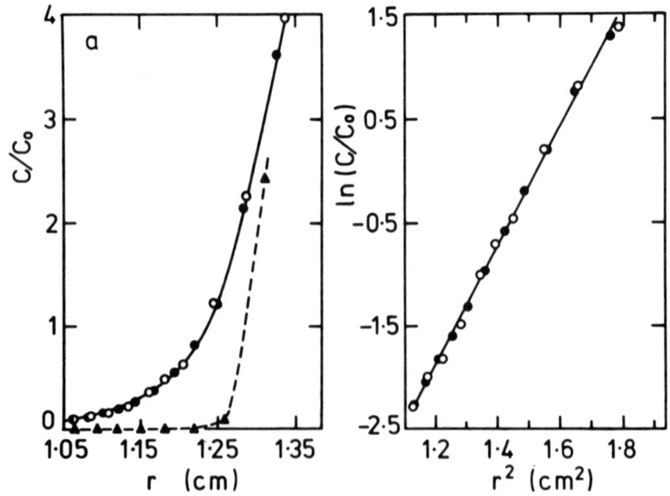

Fig. 4. *Ligand-affinity sedimentation equilibrium. Sedimentation equilibrium profiles for mixtures of palmitic acid (1×10^{-6}M; 0.4 μCi/ml) and human serum albumin (1.3×10^{-5}M) in phosphate buffered saline. Duplicate samples of 150 μl (○,●) were centrifuged in the air-driven ultracentrifuge at 37,360 rpm for 18 h at 8°C. Each sample contained Dextran T40 (5 mg/ml) for density stabilization. The concentration (c) refers to the concentration of radioactive ligand at radial position (r); c_o refers to the initial concentration before centrifugation. The same sample centrifuged at 100,000 rpm for 18 h (▲).*

ANALYSIS OF MACROMOLECULAR INTERACTIONS

Self-Association

The results of Pollet et al [1979a] indicate that the self-association of ^{125}I-labeled insulin can be readily detected by sedimentation equilibrium using an air-driven ultracentrifuge. Thus, at insulin concentrations of 1×10^{-10}M, 2×10^{-5}M and 2×10^{-4}M the weight average molecular weights were 5,600, 9,800, and 22,000 respectively.

The plant lectin, concanavalin A, provides another example of a self-associating solute [Senear and Teller, 1981]. The results in Figure 5a refer to the sedimentation equilibrium behavior of ^{125}I-labeled concanavalin A at pH 7.0. Analysis of the data close to the meniscus yields a value for the molecular weight (51,000) which is close to that expected for the dimer of concanavalin A. The data suggest that under these conditions concanavalin A exists as a dimer-tetramer mixture as described by Senear and Teller [1981]. The data in Figure 5b were obtained for the sedimentation equilibrium behavior of chorismate mutase/prephenate dehydrogenase of *E. coli.* In the absence of added ligands the enzyme behaves as a single species (dimer) of molecular weight approximately 76,700. In the presence of nicotinamide adenine dinucleotide (NAD^+) (0.5 mM) the weight-average molecular weight increases as shown by the increased slope of the line drawn through the

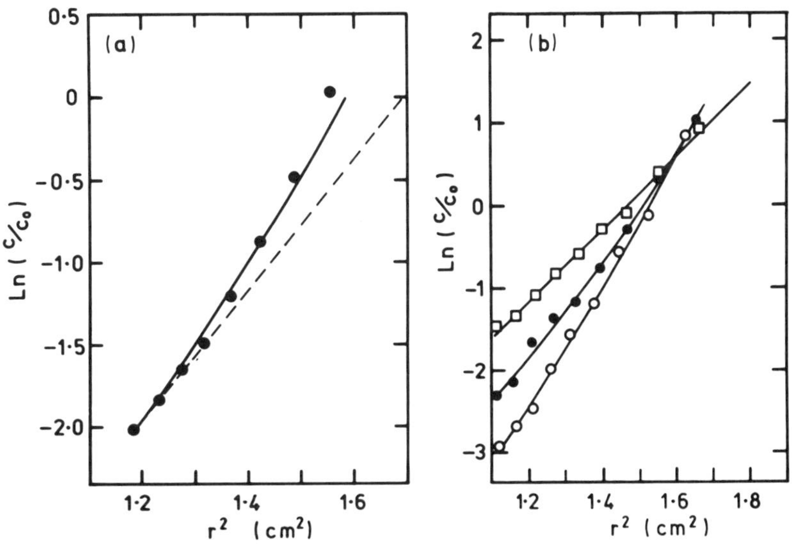

Fig. 5. *a). The sedimentation equilibrium behavior of ^{125}I-labeled concanavalin A. The buffer used was 0.26 M potassium phosphate, pH 7.0, 5 mg/ml bovine serum albumin. The initial concentration of concanavalin A was 2 mg/ml. The sample was centrifuged for 22 h at 36,670 rpm and 10°C. The broken line was calculated using Equation 4 and values for the molecular weight (51,000) and partial specific volume (0.731 ml/g) appropriate for dimeric concanavalin A [Senear and Teller, 1981]. The solid line was calculated using Equations 5–8 assuming a dimer-tetramer equilibrium constant of 1.6×10^4 M^{-1}. (b) The effects of NAD^+ and tyrosine on the sedimentation equilibrium behavior of chorismate mutase/prephenate dehydrogenase of E. coli. The buffer used was 50 mM Tris-HC1, pH 7.5 containing 0.5 mM EDTA, 10 mM 2-mercaptoethanol, 5 mg/ml dextran T40, and 5% v/v glycerol. The initial concentration of enzyme was 0.86 mg/ml, and the sample was centrifuged for 22 h at 31,440 rpm and 10°C. In the absence of ligands (□) the slope of the ln c/c_o vs r^2 plot yields a molecular weight of 76,700 for the enzyme. The data obtained in the presence of 0.5 mM NAD^+ (●) and in the presence of 0.05 mM NAD^+, 0.05 mM tyrosine (○) have been analyzed in terms of a dimer-tetramer equilibrium. The solid lines were calculated using values of 0.1×10^6 M^{-1} and 2.3×10^6 M^{-1} for the apparent equilibrium constants, respectively.*

data in Figure 5b. Similar results are obtained when 0.5 mM tyrosine is used instead of NAD^+. In the presence of both NAD^+ and tyrosine the weight average molecular weight increases still further to 137,000. This suggests that NAD^+ and tyrosine bind preferentially to the tetramer, causing a change in the equilibrium position in favour of tetramer.

Conventional procedures for the analysis of sedimentation equilibrium distribution to obtain the equilibrium constants for self-associating solutes rely on the use of point-weight average molecular weight data obtained as a function of total concentration [eg, Teller, 1973; Van Holde, 1975]. This approach requires the determination of the slope of the ln c/c_o vs r^2 plots at different radial positions. An alternative approach is to use curve fitting procedures and to fit the ln c/c_o vs r^2 plots directly using assumed values for the association equilibrium constant. Such curves may be constructed as follows: The total concentration (in g per unit volume) at the meniscus $\bar{c}(r_m)$ is equal to the sum of the concentrations of dimer $c_D(r_m)$ and tetramer $c_T(r_m)$:

$$\bar{c}(r_m) = c_D(r_m) + c_T(r_m). \qquad (5)$$

The equilibrium constant is defined as

$$K = c_T(r_m)/c_D^2\,(r_m). \qquad (6)$$

Combining Equations 5 and 6 yields

$$\bar{c}(r_m) - c_D(r_m) - Kc_D^2(r_m) = 0. \qquad (7)$$

Taking the experimental value for $\bar{c}(r_m)$ and an assumed value of K, it is therefore possible to solve equation 7 to determine $c_D(r_m)$ and then to obtain $c_T(r_m)$ using Equation 6. The entire concentration distribution can then be determined using conventional sedimentation equilibrium theory [Svedberg and Petersen, 1940].

$$\bar{c}(r) = c_D(r_m)e^{\sigma_D(r^2 - r_m^2)} + c_T(r_m)e^{\sigma_T(r^2 - r_m^2)}, \qquad (8)$$

where σ is defined as in Equation 4. This procedure can be repeated using different values of K until the theoretical curve provides the best fit to the experimental data. The solid line in Figure 5a was calculated using a value of $K = 1.6 \times 10^4\ M^{-1}$ for the dimer-tetramer equilibrium constant for concanavalin A. This value is in reasonable agreement with a value of $5 \times 10^4\ M^{-1}$ reported by Senear and Teller [1981]. Analysis of the data in Figure 5b for chorismate mutase/prephenate dehydrogenase in the presence of NAD^+ (solid circles) and NAD^+ and tyrosine (open circles) yielded values for the apparent dimer-tetramer equilibrium constant of $0.1 \times 10^6\ M^{-1}$ and $2.3 \times 10^6\ M^{-1}$ respectively. The effects of these ligands on the equilibrium between the different forms of the enzyme provide a quantitative description of the feedback inhibition of the enzyme by tyrosine [Hudson et al, 1983].

Protein-Protein Interactions

The interactions between two different macromolecules such as antibodies and antigens or lectins and glycoproteins are difficult to study using the analytical ultracentrifuge because of the inability of the optical systems to distinguish the different components. Using the air-driven ultracentrifuge it is possible to determine the effect of a nonlabeled component on the sedimentation equilibrium behavior of another component which is labeled with ^{125}I. The results presented in Figure 6 show the sedimentation equilibrium behavior of ^{125}I-labeled α_1-acid glycoprotein in the presence and absence of nonlabeled concanavalin A. The increased sedimentation in the presence of concanavalin A is a clear indication of an interaction. The weight average molecular weight of the ^{125}I-labeled component in the region close to the meniscus changes from 40,900 to 52,000 on the addition of concanavalin A. The competitive inhibitor, α-methyl mannoside (10 mM) prevents the interaction. For the electrostatic interaction between lysozyme and ovalbumin, measurements of the concentrations of each protein near the meniscus at sedimentation

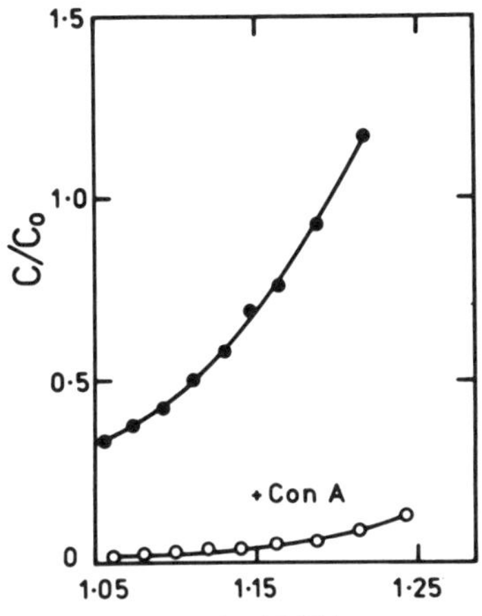

Fig. 6. *The sedimentation equilibrium distribution of* ^{125}*I-labeled rat serum* α_1*-acid glycoprotein in the presence (●) and absence (○) of concanavalin A. The solution conditions were 20 mM Tricine, pH 8.0, and 5 mg/ml bovine serum albumin. The initial concentration of* α_1*-acid glycoprotein was 5 μg/ml, and the concentration of concanavalin A was 1 mg/ml.*

equilibrium have been used to provide estimates of the equilibrium constant for the interaction [Clarke and Howlett, 1979]. This approach has also been used to determine the equilibrium constant and stoichiometry for the interaction between a monoclonal Fab fragment and an influenza virus neuraminidase [Jackson et al, 1983]

Nonideality

A third form of protein-protein interaction which occurs at high concentrations of protein is due to the effects of excluded volume or nonideality [Adams and Fujita, 1963]. These effects are a significant factor in the determination of colloidal osmotic pressure, solubility, and chemical equilibria and must be considered when interpreting the sedimentation equilibrium behavior of concentrated solutions of macromolecules. Recent studies [Howlett et al, 1982a] show significant nonideality effects for ^{125}I-bovine serum albumin in solutions of bovine serum albumin ranging in concentration from 20–80 mg/ml. The effects of nonideality are to reduce the slope of the ln c/c_c vs r^2 plot and to cause increased curvature at the higher protein concentrations. Using procedures similar to those described for the analysis of dimer-tetramer equilibria (Eg. 5—8) it is possible to analyse such non-ideality to obtain the appropriate nonideality constants [Howlett et al, 1982a]. When the total concentration of protein in solution is 5 mg/ml or less the effects of nonideality on sedimentation equilibrium results can generally be neglected.

The Interaction of Proteins With Membranes

Interactions of proteins at the cell surface are of great biological importance. In the case of erythrocytes, the cytoskeletal proteins (spectin, actin, band 2.1 and band 4.1) which interact with the membrane are major determinants of cell shape and flexibility [for reviews see Branton et al, 1981; Gratzer, 1981]. To study the rate of association or dissociation of proteins from the membrane, it is desirable to use high angular velocities in order to reduce the time taken for the separation of membrane bound and free protein. The air-driven ultracentrifuge is well suited to such studies, since it can sediment membrane-bound protein within minutes. The results shown in Figure 7 show that the cytoskeletal proteins of erythrocytes are rapidly released from the membrane at low-ionic strength and in the presence of 0.1 mM ethylenediamine tetraacetic acid (EDTA). The rate of release is decreased by adjusting the temperature from 37°C to 23°C and the ionic strength to 1.4 mM. The rate of release of spectrin/actin is even less for membranes prepared from patients with hereditary spherocytosis (Fig. 7). This disease affects both the shape and osmotic fragility of erythrocytes. Studies on the rate of reassociation at higher ionic strength of spectrin/actin with spectrin/actin-depleted vesicles indicated that the defect in hereditary spherocytosis resides not in a structurally altered spectrin or actin, but in an abnormality of the spectrin/actin-depleted vesicles [Hill et al, 1982]. The air-driven ultracentrifuge has been used in a similar way to determine the effect of ATP and 2,3-diphosphoglycerate on the structural stability of erythrocyte cytoskeletons prepared by detergent extraction of

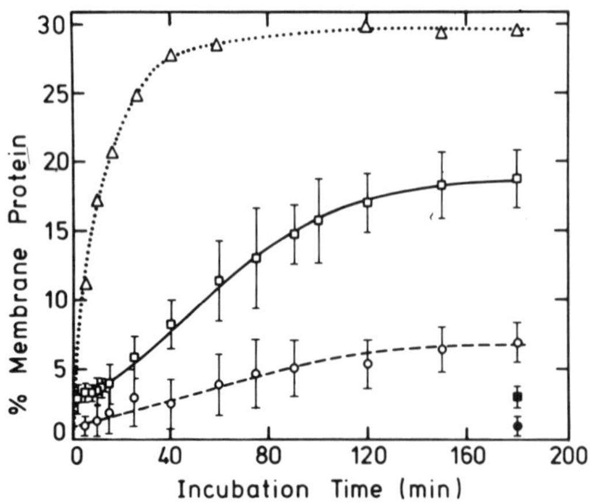

Fig. 7. *Kinetics of the dissociation of protein from erythrocyte cell membranes at low ionic strength. The percentage of total protein solubilized represents the fraction that remains in the supernatant after centrifugation at 150,000g for 30 s (total centrifugation time including acceleration and deceleration, 4 min). Results for normal ghosts in 0.1 mM-imidazole + 0.1 mM EDTA, pH 7.6 at 37°C are indicated by open triangles (△). Results for normal ghosts (□) and ghosts prepared from red blood cells of hereditary spherocytosis(HS) patients (○) in 1.2 mM-KC1/0.1 mM-EDTA/0.2 mM-imidazole (pH 7.6) at 23°C are also shown. The error bars represent means (± SD) for membranes obtained from seven different HS patients and from paired normal subjects. The filled symbols represent control experiments in which membranes were maintained in a solution containing 6 mM KC1, 0.1 mM imidazole (pH 7.6) at 23°C.*

erythrocyte membranes. In this case, changes in erythrocyte membrane morphology correlated with an increased tendency for dissociation of the cytoskeletal network [Sheetz and Casaly, 1980].

BUOYANT DENSITY STUDIES

Phospholipid Vesicles

The sedimentation equilibrium behavior of macromolecules depends on both molecular weight and partial specific volume (Eq. 4). The studies described so far used published values for partial specific volumes which were obtained from separate densitometric measurements. An alternative procedure which has been used to determine the partial specific volume of proteins involves sedimentation studies in solutions of different densities [Edelstein and Schachman, 1967]. In the case of phospholipid vesicles where the partial specific volume is close to unity it is possible to vary the solution density sufficiently such that the vesicles either sediment or float [Huang and Charlton, 1971]. By interpolation of the data it is possible to determine the solution density for zero sedimentation and to deduce the appropriate value for the partial specific volume. Such a procedure has been applied to sedimentation equilibrium experiments using the air-driven ultracentrifuge [Clarke et al, 1979]. The value obtained for the effective specific volume of the vesicles (0.9885 ml/g) was similar to the value obtained by interpolation of sedimentation velocity results. Comparison of the sedimentation behavior of the vesicles with that of standard proteins and correction for the effects of the different partial specific volumes yielded a value of (1.7–2.3) $\times$ 10^6 for the molecular weight of the vesicles. With the development of techniques to determine complete sedimentation equilibrium profiles [Pollet et al, 1979a], it may be possible to extend such studies to determine the degree of polydispersity of the vesicles and to study the interaction of vesicles with other macromolecules.

Detergent-Solubilized Proteins

Recent studies have shown that it is also possible to characterize detergent-solubilized proteins using the air-driven ultracentrifuge [Pollet et al, 1981; Howlett et al, 1982b]. The sedimentation equilibrium behavior of such systems depends on both the molecular weight of the protein and the amount of bound detergent. To analyse such systems it is convenient to define the sedimenting species (component 2*) on the basis of a hypothetical dialysis experiment such that any excess or deficiency of component 3 (electrolyte, buffer ions, detergent, and other low molecular weight solutes) on the solution side is included in component 2* [Casassa and Eisenberg, 1964]. The sedimentation equilibrium behavior of the detergent-protein complex may then be written as:

$$\sigma_2^* = \frac{d\ln c_2^*}{dr^2} = M_2^* (1 - \Phi_2^* \rho)\, \omega^2/2RT, \quad (9)$$

where Φ_2^* is the apparent specific volume of component 2* and M_2^* is the molecular weight.

In studies of detergent-solubilized mixtures it is not possible to determine Φ_2^* from densitometric studies. This parameter can be determined indirectly by measuring σ^* in solutions of different densities [Edelstein and Schachman, 1967]. This approach involves the assumptions that other macromolecules in solution do not interact with the macromolecule of interest and that preferential interactions of component 2* with components 1 (water) and 3 are identical in the different solvents. H_2O and D_2O solutions have been widely used for such studies.

$$\Phi_2^* = \frac{k - (\sigma_2^*)\, D_2O/(\sigma_2^*)\, H_2O}{\rho_{D_2O} - \rho_{H_2O}\, (\sigma_2^*)D_2O/(\sigma_2^*)H_2O}, \quad (10)$$

where k is a measure of the effect of deuteration on the molecular weight of component 2*. Determination of Φ_2^* permits determination of M_2^* using Equation 9.

Analysis of results obtained in D_2O and H_2O for ^{125}I-labeled insulin covalently bound to its receptor and solubilized by Triton X-100 yielded values of 0.78 ml/g and 475,000 for Φ_2^* and M_2^* [Pollet et al, 1981]. These authors also used D_2O^{18} as a method of increasing the density difference between the two solutions. In studies on γ-glutamyl transpeptidase [Howlett et al, 1982b], small plastic caps were fitted to the Airfuge tubes to prevent transfer of D_2O and H_2O between tubes. This procedure permitted the D_2O and H_2O samples to be centrifuged in the same rotor and resulted in a better comparison of

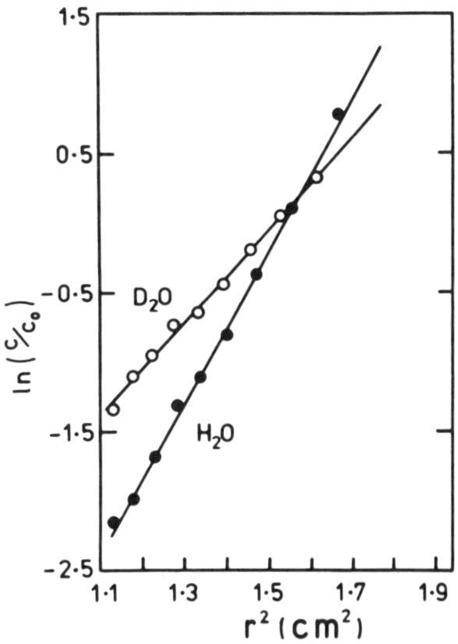

Fig. 8. *Sedimentation equilibrium profiles for rat kidney γ-glutamyl transpeptidase solubilized in 1% Triton X-100. The buffer used was 40 mM potassium phosphate pH 6.75, 0.1M NaCl containing 5 mg/ml bovine serum albumin. The solvents used were H_2O (●) and 90% D_2O (○). The samples were centrifuged for 30 h at 33,720 rpm.*

the effects of density on sedimentation. A typical result for this system is presented in Figure 8. The linearity of the plots indicate that the enzyme is relatively homogeneous with respect to molecular weight. Values of 0.819 ml/g and 175,000 were obtained for Φ_2^* and M_2^*, respectively.

Further analysis of the value obtained for Φ_2^* permits an estimation of the detergent protein ratio. The amount of detergent bound per gram of macromolecule (δ) is given by

$$\Phi_2^* = \frac{\bar{v}_2 + \delta \bar{v}_3}{1 + \delta}, \quad (11)$$

where $\bar{v}_2$ is the partial specific volume of the protein and $\bar{v}_3$ is the partial specific volume of detergent. The use of Equation 11 implies that there are no preferential interactions between the protein and other components in solution. Determination of δ permits the evaluation of

$$M_2 = M_2^*/(1 + \delta), \quad (12)$$

where M_2 is the molecular weight of the protein alone. For the case mentioned above a value of $M_2 = 79{,}000$ was obtained for the molecular weight of Triton X-100 solubilized γ-glutamyl transpeptidase. This value compares well with values of 87,000 obtained by sucrose gradient centrifugation and a value of 81,000 obtained from an analysis of the separate subunits of γ-glutamyl transpeptidase by gel electrophoresis [Hughey and Curthoys, 1976]. The molecular weight of γ-glutamyl transpeptidase was found to increase in the presence of convanavalin A [Howlett et al, 1982b]. This finding is consistent with studies on the purified enzyme which indicate that γ-glutamyl transpeptidase is a glycoprotein capable of binding concanavalin A [Tate and Meister, 1975]. Characterization of lectin-binding specificities of detergent-solubilized enzymes in heterogeneous solutions is important for purification procedures based on lectin affinity chromatography. Furthermore, the results indicate the feasibility of analyzing the interactions between the components of detergent-solubilized membrane preparations.

CONCLUSIONS

The air-driven ultracentrifuge provides a simple and relatively inexpensive method for the analysis of molecular interactions in solution and for the characterization of macromolecular size. At maximum angular velocity the gravitational fields generated are sufficient to sediment even small proteins such as ribonuclease to the base of the centrifuge tube. Ligand-binding studies can be performed by selectively removing ligand-acceptor complexes from the meniscus and determining the concentration of free ligand directly. Effects due to reequilibration of the ligand with the receptor during centrifugation are small for low molecular weight ligands. The molecular weight of macromolecules can be determined either by measuring the concentration near the meniscus at sedimentation equilibrium and comparing this value to those obtained using standard proteins or by sequentially fractionating the centrifuge tube contents and determining the complete sedimentation equilibrium profile. Molecular interactions due to protein-protein interactions, and nonideality can readily be detected and methods are available to analyze such data to obtain the appropriate thermodynamic parameters. The accuracy and reproducibility of data obtained using the air-driven ultracentrifuge have now improved to the extent that it is possible to determine both the molecular weight and density of detergent-solubilized proteins by determining the effect of solution density on sedimentation behavior. Such methods may ultimately be useful for the analysis of the interactions of detergent-solubilized membrane proteins in solution.

REFERENCES

Adams ET, Fujita H (1963): Sedimentation equilibrium of reacting systems. In Williams JW (ed): "Ultracentrifugal Analysis in Theory and Experiment." New York and London: Academic Press, pp 119–129.

Albers RW, Krishnan N (1979): Application of the miniature ultracentrifuge in receptor-binding assays. Anal Biochem 96:395–402.

Behrens PQ, Spiekerman AM, Brown JR (1975): Structure of human serum albumin. Fed Proc 34:591.

Bothwell MA, Shooter EM (1977): Dissociation equilibrium constant of β nerve growth factor. J Biol Chem 252:8532–8536.

Bothwell MA, Howlett GJ, Schachman HK (1978): A sedimentation equilibrium method for determining molecular weights of proteins with a tabletop high speed air turbine centrifuge. J Biol Chem 253:2073–2077.

Branton D, Cohen CM, Tyler J (1981): Interaction of cytoskeletal proteins on the human erythrocyte membrane. Cell 24:24–32.

Casassa EF, Eisenberg H (1964): Thermodynamic analysis of multicomponent solutions. In Anfinsen CB, Anson ML, Edsell JT, Richards FM (eds): "Advances in Protein Chemistry." New York and London: Academic Press, pp 287–395.

Clarke RG, Howlett GJ (1979): Determination of the molecular weight of proteins in heterogeneous mixtures: Use of an air-driven ultracentrifuge for the analysis of protein-protein interactions. Arch Biochem Biophys 195:235–242.

Clarke RG, Eagle GR, Howlett GJ (1979): Studies on polydisperse systems using an air-driven ultracentrifuge: Application to phosphatidylcholine vesicles. Aust J Biol Sci 33:187–196.

Duck-Chong CG (1979): Lamellar body phospholipid content of amniotic fluid: A possible index of fetal lung maturity. Ann Clin Biochem 16:191–196.

Edelstein SJ, Schachman HK (1967): The simultaneous determination of partial specific volumes and molecular weights with microgram quantities. J Biol Chem 242:306–311.

Garwin JL, Kages AL, Cronan JE (1980): Structural, enzymatic and genetic studies of β-ketoacyl-acyl carrier protein synthases I and II of *Escherichia coli*. J Biol Chem 255:11949–11956.

Gerhart JC, Schachman HK (1968): Allosteric interactions in aspartate transcarbamylase. II. Evidence for different conformational states of the protein in the presence and absence of specific ligands. Biochemistry 7:538–552.

Gratzer WB (1981): The red cell membrane and its cytoskeleton. Biochem J 198:1–8.

Heidemann PH, Wittliff JL (1979): Preparation of cytosol for steroid receptor determinations with the Beckman Airfuge. Clin Chem 25:622–625.

Hill JS, Sawyer WH, Howlett GJ, Wiley JS (1982): Hereditary spherocytosis of man. Altered binding of cytoskeletal components to the erythrocyte membrane. Biochem J 201:259–266.

Howlett GJ, Markov M (1980): Ligand-affinity sedimentation equilibrium using an air-driven ultracentrifuge. Arch Biochem Biophys 202:507–514.

Howlett GJ, Yeh E, Schachman HK (1978): Protein-ligand binding studies with a table-top, air-driven, high-speed centrifuge. Arch Biochem Biophys 190:809–819.

Howlett GJ, Dickson PW, Birch H, Schreiber G (1982a): Studies on ^{125}I-labeled proteins in rat plasma using an air-driven ultracentrifuge: Protein-protein interactions and non-ideality. Arch Biochem Biophys 215:309–318.

Howlett GJ, Birch H, Dickson PW, Schreiber G (1982b): Determination of the molecular weight of detergent solubilized enzymes by sedimentation equilibrium in an air-driven ultracentrifuge. Biochem Biophys Res Commun 105:895–901.

Huang CH, Charlton JP (1971): Studies on phospholipid vesicles. Determination of partial specific volumes by sedimentation velocity method. J Biol Chem 246:2555–2560.

Hudson GS, Howlett GJ, Davidson BE (1983): The binding of tyrosine and NAD^+ to chorismate mutase/prephenate dehydrogenase from *Escherichia coli* and the effects of these ligands on the activity and self-association of the enzyme. J Biol Chem 258:3114–3120.

Hughey RP, Curthoys NP (1976): Comparison of the size and physical properties of γ-glutamyltranspeptidase purified from rat kidney following solubilization with papain or with Triton X-100. J Biol Chem 251:7863–7870.

Jackson DC, Howlett GJ, Nestorowicz A, Webster RG (1983): The equilibrium constant for the interaction between a monoclonal Fab fragment and an influenza virus neuraminidase. J Immunol 130:1313–1319.

Klotz IM, Hunston DL (1975): Protein interactions with small molecules. Relationships between stoichiometric binding constants, site binding constants and empirical binding parameters. J Biol Chem 250:3001–3009.

Moulin S, Dewailly P, Sezille G, Fruchart JC (1979): Diagnosis of type III hyperlipoproteinemia: Contribution of an air-driven ultracentrifuge associated with an original tube-slicer. Ann Biol Clin (Paris) 37:103–105.

Pollet RJ, Haase BA, Standaert ML (1979a): Macromolecular characterization by sedimentation equilibrium in the preparative ultracentrifuge. J Biol Chem 254:30–33.

Pollet RJ, Haase BA, Standaert ML (1979b): Macromolecular characterization by sedimentation equilibrium in the preparative ultracentrifuge. J Biol Chem 254:3679.

Pollet RJ, Haase BA, Standaert ML (1981): Characterization of detergent-solubilized membrane proteins. Hydrodynamic and sedimentation equilibrium properties of the insulin receptor of the human lymphoblastoid cell. J Biol Chem 256:12118–12126.

Scatchard G (1949): The attractions of proteins for small molecules and ions. Ann NY Acad Sci 51:660–672.

Schachman HK (1959): "Ultracentrifugation in Biochemistry." New York and London: Academic Press.

Senear DF, Teller DC (1981): Thermodynamics of concanavalin A dimer-tetramer self-association: sedimentation equilibrium studies. Biochemistry 20:3076–3083.

Scheetz MP, Casaly J (1980): 2,3-Diphosphoglycerate and ATP dissociate erythrocyte membrane skeletons. J Biol Chem 255:9955–9960.

Steinberg IZ, Schachman HK (1966): Ultracentrifugation studies with absorbtion optics. V. Analysis of interacting systems involving macromolecules and small molecules. Biochemistry 5:3728–3747.

Svedberg T, Petersen K (1940): "The Ultracentrifuge." London: Oxford University Press.

Tate SS, Meister A (1975): Identity of maleate-stimulated glutaminase with γ-glutamyl transpeptidase in rat kidney. J Biol Chem 250:4619–4627.

Teller DC (1973): Characterization of proteins by sedimentation equilibrium in the analytical ultracentrifuge. In Hirs CHW, Timasheff SN (eds): "Methods in Enzymology." New York and London: Academic Press, pp 346–441.

Van Holde KE (1975): Sedimentation analysis of proteins. In Neurath H, Hill RL (eds): "The Proteins." New York: Academic Press, pp 225–291.

Molecular and Chemical Characterization of Membrane Receptors, pages 193–208

8

Molecular Weight Determination by Radiation Inactivation

Chan Y. Jung

Department of Biophysical Sciences, SUNYAB, and Biophysics Laboratory, VA Medical Center, Buffalo, New York 14215

INTRODUCTION

Membrane-bound proteins are known to play a key role in many cell functions involving the transfer of information, energy, and mass across the membrane [Mullins, 1981]. Knowledge of the size (or mass) and assembly state in situ of these proteins is important in understanding the molecular basis of these functions. Conventional physicochemical means such as hydrodynamic and electrophoretic measurements may be used for this purpose. However, the fact that these measurements require the solubilization and purification of the proteins in detergents make the obtained molecular weights less reliable. Furthermore, the molecular weight obtained in such measurements has no direct reference to the functionality of the protein. Information as to the quaternary structure of the protein is not expected from these conventional measurements.

For the study of membrane-bound proteins, the target-size measurement by high-energy radiation inactivation [Lea, 1946; Pollard et al, 1955] offers many distinct advantages over conventional techniques: This method does not require the solubilization or purification of the protein, thus eliminating the artifacts such as aggregation or degradation introduced by the use of detergents. Since the measurement is based upon the biological activities of the protein, the size obtained has direct bearing to the functional protein. Furthermore, this method is unique in that molecules in their native membrane environment can be studied. If the size of the macromolecule subunits is known, information on the association state (quaternary structure) in situ of the protein may be obtained. Finally, the method requires relatively small amounts of protein sample.

The concept of target size was first introduced by Crowther in 1924. In order to explain experimental data on the inhibition of cell division by x-irradiation, he assumed a single, sensitive volume (target) the size of a centromere existed in each cell, and a single hit of this target by radiation would cause the inhibition. In a series of studies, Hussey and Thompson [Lea, 1946] carried out the first experiments of the radiation inactivation of enzymes, and found that the concept of target size could be equally applied for biological macromolecules. In these experiments, enzymes in solution were irradiated by x-rays and the loss of enzymatic activity was measured. The survival of activity was found to be a logarithmic function of the total radiation dose. However, it was not until the late 1940s that Lea first related the direct action of ionization to the target size or molecular weight and applied this relationship to a wide variety of biological systems [Lea, 1946]. Subsequently, target molecular weight measurement became quite popular in the 1950s and 1960s. Molecular weights of many water-soluble (cytoplasmic) enzymes were determined by this method [Pollard, 1969], which reproduced the data obtained by other conventional physicochemical means, thus establishing the general validity of the method. However, for water-soluble enzymes, the target size measurement was not always found to be superior to other more conventional methods. Only recently, after the importance of membrane-associated proteins was realized, was interest on this methodology revived.

A number of excellent reviews on the target size measurement of biological molecules are available in literature [Pollard et al, 1955; Hutchinson and Pollard, 1961; Kempner and Schlegel, 1979]. This chapter discusses the theory and methods relating to the use of radiation inactivation for target-size measurement with particular emphasis on membrane-bound proteins such as receptors and effectors.

THEORETICAL CONSIDERATIONS

The Physics and Chemistry of Radiation Inactivation

High-energy radiation absorbed in a biological system initiates a complex series of reactions which may result in the inactivation of biological functions [Dartinger and Jung, 1970; Andrews, 1974]. The very first event, which takes place within less than picosecond, is the physical absorption of radiation, namely, the ionization and/or excitation of atoms. This is followed by a second step known as the "initial chemical reaction," which may last up to a few milliseconds. The two major events here are radical formation and molecular activation. Alterations of the biological function of the molecule will follow within a period of seconds to hours.

In the ionization process, an orbital electron is completely stripped from the atom. Since forces that hold a molecule together (chemical bonds) are usually made of electrons shared between atoms (valence electrons), the ejection of these electrons would lead to the dissociation of atoms (bond rupture) within the molecule. In the excitation process, on the other hand, the electrons are not completely removed from the atom but are raised to a higher energy level within the atom. In dissipating this excess energy, the excited molecule occasionally undergoes reorganization of its valence electrons, which leads to bond rupture [Augenstein et al, 1964].

Ionization is nearly exclusively produced by fast charged particles, eg, electrons and protons. High-energy radiation, such as x-rays, α-, β-, and γ-radiation, protons and neutrons, can produce fast charged particles when they traverse matter, and thus are called ionizing radiations. On the other hand, both low-energy radiation (such as UV light) and high-energy ionizing radiation can produce excitation. In the case of ionizing radiation, ionization and excitation contribute nearly equally to the energy transfer. Ionization directly due to the pass of a fast charged particle (known as primary ionization) occurs randomly and discretely. The average value of energy release is estimated to be 110 eV. The electron ejected at primary ionization often has enough to produce a few ionizations on its own account. These are called "secondary ionization." They are produced near the primary ionization, usually within 20–30 Å (less than the radius of a typical protein molecule), and form a cluster of two or three ionizations. If both primary and secondary ionizations are considered, the average release of energy per ionization is approximately 32 eV.

Electrons, α-particles, and protons are three well-known charged particles. The biochemical effects of ionizing radiation are predominantly due to the action of fast electrons. Fast electrons are produced in the interaction of x-rays and γ-rays with biological samples. Fast protons, on the other hand, are produced when neutrons interact with matter. Fast electron beams are also produced directly by acceleration of slow electrons in an appropriate electrostatic field.

The rate of energy loss of any charged particle traversing biological samples is linearly proportional to the square of the charge and inversely proportional to the velocity. The rate of energy loss along the track is commonly referred to as linear energy transfer (LET) and is expressed in units of keV/μm. For a given kinetic energy (eg, 1 MeV), the heavier the particle, the lower the velocity. Thus, protons and α-particles are 40 times more effective ionizers than electrons of the same kinetic energy. The LET of a 1-MeV electron is approximately 1 keV/μm. The LETs of a 1 MeV proton and α-particle are about 25 and 260 keV/μm, respectively. A particle with higher LET will produce ionizations with higher density along its track; the densities of ionizations along the track of a 1-MeV electron, proton, and α-particle are approximatley 30, 750, and 5,000 per μm, respectively. These values correspond to a primary ionization for every 1,000, 25, and 2 Å, respectively. As an electron travels along the track losing energy (by 32 eV) for every ionization produced, its velocity is reduced, and it will eventually stop. Since LET is inversely proportional to the velocity, the density of ionization produced near the end of the track is much higher than the rest of the track; thus the spatial distribution of ionization along the whole track is not exactly even. The distance a charged particle can penetrate into a sample before being completely stopped is called its "range." The range of a 1-MeV electron in biological sample is approximately 4 mm, whereas that of a 1-MeV proton is approximately 20 μm, or two orders of magnitude shorter. For α-particles, the range is even shorter, ie, approximately 5 μm for a 1-MeV particle.

The ionization process leads into the initial chemical reaction, which is characterized by the production of free radicals. Free radicals, having unpaired or odd electrons, are highly reactive, and are able to initiate further covalent reactions within the molecule. The disruption of chemical bonds and/or covalent change causes alteration in the secondary or tertiary structure of the macromolecule, which ultimately leads to the loss of function. This is the direct action of radiation, as the events occur strictly within the molecule to which the energy is directed.

Energy transfer via ionization is a nonspecific process transferring energy to any molecule in a biological system with little regard to the chemical status of the individual atoms. Most biological samples include excess amounts of water molecules as solvent; these water molecules also undergo ionization, yielding diffusible, highly reactive radicals (H^+, OH^-) and oxidants (H_2O_2). The radicals can diffuse an average of 30 Å within their lifetime, and react with proteins encountered within this radius, inactivating their function. This is an indirect action of radiation.

In a dilute protein solution such as in most biological samples, direct action accounts only for a minor fraction of the overall radiation effect; indirect action predominates as most of the energy transfer occurs to the solvent molecules which are present in abundance. Indirect action may be eliminated by irradiating the dried samples which lack the solvent water molecules. The indirect action may also be removed by irradiating frozen samples. The diffusion of free radicals is effectively hindered in ice.

Basic Idea of Target Theory

Ionizing radiation inactivates many biological macromolecules by direct action. Dose-dependency of this inactivation appears to be intimately related to the size of the macromolecule, and it is the treatment of this relationship which is the target theory. In this theory, we speak of the molecule or structure in which ionization is produced as the "target" and the production of ionization within it as the "hit." The basic idea of the treatment is a statistical calculation of the hit probability of a target of volume V, for a given radiation dose I, equivalent to the production of I ionizations per unit volume (cm^3).

Under proper experimental conditions

where only direct action is effected, primary ionization (or its related clusters) is spatially localized in a sphere of radius of 20 Å, a volume smaller than any obtainable by other physical means. If this volume is occupied by a protein molecule, ionization releases an amount of energy so enormous (66 eV or 1,518 kcal/mol, enough to break fifteen C-H bonds) that derangement certainly results; this in turn causes loss of the protein function. With any reasonably sized biological sample and radiation dose, ionizations produced in the sample are sparse although very large in total number, and only a small fraction of the total protein targets would be hit. These facts justify the use of Poisson statistics in calculating the expected number of hits to the target for a given radiation dose. Since the average number of hits per target molecule is the product of the volume, V, of that molecule and the number of hits per cm^3, I (ie, VI), the probability of n hits occurring in a volume V is given by the formula

$$P(n) = \frac{e^{-IV}(IV)^n}{n!} . \quad (1)$$

The simplest form of the theory deals with the "total escape" probability, where it is assumed that survival occurs only if ionization has not taken place in the target molecule, and a single hit is enough to inactivate the molecule (single-hit theory). In this case, the probability of survival activity is the probability of a zero hit [$P_{(n)}$ where $n = 0$] in the Poisson formula; thus,

$$S = e^{-IV} \text{ or } \ln S = -IV, \quad (2)$$

where S is the surviving fraction of the activity of the macromolecule after irradiation at a dose I. When $S = 1/e$ (or 0.37), it is evident that

$$V = \frac{1}{I_0} , \quad (3)$$

where I_0 is the radiation dose which reduced the activity of the molecule to 37% of that of an unirradiated control. It follows that if I_0 is known, the target volume, V, may be obtained from this simple relationship.

The theoretical prediction of Equation 2 was borne out in many experiments in which enzymes or proteins in either a dehydrated or frozen state were irradiated with ionizing radiation. Surviving activities were found to decrease as a simple exponential of the radiation dose D. The radiation dose in actual experiments is conventionally expressed in rads, the unit based on the total energy absorbed per gram of sample (1 rad is a dose of radiation sufficient to result in 100 ergs of total energy absorbed in 1 g of sample). In order to use Equation 3 in calculating the target size from the experimental data of radiation inactivation, it is necessary to convert the conventional units of radiation dose measured, D, to the unit of primary ionizations per unit volume, I. Although precise numerical treatment of this conversion is not yet available, a general relationship is [Kepner and Macey, 1968]

$$I = \left(6.24 \times 10^{13} \frac{\rho}{Q}\right) \times D, \quad (4)$$

where Q denotes the energy in eV per inactivating event, and ρ is the density of the sample. Combining Equations 3 and 4, and using Avogadro's number (6.02×10^{23}) the sample molecular weight, Mr, is

$$Mr = 0.96 \times 10^{10} \frac{Q}{D_0} , \quad (5)$$

where D_0 is the dose in rads which reduces activity to 37% of the control. Using the value of 60 eV for Q, Pollard [1959] showed that, for a number of enzymes, the relationship of Equation 5 was capable of reproducing the molecular weights obtained by conventional physicochemical methods. However, for many other enzymes, he also noted the deviation from this relationship was much larger than experimental error could account for, and this was particularly so where D_0 was measured at lower temperatures. Subsequently, Kepner and Macey [1968] took an empirical approach, using selected D_0 data obtained only at room temperature (25°C). For more than 20 different enzymes and macromolecules, a comparison of D_0 with their Mr determined by physicochemical methods gave them an empirical relationship

$$Mr = 6.4 \times \frac{10^{11}}{D_{0\ (rads)}} . \qquad (6)$$

This expression has been used successfully for calculations of target molecular weights. Incidentally, Equation 6 implies that Q is equal to 66 eV.

METHODOLOGICAL CONSIDERATIONS

The experimental part of radiation target-size measurements involves two steps: irradiation of samples at varying radiation doses under well-controlled experimental conditions, and quantitation of resulting inactivation of the macromolecules after irradiation.

Quality of Radiation

In principle, any type of radiation which produces ionization randomly and sparsely throughout the volume of a sample may be used for target-size measurements. Target theory assumes that the radiation sensitivity of a protein is independent of its chemical or atomic state. The energy transfer by ionization is practically independent of the chemistry of the molecule. The energy adsorption by excitation on the other hand is sensitive to the chemical state (eg, bond energies) of molecules. This fact rules out the use of any low-energy radiation (such as UV) for target-size measurements, as such radiation produces only excitation.

The most commonly used radiation in target-size measurements are high-energy electrons. High-energy x-rays and γ-rays are also used. Either of these electromagnetic forms of radiation produces fast electrons (either Compton electrons or photoelectrons) when they interact with samples, and it is essentially these high-energy electrons which eventually dissipate energy through ionization. As electrons with kinetic energies of 1 MeV or more produce primary ionizations separated by at least 1,000 Å in space, these ionizations would be random and sparse. Let us consider a sample of proteins with a Mr of 10^6. This protein would have a radius, r, of 67 Å (assuming a spherical shape and a density of 1.3 g/cm^3), and the average path length through this molecule would be 1.3 r or 90 Å; only a small fraction (less than one tenth) of the total number of molecules would be hit by these fast (low LET) electrons.

If inactivation of a molecule can occur as the result of a single hit (single primary ionization), the occurrence of several primary ionizations per molecule would be wasteful. Indeed, many experimental results demonstrate that radiation of high LET which produces ionizations close together would be less effective in inactivation and would yield incorrectly a small target size. In extreme situations, high LET ionizing radiation such as protons or α-particles are sometimes purposefully used; here the probability of the protein molecule being hit is almost certain ($P_{(n)}$ = 1), and thus target cross-section rather than target size is obtained [Marshall et al, 1970]. An opposite type of complication may also occur in the use of α-particles, which would give incorrectly a large target size, and this is due to the fact that α-particles produce many high-energy secondary electrons (δ-rays), whose total track length is even greater than the track of the α-particle itself.

Another factor important in selecting radiation sources for target analysis is the range of radiation, as it is necessary that the radiation penetrate the total thickness of the sample. For a sample thickness of up to 2 mm, 1-MeV electrons may be used. Such electron beams can be readily provided by a Van de Graaff electron generator [Blum and Alpher, 1971], the γ-rays of cobalt-60 [Beauregard and Potier, 1982], or 1,000-kVcp x-rays. For thicker samples, higher MeV electron beams produced by a linear electron accelerator may be used [Harmon et al, 1980].

For proteins of smaller molecular weight, inactivation analysis sometimes requires radiation doses as high as 30 Mrads. With a Van de Graaff electron generator having a dose rate of 0.5 Mrad/min, total irradiation time could take up to 1 h.

Sample Preparation and Irradiation

A wide spectrum of sample preparations may be used for target-size studies. These include intact cells, subcellular fractions, isolated membranes, reconstituted membranes, and purified proteins. When intact cells are irradiated and the assay of the function of the molecules studied requires cell viability, cryoprotective techniques can be used [Cuppoletti et al, 1981]. Typically, a volume of packed cells is mixed with an equal volume of a freezing solution and equilibrated for 30 min at 0–4°C before freezing. The freezing solution is a mixture of 280 g of glycerol, 28 g of sorbitol, and 720 ml of 150 mM NaCl.

Since radiation target-size analysis is based on the direct action only, it is necessary to eliminate the indirect action of radiation. This is done by irradiating samples in either a dry or frozen state. Many experimental data support that both of these methods work. There are some proteins, however, which are known to be inactivated by freeze-drying and subsequent resolvation in an aqueous buffer, eg, some receptor and ion transporters [Fraser and Venter, 1982; Lilly et al, 1983; Venter et al, 1983]. For such proteins there is no choice but to irradiate in the frozen state. Furthermore, there are many membrane-associated functions whose assay requires an intact membrane diffusion barrier. A typical example is transmembrane solute transport mediated by specific transport proteins. Membrane diffusion barriers, particularly that of natural membranes, are usually not restored by rehydration once freeze-dried. However, by the use of cryoprotective agents many cells may be frozen and thawed without losing their membrane diffusion barrier. One well-known complication of using frozen-state irradiation, however, is the need for temperature-correction in dose-activity analysis, which will be discussed later.

For dry-state irradiation, samples may be lyophilized in a hard glass tube, then sealed under high vacuum [Schlegel et al, 1979]. Alternatively, for both dry and frozen irradiation, samples may be prepared on a shallow aluminum tray [Saccomani et al, 1981]. Aluminum is used here because of its superior heat conductance. Typically, the sample tray (Fig. 1) has the following dimensions: 80 mm long, 10 mm wide, and 3 mm deep. Approximately 1 ml of sample solution or suspension is evenly spread in the tray and dried in a vacuum dessicator having a close-coupled side-arm trap packed in dry ice. An overnight dessication in this device gives a dry film of uniform thickness. For frozen irradiation, the samples are flushed with nitrogen gas before being applied on the sample tray. The trays are first suspended horizontally near the surface of a liquid nitrogen bath for 15 to 20 s, then plunged into the bath for 5 to 10 s. This procedure produces uniformly thin (less than 2 mm) frozen samples on the tray.

Actual irradiation of the samples is done in a closed chamber (target chamber) (Fig. 1) to eliminate oxygen and to control temperature. The target chamber consists of a base and a cover, both made of aluminum. Typically, the base is a 6 × 22-in block with a 0.25-in-deep gasket encircling a channel (sample channel) which accommodates sample trays. An aluminum foil window (0.005 in thick) is sealed to the cover, providing an airtight unit when in place over the gasket, and bolted along the perimeter of the base. During irradiation, the temperature of the target chamber is maintained at a preset value. When frozen samples are used, this is done by flushing liquid nitrogen through two ports at opposite ends of the base. As nitrogen changes states it purges the channel of air through the exit port, providing

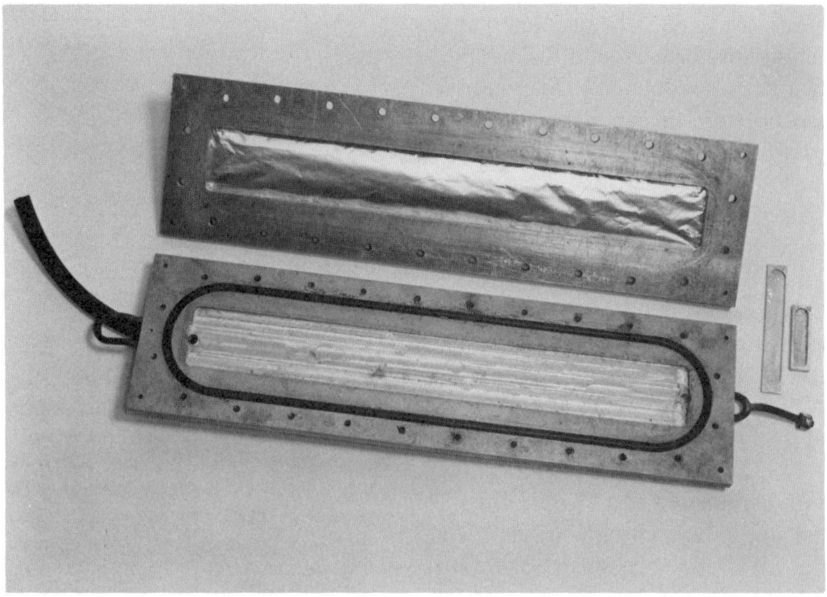

Fig. 1. *Unassembled view of the target chamber and two sample trays of different sizes.*

an oxygen-free atmosphere. When irradiation is done at room temperature, a suitable coolant may be used which circulates through the copper tubing encased beneath the sample channel in the base. The precise measurement of the temperature of the sample during irradiation is difficult if not impossible. It may be approximated using the temperature of the target chamber. A temperature-sensitive probe inserted through the wall of the target chamber may be used for this purpose. The irradiation room should be kept well ventilated to prevent possible frost formation on the target chamber window.

In the case of the Van de Graaff electron accelerator, the target chamber is placed on a chain conveyor system which is positioned underneath the drift tube. The distance between the drift tube window and the target chamber, usually 6.5 in, may be adjusted by changing the elevation of the conveyor. Through the use of electromechanical switches, the target chamber can be made to traverse beneath the beam and return, registering each "pass" on a control room counter. In normal operation, the terminal potential is adjusted to its rated maximum, typically 1.5 million V, with its beam currents between 0.2 and 1.0 mA, depending on the degree of irradiation required. The beam current is scanned, providing a lateral beam spread of about 16 in to encompass the entire sample channel. The beam exits the drift tube with a fairly narrow width; Ozalid paper affixed across the length of this window displays a uniform burn pattern of typically ¼ in when exposed for 1 min at the rated voltage and at 100 μA of beam current. This band widens to extend a few inches on either side of the drift tube at the target chamber level. When a linear accelerator of 13 to 15 MeV is used, a uniform beam field of 10–17-cm diameter with a dose rate of approximately 2 Mrad/min may be achieved. In this example, the radiation dose may be varied by irradiating the samples in the target chamber for a varying length of time.

Determination of Radiation Doses

Relative dose of radiation can be conveniently expressed in terms of "pass" numbers or exposure time, provided that other irradia-

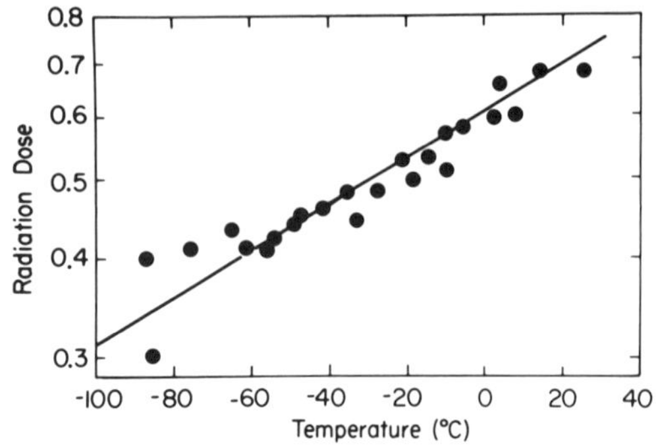

Fig. 2. *Radiation sensitivity of Blue Cellophane (du Pont) as a function of irradiation temperature. Blue Cellophane films were irradiated with a 1.5-MeV electron beam of Van de Graaff with a beam current of 0.6 mA for a fixed time interval of 20 s at each temperature. The changes in light transmittance of the film due to irradiation were measured at 655 nm with a Beckman DU spectrophotometer. Radiation doses corresponding to the bleaching were read in rads from the calibration curve, which was constructed from the data of the film bleaching at known radiation doses measured by Fricke's dosimetry. Each data point represents a single measurement. The straight line represents the least-squares line fitted by Kempner and Haigler [1982] to the composite data of radiation sensitivity for five different enzymes. It is apparent that the Blue Cellophane reproduces the change in the radiation sensitivity of enzymes as a function of temperature between −60 and 20°C.*

tion conditions are fixed. The absolute value of the radiation dose delivered to the sample, however, is difficult to obtain. This value is usually deduced from the physical or chemical changes produced in a well-defined reference system subjected to the same irradiation. One example of such a system is the Fricke dosimeter [ICRU Report 17, 1970]. This system is based on the fact that radiation oxidizes ferrous iron (Fe^{2+}) to the ferric state (Fe^{3+}) with a well-defined chemical yield [Fricke and Morse, 1927]. This system is suitable for medium to high dose radiations (up to 0.05 Mrads) and linear extrapolation may be used for radiation doses mostly encountered in the target-size measurement. Many other chemical systems (eg, radiochrome films) are known which are used for dosimetry based on similar principles.

Many of these chemical systems, however, show somewhat different responses to temperature [Miller et al, 1975; Levine et al, 1979]. Figure 2 illustrates the temperature dependence of Blue Cellophane (du Pont). This chemical system is a cellophane sheet of uniform optical and physical characteristics containing dimethoxy-diphenyl-bisazo-bis-8-amino-1-naphthol-5,7-disulphonic acid dye, which has an absorption peak at 655 nm [Henley, 1954]. Irradiations are carried out in the aluminum target chamber referred to above under the conditions identical to those for sample irradiations using 1.5-MeV electrons generated by a Van de Graaff. The radiation doses are measured by the Fricke dosimetry. For the fixed temperature of 25°C, the change in percent transmission of the cellophane sheet is a linear function of radiation dose up to 5 Mrads. In the case of irradiation at varying temperatures, the dose rate decreases almost linearly as the temperature is reduced. There is at present no adequate explanation for such a temperature effect of ionizing radiation. It may be suggested that the decrease in bond lengths at low temperatures is sufficient to impart extra stability to the molecule. The excitation energy, which spreads through the molecule, is usually insufficient to rupture even weak bonds, and deexcitation is delayed until an energy fluctuation raises bonds over the dissociation energy. It may be possible that this type of energy fluctuation is temperature sensitive. In solids, this energy fluctuation is practically prohibited, and heating or some other stimulation is needed to induce the release of the trapped energy. This physical system is put to practical use in radiation dosimetry (known as thermoluminescent dosimetry). Thermoluminescent dosimetry is known to be independent of temperature at 30°C or below [Kempner and Haigler, 1982]. This is easily understood if the temperature effect observed in these chemical reference systems is indeed due to energy fluctuation.

Assay of Biological Function of Irradiated Samples

After irradiation the samples are usually stored at liquid nitrogen temperatures in the dark for 24–48 h prior to the assay for their biological functions. Irradiated samples are either resuspended in a suitable aqueous buffer (in the case of dry irradiation) or gently thawed (in the case of frozen irradiation) to prepare assay samples. Various membrane-associated functions have been studied after irradiation. They are typically ligand-binding activities, enzymatic activities, and transport activities. Ligand-binding activities studied include those of physiological ligands such as hormones [Harmon et al, 1979; Venter, 1983], antibodies [Folkerd et al, 1977], and enzymes [Steer et al, 1981], and various pharmacological ligands [Jung et al, 1980]. When intact cells are studied, some complications in binding assays have been experienced because of unexpected increases in nonspecific binding after irradiation [Green et al, 1983]. The assay for enzymatic activities [McIntyre et al, 1983] is not in general problematic with the exception of those membrane-bound enzymes that exhibit vectoral asymmetry for substrates and activators. One such example is Na^+-K^+-ATPase (Hah, unpublished data). When transport proteins are studied and fluxes of the substrates across the membrane are to be measured, the diffusion barrier should be restored after irradiation [Cuppoletti et al, 1981].

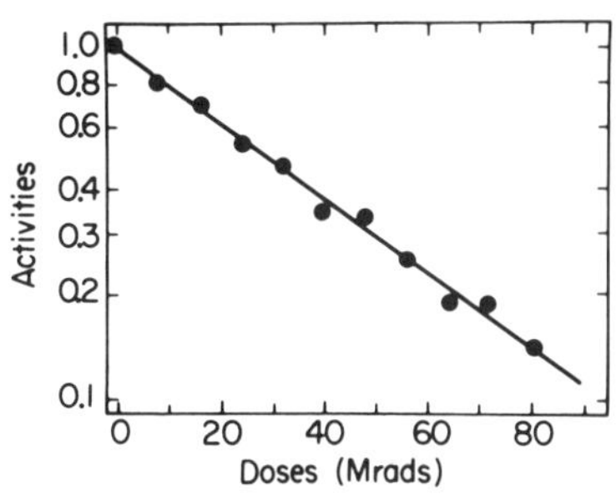

Fig. 3. *Lysozyme activity remaining after irradiation with a 1.5-MeV electron beam of varying doses. The enzyme activity is expressed as a fraction of that of unirradiated control. Radiation doses were measured by Blue Cellophane film method.*

Along with the assay of the biological activities of the molecules studied, it is often informative to characterize the chemical changes produced by irradiation. This is usually done through sodium dodecyl sulfate (SDS) gel electrophoresis of the irradiated samples [Saccomani et al, 1981]. Any change (breakdown or aggregation) observed in the sample should be noted and correlated with the results of the functional assay to determine the relevance.

ANALYSIS OF INACTIVATION DATA

The first step in the analysis of inactivation data for target-size measurement is to establish the relationship of fractional survival of activity or related parameters of the macromolecule to radiation doses. The simplest of such relationships is where survival is linearly related to radiation dose. This is typically found in simple chemical reactions such as the oxidation of 0.01 M $FeSO_4$ in 0.8 N H_2SO_4 solution (Fricke's ferrous sulfate dosimeter), provided that the radiation dosage is kept within a certain range [Weiss et al, 1956].

Exponential Survival Curves

The most common survival relationship found in biological systems, however, is where the logarithm of survival is linearly related to radiation dosage [Pollard, 1969]. This relationship conforms to the theoretical prediction of Equation 2 and represents the single-target, single-hit model of the classic target theory [Fowler, 1964]. Figure 3 illustrates, as an example of such a relationship, the inactivation of lysozyme. This enzyme is a single polypeptide chain of 129 amino acids with a molecular weight of 14,600. It is a relatively compact protein of roughly ellipsoidal shape with dimensions 45 × 30 × 30 Å. In this experiment, lyophilized lysozymes (Worthington Biochemical Corp.) of hen egg whites were irradiated at 2O°C with an increasing dose of 1.5-MeV electron beams and subsequently assayed for enzyme activity [Fluke, 1966]. In the plot of Figure 3, the ordinate represents the fractional survival activities in logarithmic scale and the abscissa represents radiation dosage in Mrads in linear scale. The straight line represents the linear regression calculated via the least squares method. A D_0 of 42.4 Mrads was obtained from the reciprocal slope of this semilog plot. This D_0 value and Equation 6 gave a Mr of 15,100, which is slightly larger than the value of 14,600 obtained by conventional biochemical means.

Exponential Survival Curves With a Shoulder

Although radiation inactivation of most biological macromolecules studied follow the exponential survival-dose relationship described above, exceptions to this simple situation can occur in theory in target-size measurements [Fowler, 1964]. One example is where the log survival-versus-dose plot shows a shoulder in the low dose range [Andrews, 1974]. This is known as a sigmoidal curve because of the shape it assumes in a linear plot. This particular mode of dose-response curve is rather common in the radiation-induced killing of mammalian cells in culture, but little data exist in which this mode operates in the inactivation of well-defined molecular systems such as enzymes and receptors. This mode in the semilog plot is illustrated in Figure 4. Up to a certain radiation dose there is only a slight decrease in the surviving fractions, whereas at higher doses the number decreases exponentially. A linear regression line may be ascertained describing the exponential fall of the survival fraction at high doses, from which the slope, R, and its Y-intercept, N_e (extrapolation number), can be obtained. Some theoretical treatment of this mode of dose-effect curve is available in literature [Oliver and Shepstone, 1964]. A more general expression of the classical target theory follows:

$$\frac{A}{A_0} = 1 - \left\{1 - e^{-D/D_0} \sum_{n=0}^{n-1} \frac{(D/D_0)^{n-1}}{(n-1)!}\right\}^m, \quad (7)$$

where A is the activity surviving a dose D, A_0 is the activity at no dose (control), and D_0 is the dose causing a single event in the target molecule as defined previously. In this treatment, it is assumed that the biological unit in concern consists of m targets of equal volume, and that each of these m targets must receive n hits to inactivate the unit (multiple target-multiple hit model). In a special case where only one hit is required in each of m targets to inactivate the unit (n = 1), Equation 7 is reduced to the form

$$\frac{A}{A_0} = 1 - \left\{1 - e^{-D/D_0}\right\}^m. \quad (8)$$

This is known as the single-hit, multiple target model. Here, the semilog plot gives a straight line at high doses, m coincides with N_e (the extrapolation number), and R is equal to $1/D_0$. Another special case is where a sin-

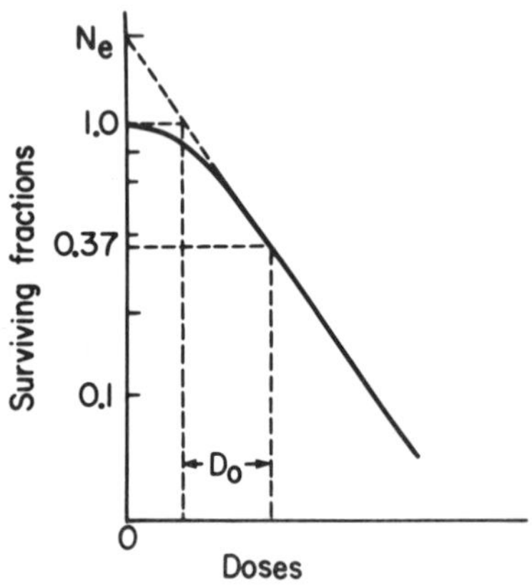

Fig. 4. *Illustration of a typical exponential survival curve with a shoulder.*

gle target in a given unit requires n hits for inactivation. In this case (m = 1), Equation 7 reduces to

$$\frac{A}{A_0} = e^{-D/D_0} \sum_{n=0}^{n-1} \frac{(D/D_0')^{n-1}}{(n-1)!}, \quad (9)$$

where D_0' is the D_0 for the special case of the single hit (n = 1) of this model. In this case, there is no limiting slope at high doses in the semilog plot, and the relationships between n and N_e, and R and D_0 become rather complex [Oliver and Shepstone, 1964].

In applying these models for analysis of dose-effect curves showing a distinct shoulder in the semilog plot, one may first compute the theoretical curves of $\log(A/A_0)$ as a function of D/D_0 from Equations 8 and 9 for a variety of values of m and n, respectively. Once this is done, one can superimpose the experimental survival data on the theoretical curves, adopting a range of numerical values for D_0 bracketing $-0.567/R$. This allows one to determine the value of D_0 that gives the best fit of data to a theoretical curve. The molecular weight can then be calculated using this value of D in Mrads according to Equation 5. One example of such analyses is shown in Figure 5. Irradiation of human erythrocytes with an increasing dose of electron beam abolishes hypertonic cryohemolysis, a type of lysis induced by cold temperatures and high tonicity [Green et al, 1981]. The semilog plot for radiation inactivation of cryohemolysis shows a distinct shoulder followed by a steep, almost linear fall at high doses. Analysis of the steep fall of the curve gives a R value of -5.35 Mrads^{-1} and an E value of 2×10^2. For the multiple-target model, this would imply that each of the 200 individual-targets with a molecular weight of 8×10^6 would need to be hit for inactivation. On the other hand, for the multiple-hits model, this same set of parameters implies that the inactivation involves a single target with a molecular weight of approximately 15×10^6 which would receive 10 to 12 hits. It should be noted that neither of these models precisely simulates the behavior of the observed survival data, and it is rather arbitrary which value of molecular weight is to be taken. Nevertheless, it may be safe to conclude that a macromolecular structure of a size commensurate with a molecular weight of 8–15 $\times$ 10^6 is involved in this hypertonic hemolysis. The presence of large oligomeric species (up to 11 dimers) of cytoskeletal proteins, spectrins, in erythrocytes is well known. Based on independent evidence this spectrin structure has been implicated in hypertonic cryohemolysis [Green et al, 1981], suggesting that the target size obtained represents such a spectrin network.

SPECIFIC COMMENTS

Temperature Effects on Target-Size Measurement

Although it is not theoretically predicted, the action of ionizing radiation on most biological systems show a great sensitivity to temperature during irradiation [Setlow and Doyle, 1953; Fluke, 1966, 1972]. The inactivation due to a given dose of radiation is low at low temperatures and increases with increasing temperatures. This temperature effect is observed with the samples in frozen

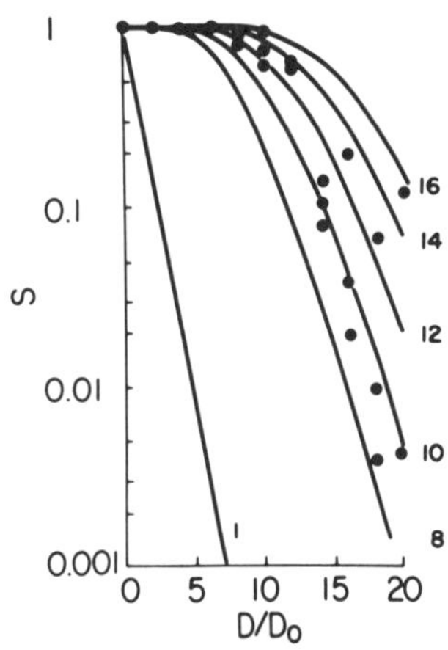

Fig. 5. *Hypertonic cryohemolysis of human erythrocytes after irradiation with a 1.5-MeV electron beam at varying doses [Green et al, 1981].*

and dry states, the conditions where radiation inactivations are measured for the target size analysis. The temperature effects for many well-characterized enzymes show three components in Arrhenius analysis, one of which is temperature independent, while the other two have activation energies of approximately 1 kcal/mol and 4–6 kcal/mol, respectively. Which particular molecular processes of radiation inactivation are associated with these activation energies are not well understood at present.

The primary concern in this chapter with regard to temperature effect is its effect on target-size measurements. Theoretically, in calculations of target size, molecular weight based on Equations 1 and 2, the only parameter which could be significantly temperature dependent is I, the number of ionizing events in a unit volume. Q, the energy per an inactivating event or primary ionization, and D, the total energy absorbed per gram of sample, in turn, determine the parameter I. (See Equation 4; ρ is relatively temperature insensitive in practical situation.) The parameter D, the total energy of radiation absorbed per gram of sample, as defined above, is expected to be temperature independent, and it is most likely the parameter Q that is primarily involved in the temperature effect. Equation 6, the most widely used relationship of molecular weight to radiation sensitivity, is derived empirically based on the inactivation data obtained at ambient temperature [Kepner and Macey, 1968] and implicitly assumes that Q is equal to 66 eV at this temperature. However, the target-size molecular weight determination of some membrane-bound proteins is possible only while irradiation at the frozen state (see above). Kempner and Schlegel found that the radiation sensitivity at $-130°C$ relative to that at $+30°C$ is similar for many different enzymes, which is estimated to be 2.57 ± 0.29. These analyses were further extended to show that the same quantitative relationship holds for 25 different enzymes. This correction factor has been used successfully in target size molecular weight calculation from the irradiation data obtained at $-130°C$ [Kempner and Schlegel, 1979]. In these experiments, radiation doses were measured using thermoluminescent dosimeters which are known to be independent of temperature.

Kempner and Schlegel's temperature correction method discussed above requires measurement of absolute radiation dose independent of temperature. Thermoluminescent dosimetry is shown to satisfy this requirement. Many chemical dosimetries, on the other hand, are shown to be sensitive to irradiation temperature (see above). Exact modes of the temperature dependency reported in literature vary with each chemical system. One well-characterized chemical system for its temperature dependence is Blue Cellophane, discussed in the preceeding section (Fig. 3). Virtual identity of the temperature dependence of Blue Cellophane with that of five representative enzymes compiled by Kempner and Haigler [1982] strongly suggests that the same molecular and atomic mechanisms are responsible for both biological macromolecules (such as enzymes) and this chemical dosimeter. It follows that if such a dosimetry is done at the same temperature at which the samples are irradiated, the target molecular weight can be calculated from Equation 6 directly without any temperature correction. This method has been used successfully in a number of laboratories [Cuppoletti et al, 1981; Saccomani et al, 1981; McIntyre et al, 1983].

More recently, Lo et al [1982] introduced a practical approach to cope with this temperature problem in target-size measurements. This method bypasses the necessity of determining the absolute radiation dose. This method makes the use of a calibration curve between molecular weight and radiation sensitivity (ie, $1/D_0$) which can be constructed from several marker proteins of known molecular weights. This is based on the fact that, according to target theory, molecular weight is directly proportional to radiation sensitivity. If a series of protein molecules of well-defined molecular weight are available as reference molecules and if relative radiation sensitivity of any unknown protein to the radiation sensitivity of the reference molecule is known, the molecular weight of this unknown can be calculated. In this case, the radiation sensitivity does not need to be expressed in absolute radiation dose (such as rads) but in any convenient, relative unit (such as "number of passes" or "irradiation time").

Target Size and True Molecular Weight

In theory, target size of a protein represents the size of radiation sensitive mass, within which the ionizing energy absorbed at any one point can destroy the function of the protein. In the case of a single polypeptide chain protein (monomeric protein), the target size is expected to be that of the polypeptide folded into a compact three-dimension. However, many proteins are made of two or more peptide chains with no covalent linkages between them. For such oligomeric proteins, the target size could be of subunit peptides or of the entire oligomer if sufficient energy transfer occurs between the subunits. Compilation of more than 30 water-soluble oligomeric enzymes, whose molecular and subunit weights are known, revealed that the target size measured coincides with the size of entire oligomer in half of the cases and that of single subunits for the rest [Blum and Alper, 1971; Kempner and Schlegel, 1979]. For some tetrameric enzymes such as β-fructofuranosidase [Darnell and Klotz, 1975], the target size [Pollard et al, 1952] corresponded to the size of two subunits rather than a single subunit, suggesting that the molecular interaction between four subunits is not equal.

It is conceivable that two or more functional units of single or oligomeric polypeptides may occur in cell membrane matrix as a cluster or complex. In such a fortuitous assembly, the member units stick together very tightly so that significant energy transfer can occur between them, and the target size may reflect the entire assembly rather than cach minimal functional unit. It is known for the Na^+-K^+-ATPase of human erythrocytes that the target size in intact cells is 610,000 daltons, whereas that in isolated ghosts is 310,000 daltons (Hah and Jung, unpublished observation). This observation is most readily understood if one assumes that in intact cells the enzyme occurs as a fortuitous assembly of two functional units which dissociate during membrane isolation procedure. Alternatively, the large target size of the intact cell system may reflect putative supramolecular assembly which consists of one functional enzyme in complex with a particular set of glycolytic enzymes [Shrier et al, 1966]. The target-size molecular weight thus may not always give the size of the *minimal* assembly of structures required for the measured function but simply in situ assembly.

Target sizes smaller than the component peptide chain have been observed in a few instances. The receptor for immunoglobulin E (IgE) of rat basophile is made of two noncovalently attached polypeptides: a 50,000-dalton α-chain and a 30,000-dalton β-chain. The α-chain contains the binding site for IgE and has approximately 30% carbohydrate. α-Chains can bind IgE in the absence of β-chains. The target size of this receptor [Fewtrell et al, 1981] was found to be only 28,000 and 38,000 daltons in the presence and the absence of detergent, respectively. Another example of this complication is the (Na^+K^+)ATPase of pig kidney [Richards et al, 1981]. The target size of the enzyme measured based on the Rb^+ occlusion ability and p-nitrophenylphosphatase activity are approximately 39,000 and 164,000 daltons, respectively. If (Na^+K^+)ATPase activity was used instead, the target size of this enzyme was approximately 300,000 daltons [Kepner and Macey, 1968; Ellory et al, 1979; Peters et al, 1981]. Based on extensive biochemical studies, this enzyme is known to be made of two copies of a 120,000-dalton α-peptide and two copies of a 40,000–60,000-dalton (protein part only) β-glycopeptide with the total molecular weight of approximately 300,000 daltons. The target size of the ATPase activity thus corresponds to that of $\alpha_2\beta_2$, whereas the target size of the Rb^+ occulation is grossly smaller than α or β. With water-soluble proteins for which a large body of information is available, no target size which is significantly smaller than the size of subunits has been reported, and this type of complication may be unique to membrane-associated proteins. It is conceivable that the lipid bilayer may stabilize the intrinsic membrane protein conformation even after the protein is damaged by ionization, thus allow it to carry out its partial function.

Target-Size Analysis on Complex Systems

The modes of survival-radiation dose relationship discussed above are those developed from studies on simple, homogeneous, noninteracting systems in which any one individual member is independent of events in other members. Many membrane-associated biological functions, however, involve a macromolecular assembly or oligomeric structure consisting of two or more mutually interacting molecular species. A well-known example is adenylate cyclase system. This system includes at least three interacting macromolecules, namely, hormone receptor (R), GTP-binding protein (G), and adenylate cyclase catalytic unit (C), which exist in various assembly states in equilibria where a certain assembly state dictates function. Hormones (such as glucagon) binding to the receptor, induce a shift in the equilibria of molecular assembly and modified function. Thus, the adenylate cyclase activity as measured in the presence of hormone is the consequence of the interactions between these three components and is thus dependent on the concentration of these components. As all of these component proteins are inactivated by radiation depending on their respective molecular size, irradiation would induce shifts in equilibria in interactions and thus assembly states. The survival curves of such a mutually interacting system are expected to be complex, and little theoretical treatment is currently available.

Target-size analysis [Houslay et al, 1977; Martin et al, 1979; Schlegel et al, 1979; Nielsen et al, 1981] applied to such a complex system is at best arbitrary because of the high degree of freedom, and often leads to overinterpretation even on the basis of a simple model [Swillens and Dumont, 1981]. This has been well exemplified in the theoretical analysis [Simon et al, 1982] of a simple equilibrium system consisting of a catalytic unit, C, and one regulatory unit, R, where R and C interact reversibly to form the complex, RC with an equilibrium-dissociation constant; K_d and only RC expresses enzymatic activity. Even in this rather simple model, the target-size calculation is sensitively affected by equilibrium parameters, namely, the ratio of K_d to the total concentration of C, the ratio of the total concentration of R to that of C, and the relative sizes (molecular weights) of R and C. Furthermore, transfer of radiation energy between the subunits of the complex RC may or may not occur, and this affects the calculated target size. Fortunately, information for the equilibrium parameters can be obtained experimentally and can be used for the target size analysis. However, the semilog plot of the survival-dose data of such a system is shown to be nonlinear, and the target-size analysis requires nonlinear regression analysis of data covering inactivation of at least two or three common log units. If insufficient inactivation data were used, the nonlinearity might not be obvious, and the data could be fitted to a straight line which would falsely give a large target size.

Another interesting mode of the survival-to-dose relationship that has been reported is the one where activation, rather than inactivation, was indicated at low doses of radiation. This mode shows an initial hump rather than a shoulder on the usual log survival-dose plot, which is followed by a simple exponential decay. This type of radiation inactivation was observed with the specific binding of insulin to its receptors in hepatocytes [Harmon et al, 1980] and also with the calmodulin-independent phosphodiesterase activity of bovine brain [Kincaid et al, 1981]. In the latter case, the data were analyzed based on the assumption that phosphodiesterase exists in two forms, enzymatically active monomers and inactive dimers. This analysis further involves less realistic assumptions that there is no equilibration between monomeric and dimeric species and that the only effect of radiation on dimers is its dissociation to generate two monomers without inactivating their enzymatic activity. For the case of insulin receptor studies, it was assumed that the system included two molecular species, an insulin receptor and an inhibitor where the binding of the inhibitor molecule to receptor reduces the affinity of the receptor to insulin. Here, again, the analysis was simplified by further assuming that there is no dynamic equilibration among inhibitors, receptors, and their complexes. It seems quite desirable in both of these cases to work out analyses which account for the equilibria of mutually interacting molecular species.

REFERENCES

Andrews HL (1974): "Radiation Biophysics." Englewood Cliffs, New Jersey: Prentice Hall Inc.

Augenstein LG, Brustad T, Mason R (1964): The relative roles of ionization and excitation processes in the radiation inactivation of enzymes. Adv Radiat Biol 1:228–266.

Beauregard G, Potier M (1982): Radiation inactivation of enzymes at lower dose rates: Identical molecular weights of rat liver cytosolic and lysosomal neurominidases. Anal Biochem 122:379–384.

Blum E, Alpher T (1971): Radiation target molecular weights of urease and L-glutamate dehydrogenase and their relevance to the size of the functional subunits. Biochem J 122:677–680.

Crowther JA (1924): Actions of X-rays on tissue cells. Proc R Soc Lond [Biol] 96:207–211.

Cuppoletti J, Jung CY, Green FA (1981): Glucose transport carrier of human erythrocytes. J Biol Chem 256:1305–1306.

Darnell DW, Klotz IM (1975): Subunit constitution of proteins: A Table. Biochim Biophys Acta 166:651–682.

Dartinger H, Jung H (1970): "Molecular Radiation Biology. The Action of Ionizing Radiation on Elementary Biological Objects." New York: Springer-Verlag.

Ellory JC, Green JR, Jarvis SM, Young JD (1979): Measurement of the apparent molecular volume of membrane-bound transport systems by radiation inactivation. J Physiol 295:10–11.

Fewtrell C, Kempner E, Poy G, Metzger H (1981): Unexpected findings from target analysis of immunoglobulin E and its receptor. Biochemistry 20:6589–6594.

Fluke DJ (1966): Temperature dependence of ionizing radiation effect on dry lysozyme and ribonuclease. Radiat Res 28:677–693.

Fluke DJ (1972): Temperature dependence of the direct action of ionizing radiation on beef heart lactate dehydrogenase: Enzyme activity, substrate and coenzyme affinities. Radiat Res 51:56–71.

Folkerd EJ, Ellory JC, Hughes-Jones NC (1977): A molecular size determination of Rh(D) antigen by radiation inactivation. Immunochemistry 14:529–531.

Fowler JF (1964): Differences in survival curve shapes for formal multi-target and multi-hit models. Phys Med Biol 9:177–188.

Fraser CM, Venter JC (1982): The size of the mammalian lung β_2-adrenergic receptor as determined by target size analysis and immunoaffinity chromatography. Biochem Biophys Res Commun 109:21–29.

Fricke H, Morse S (1927): Action of X-rays on ferrous sulphate. Am J Roentg 18:426–430.

Green FA, Jung CY, Cuppoletti J, Owens N (1981): Hypertonic cryohemolysis and the cytoskeletal system. Biochim Biophys Acta 648:225–230.

Green FA, Owen NA, Hui HL, Jung CY, Cuppoletti J (1983): Molecular size of the Rho(D) antigen of the human erythrocyte in situ by radiation inactivation. Mol Immunol 20:361–365.

Harmon JT, Kahn CR, Kempner ES, Schlegel W (1980): Characterization of the insulin receptor in its membrane environment by radiation inactivation. J Biol Chem 255:3412–3419.

Harmon JT, Kempner ES, Kahn CR (1981): Demonstration by radiation inactivation that insulin alters the structure of the insulin receptor in rat liver membranes. J Biol Chem 256:7719–7722.

Henley EJ (1954): Gamma ray dosimetry with cellophane-dye systems. Nucleonics 12:62–63.

Houslay MD, Ellory JC, Smith GA, Hesketh TR, Stein JM, Warren GB, Metcalfe JC (1977): Exchange of partners in glucagon receptor-adenylate cyclase complexes. Physical evidence for the independent, mobile receptor model. Biochim Biophys Acta 467:208–219.

Hutchinson F (1962): Radiation effects on macromolecules of biological importance. In Segre E (ed): "Annual Rev Nucl Sci," Vol 12. Palo Alto, CA: Annual Rev. Inc., pp 535–563.

Hutchinson F, Pollard E (1961): Target theory and radiation effects on biological molecules. In Errera M, Forssberg (eds): "Mechanism in Radiology," Vol 1. New York: Academic Press, p 73.

Jung CY, Hsu TL, Hah JS, Cha C, Haas MN (1980): Glucose transport carrier of human erythrocytes. J Biol Chem 255:361–364.

Kempner ES, Haigler HT (1982): The influence of low temperature on the radiation sensitivity of enzymes. J Biol Chem 257:13297–13299.

Kempner ES, Schlegel W (1979): Size determination of enzymes by radiation inactivation. Anal Biochem 92:2–10.

Kepner GR, Macey RL (1968): Membrane enzyme systems molecular size determination by radiation inactivation. Biochim Biophys Acta 163:188–203.

Kincaid RL, Kempner E, Manganiello VC, Osborne, Jr. JC, Vaughan M (1981): Calmodulin-activated cyclic nucleotide phosphodiesterase from brain. J Biol Chem 256:11351–11355.

Lea DE (1946): "Action of Radiations on Living Cells." London and New York: Cambridge University Press.

Levine H, McLaughlin WL, Miller A (1979): Temperature and humidity effects on the gamma ray response and stability of plastic and dyed plastic dosimeters. Radiat Phys Chem 14:551–574.

Lilly L, Fraser CM, Jung CY, Seeman P, Venter JC (1983): Molecular size of the canine and human brain D2 dopamine receptor as determined by radiation inactivation. Mol Pharmacol 23:611–617.

Lo MMS, Barnard EA, Dolly JO (1982): Size of acetylcholine receptors in the membrane. An improved version of the radiation inactivation method. Biochemistry 21:2210–2217.

Marshall M, Holt PD, Gibson JAB (1970): Physical parameters of radiation determined from the inactivation of ribonuclease. Int J Radiat Biol 18:139–146.

Martin BR, Stein JM, Kennedy EL, Doberska CA, Metcalfe JC (1979): Transient complexes. A new structural model for the activation of adenylate cyclase by hormone receptors. Biochem J 184:253–260.

McIntyre JO, Churchill P, Maurer A, Berenski CJ, Jung CY (1983): Target size of D-β-hydroxybutyrate dehydrogenase. J Biol Chem 258:953–959.

Miller A, Bjergbakke E, McLaughlin W (1975): Some limitations in the use of plastic and dyed plastic dosimeters. Int J Appl Radiat Isotopes 26:611–620.

Mullins LJ (1981): "Annual Reviews Reprints: Cell Membranes, 1978–1980." Palo Alto, CA: Annual Reviews Inc.

Nielsen TB, Lad PM, Preston MS, Kempner E, Schlegel W, Rodbell M (1981): Structure of the turkey erythrocyte adenylate cyclase system. Proc Natl Acad Sci USA 78:722–726.

Oliver R, Shepstone BJ (1964): Some practical considerations in determining the parameters for multi-target and multi-hit survival curves. Phys Med Biol 9:167–175.

Peters WHM, De Pont JJHHM, Koppers A, Bonting SL (1981): Studies on (Na^+ and K^+)-activated ATPase. Biochim Biophys Acta 641:55–70.

Pollard EC (1959): Rev Mod Phys 31:273.

Pollard E (1969): Radiation inactivation of enzymes, nucleic acids and phage particles. In Oncley JL (ed): "Biophysical Science—A Study Program." New York: Wiley.

Pollard EC, Guild WR, Hutchinson F, Setlow RB (1955): The direct action of ionizing radiation on enzymes and antigens. "Prog Biophys Biophys Chem." Vol 5. Long Island City, New York: Pergamon, pp 72–108.

Pollard EC, Powell WF, Reaume SH (1952): The physical inactivation of invertase. Proc Natl Acad Sci USA 38:166–172.

Richards DE, Ellory JC, Glynn IM (1981): Radiation inactivation of (Na^+-K^+)-ATPase. A small target size for the K^+ occluding mechanism. Biochim Biophys Acta 648:284–286.

Saccomani G, Sachs G, Cuppoletti J, Jung CY (1981): Target molecular weight of the gastric (H^+ + K^+)-ATPase functional and structural size. J Biol Chem 256:7727–7729.

Schlegel W, Kempner ES, Rodbell M (1979): Activation of adenylate cyclase in hepatic membranes involves interactions of the catalytic unit with multimeric complexes of regulatory proteins. J Biol Chem 254:5168–5176.

Schrier SL, Doak LS, Carr B (1966): Organization of enzymes in human erythrocyte membranes. Am J Physiol 210:139–145.

Setlow R, Doyle B (1953): The effect of temperature on the direct action of ionizing radiation of catalase. Arch Biochem Biophys 46:46–52.

Simon P, Swillens S, Dumont JE (1982): Size determination of an equilibrium enzymic system by radiation inactivation. Biochem J 205:477–483.

Steer CJ, Kempner ES, Ashwell G (1981): Molecular size of the hepatic receptor for asialoglycoproteins determined *in situ* by radiation inactivation. J Biol Chem 256:5851–5856.

Swillens S, Dumont JE (1981): A pitfall in the interpretation of data on adenylate cyclase inactivation by irradiation. FEBS Lett 134(1):29–31.

Venter JC, Fraser CM, Schaber JS, Jung CY, Bolger G, Triggle DJ (1983): Molecular properties of the slow inward calcium channel: Molecular weight determinations by radiation inactivation and covalent affinity labelling. J Biol Chem 258:9344–9348.

Venter JC (1983): Muscarinic cholinergic receptor structure. I. Receptor size, membrane orientation and absence of major phylogenetic structural diversity. J Biol Chem 258:4842–4848.

Weiss J, Allen AO, Schwarz HA (1956): In "Proceedings of the International Congress on Peaceful Uses of Atomic Energy, Geneva, 1955," Vol 14. New York: Columbia University Press pp 179–186.

Molecular and Chemical Characterization of Membrane Receptors, pages 209–239

9

Reconstitution of Membrane Receptors

Richard D. Klausner, Jos van Renswoude, Robert Blumenthal, and Benjamin Rivnay

Laboratory of Biochemistry and Metabolism (R.D.K., J.R.) and Arthritis and Rheumatism Branch (B.R.), National Institute of Arthritis, Diabetes, and Digestive and Kidney Diseases; and Laboratory of Theoretical Biology, National Cancer Institute (R.B.), National Institutes of Health, Bethesda, Maryland 20205

INTRODUCTION

It has long been recognized that receptors play a key role in conveying molecular messages from the outside world to the interior of the cell. Our knowledge of receptor physiology has grown considerably through the development of procedures in which receptors are taken out of their natural habitat (eg, a biological membrane) and put back into an artificial or semiartificial membrane environment. This process is generally known as receptor reconstitution. The most conspicuous advantage of reconstitution is undoubtedly that the receptor can be studied in a relatively defined environment. In this way, reconstitution experiments have led to some understanding of the molecular relationship between a receptor and the membrane in which it resides, eg, how receptor function depends on membrane lipid composition, and how a functioning receptor is positioned with respect to the lipid backbone of the membrane.

In this chapter we will first discuss, in detail, the general methods of receptor reconstitution. Next we will examine some of the physical-chemical principles involved in detergent solubilization of receptors from biological membranes. We will pay particular attention to the role of lipid and detergent in attaining successful reconstitution, and discuss the nature of the vesicular structures that are formed in reconstitution. These issues will be presented with particular reference to the two most well studied receptors, those for acetylcholine and immunoglobulin E (IgE). The next section deals with the various techniques employed in reconstitution of other receptors including those for asialoglycoproteins, insulin, β-adrenergic ligands (β-sympathomimetics) and thyrotropin. The chapter is closed by a review on the use of planar lipid bilayers in receptor reconstitution.

GENERAL METHODS OF MEMBRANE PROTEIN RECONSTITUTION

There are many methods that can be used to reconstitute receptors. For each general method, virtually all experimental parameters can and have been varied to yield a functioning reconstituted membrane. Underlying a gradually emerging science of reconstitution is an apparent caprice and serendipity that forces the experimenter to try a seemingly inexhaustible list of variations to find the recipe that works. Despite this there are general approaches to reconstitution and some general rules to follow. Before examining specific examples of receptor reconstitution, we will briefly explore some of these general approaches. It is important to point out that the techniques for receptor reconstitution were developed for nonreceptor integral membrane proteins, especially membrane-bound enzymes. They, however, have provided the bases for reconstitution techniques for a variety of membrane proteins [Racker, 1974].

Detergent Removal

Most reconstitutions begin with the isolation of integral membrane proteins from their native membrane environment. This most frequently (but not always) requires detergent solubilization of the membrane. The membrane proteins are now present as mixed micelles of protein, detergent, and perhaps some of the original membrane lipid. It is generally advantageous to purify the protein to be reconstituted, but this is by no means required. In general, a suitable source for the particular protein is a cell whose membranes are rich in that protein. An example is the electric organ of *Torpedo californica*, which is rich in acetylcholine receptors. Fractionation of the membrane or membranes containing the protein is the recommended step before detergent solubilization. Solubilization presents the first major stumbling block. It is important that the detergent solubilizes the protein without irreversibly denaturing it. For this reason the highly denaturing detergent sodium dodecyl sulfate (SDS), although an excellent solubilizer, is rarely useful for reconstitution [MacLennan, et al, 1976]. The most popular detergents are cholate, deoxycholate, octyl glucoside, and the new detergent CHAPS 3- [3′-cholamidopropyl] dimethylammonio - 1-propane-sulfonate). Triton X-100 is a mild detergent that is effective at solubilizing membrane proteins. It is generally avoided because it is poorly suited for the next step in reconstitution: detergent removal. Often detergents are used in combination. For example, a mixture of octyl glucoside with cholate was useful for the reconstitution of the α-aminobutyric acid transporter [Kanner, 1970]. In certain systems optimal reconstitution was achieved after solubilization in a mixture of cholate and deoxycholate. The Ca^{++} pump from sarcoplasmic reticulum required this mixture to display pump activity in the reconstituted state [Knowles and Racker, 1974].

The next step in this most common reconstitution procedure is the addition of phospholipid, which will be readily solubilized by the detergent, with the resultant formation of mixed lipid-protein-detergent micelles. The exact point at which phospholipid is added can be varied. It may be present during the initial detergent solubilization of the protein. Alternatively the phospholipid may be dried out of a solvent onto the walls of a glass vessel and then combined with the detergent solubilized protein [Kagawa and Racker, 1971].

Although one should avoid "denaturing" detergents such as SDS, it is important to remember that any detergent may denature particular membrane proteins. It is a good rule of thumb to minimize the amount of time that the protein is maintained in a detergent suspension [Froman et al, 1980]. Various maneuvers have proven successful in protecting proteins against detergent denaturation. The presence of succinate during the reconstitution of succinate dehydrogenase was found to be required for an active reconstituted enzyme, presumably because the substrate protects the enzyme from denaturation during solubilization [King, 1963]. Certain enzymes can be protected from detergent denaturation through the continuous presence of phospholipids throughout the solubilization procedure. These phospholipids may be required to be bound to the membrane proteins to protect them from detergent denaturation as was found for the *Escherichia coli* lactose transport system [Newman and Wilson, 1980]. A number of receptors have been shown to exhibit normal binding activity when in a detergent-solubilized form. These include the insulin receptor, the low-density lipoprotein (LDL) receptor, the asialoglycoprotein receptor, and the transferrin receptor. The knowledge that a particular detergent does not interfere with binding activity can be used in choosing the type of detergent from which to reconstitute the receptor into a phospholipid membrane.

After solubilization with detergent, reconstitution is achieved by removal of the detergent in the presence of phospholipid (Fig. 1). As the detergent is removed two events occur: (1) The lipid is no longer solubilized in detergent micelles and spontaneously forms vesicles in which it takes the form of a bilayer; and (2) the protein no longer finds itself in detergent micelles but rather inserts in the nascent lipid bilayer. The exact molecular details of this process are unknown. In other words, whether a mixed proteoliposome forms out of a lipid-protein-detergent micelle or whether a pure lipid vesicle forms first with subsequent insertion of the protein, is unclear. Recent evidence from some systems points to the latter process (see below). In either event the removal of detergent leads to successful reconstitution in a great number of systems. This technique, with cholate as the detergent, was first used by Racker for the successful reconstitution of the enzymes of oxidative phosphorylation [Kagawa et al, 1973; Kagawa and Racker, 1971]. In these pioneering studies, phospholipid was mixed with sodium cholate by sonication. Next, the isolated protein in cholate (about 2%) was added and the mixture dialyzed against buffer for 20 h. The removal of cholate led to the formation of proteoliposones containing the protein(s) in a reconstituted form. Cholate dialysis remains one of the most common and useful techniques for reconstitution. Cholate is readily dialyzed out of the system because of its high critical micelle concentration. This fact is essential to the success of the technique. Deoxycholate and octylglucoside share this property with cholate and are therefore useful agents for reconstitution. Triton X-100, on the other hand, forms large micelles and has a low critical micelle concentration [Tanford and Reynolds, 1976]; it is very difficult to remove by simple dialysis. For this reason it is much less useful as a detergent for reconstitution. Recently the applicablility of Triton X-100 in this respect has been greatly improved by the finding that certain resins such as SM-2 biobeads [Holloway, 1973] can remove Triton X-100 across dialysis membranes.

The detergent-removal technique has been expanded to include modifications in the way the detergent is removed. First of all, the speed of removal can be critical to the success of the reconstitution. Because of the rule which tells us to minimize the amount of time that the protein is exposed to detergent, attempts have been made to find the minimum time required for removal of the detergent by

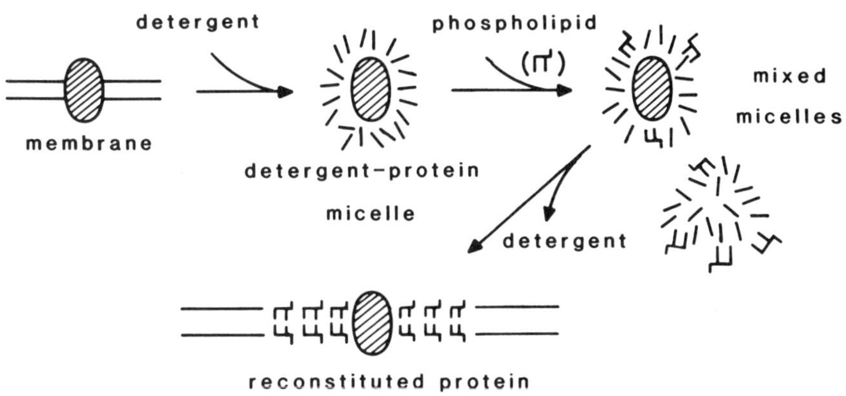

Fig. 1. *Schematic representation of membrane protein solubilization and reconstitution.*

dialysis. Using a hollow-fiber dialysis apparatus, efficient and rapid removal of detergent has allowed for successful reconstitution [Goldin, 1977]. Nevertheless, certain proteins can only be properly reconstituted by prolonged dialysis. An example is the sodium pump, which requires four days of gradual dialysis [Goldin and Tong, 1974]. If, in the case of bacteriorhodopsin solubilized in octylglucoside, the detergent was too rapidly removed in the presence of phospholipid, pure lipid vesicles would be formed, and aggregated, nonreconstituted bacteriorhodopsin would settle out of solution [Klausner et al, 1980]. Dialysis is not the only technique for removing detergent. Cholate removal can be achieved by passing the lipid-protein-detergent mixture over a Sephadex column. Proteoliposomes are formed as the mixture passes through the column; they separate from and elute before the detergent [Padan et al, 1979]. Cholate can also be preferentially absorbed to Sephasorb beads and removed from a mixed micelle, leaving lipid and protein to rearrange into a reconstituted proteoliposome [Rogner et al, 1979].

In a still more marked variation on this technique, actual removal of the detergent is not required; it is sufficient to merely reduce the detergent concentration far enough below the critical micelle concentration (CMC) to allow the formation of proteoliposomes. This was first observed by Racker in the reconstitution of cytochrome oxidase [Racker, 1972]. Cholate, again, has been the most widely used detergent for this procedure. The initial concentration of cholate in the lipid-protein-detergent mixture is lower than in the usual detergent dialysis scheme. In this case it is generally between 0.5% and 0.8%. The mixture is incubated for a short time (less than 30 min) and then diluted 20-fold or more. This procedure has resulted in the successful reconstitution of a number of membrane proteins [Wickner, 1976]. The striking advantage of this technique is its speed. This not only makes this procedure more convenient, it may enhance the likeliness of success as well, because it minimizes the time that the protein spends in the presence of detergent. Perhaps for this reason the dilution technique may be applied successfully where the dialysis technique has failed. This is the case for the DCCD-sensitive ATPase from spinach chloroplast [Winget et al, 1977]. On the other hand, several systems have been reported in which the dialysis technique is successful while the dilution technique is not. Unfortunately the reasons for success or failure are, in most cases, unclear.

Sonication and Freeze-Thawing

Another class of techniques for reconstitution involves mechanical means of forming proteoliposomes. The most common of these approaches is sonication. The advantage of this procedure is that it provides a way to avoid detergent. Of course, the membrane protein in question should withstand removal of detergent without denaturation or irreversible aggregation. Such detergent-free preparations of membrane proteins have been produced for a number of proteins such as the Ca^{+}-ATPase from sarcoplasmic reticulum [Racker, 1973]. In this procedure the detergent-free protein is mixed with an aqueous dispersion of lipids, and the mixture is subjected to sonic oscillation. This leads to proteoliposome formation. The technique is rapid and has been successfully applied in the case of certain proteins that could not be reconstituted via detergent removal techniques. Its problems are twofold: First and foremost is that many membrane proteins are very sensitive to sonication; they are destroyed by even short treatments at low energy levels [Tanford and Reynolds, 1976]. The damage can be minimized by attempting to sonicate for as short time periods as possible [Racker, 1979]. One must titrate each protein to determine the sensitivity to sonic denaturation. The second drawback to sonication is the "apparent" irreproducibility of results. This arises out of the nature of the sonication itself. The exact energy input into a solution depends upon the size, shape, and composition of the vessel;

the volume of the solution; the instrument; and the energy output. Temperature control is important, especially when probe rather than bath sonicators are used. If sonication is to result in reproducible preparations, great care must be taken to ensure exact and reproducible experimental conditions.

A recent modification of the sonication technique was developed for the erythrocyte glucose carrier [Kasahara and Hinkle, 1976]. In this study the detergent-free protein and preformed liposomes were mixed and rapidly frozen. After thawing, the mixture was sonicated, and successfully reconstituted proteoliposomes were formed. The subsequent sonication time was greatly reduced by the initial freeze-thawing. For example, the sonication time for the glucose transporter is only 20 s. This technique then allows for minimal sonic denaturation as well as minimal detergent denaturation. If small amounts of solvents are present in the lipid-protein mixture, the sonication time can be reduced even more.

Incorporation into Preformed Vesicles

Another technique for reconstitution involves the direct incorporation of membrane proteins into preformed lipid vesicles. In some circumstances this will occur in the absence of any detergent or detergentlike catalyst. For example, cytochrome b5 can be inserted into preformed liposomes in the absence of any detergent. Recently we have described the spontaneous insertion of the hepatic asialoglycoprotein receptor into preformed unilamellar liposomes [Klausner et al, 1980]. It is frequently observed that this technique requires specific lipids and a specific lipid composition to be successful. For example, the presence of both phosphatidylethanolamine and an acidic lipid such as cardiolipin is often required. The use of phosphatidylinositol and phosphotidylserine is also frequently of help. Why is this? These lipids, in contrast to phosphatidylcholine, tend to destabilize bilayers. Phosphatidylethanolamine tends to form inverted structures [DeKruijff et al, 1979]. It is not surprising that the presence of these lipids in pure lipid vesicles results in vesicle-vesicle fusion [Verkleij et al, 1979]. Most likely, defects in the structure of the bilayer are a prerequisite for insertion of the protein. The lipids mentioned above probably introduce such defects. If the liposomes used for spontaneous insertion are highly curved (ie, very small) they are more likely candidates for insertion than are larger vesicles. Lipid bilayers tend to be less stable when they are highly curved. A variation on this technique includes the addition of very low concentrations of detergents such as cholate (0.1% or less), octylglucoside, or lysolecithin. Once again, these agents promote vesicle-vesicle fusion. It is likely that they may facilitate insertion of proteins by several mechanisms, eg, by destabilizing the membrane, causing structural defects into which the protein inserts. Alternatively, they may induce fusion of vesicles with nonvesicular lipid-protein micelles. Thus the protein in solution would bind lipid and the resulting complex would bind to the surface of the preformed vesicle. The addition of a small amount of detergent would then lead to reorientation and insertion of the absorbed protein into the bilayer.

It is clear that there are advantages and disadvantages to each method of reconstitution. It cannot be stressed enough that it is quite unpredictable which of the techniques may be successfully applied for any given membrane protein. It is probably reasonable to begin any new reconstitution with the detergent dialysis method. This is the most widely used and most frequently successful approach. If this does not result in a satisfactory reconstitution the first question to be asked is whether the protein has been irreversibly denatured. To avoid possible denaturation several options should be tried: (1) addition of phospholipid to stabilize the protein, (2) changing the detergent, (3) minimizing the time in detergent, and (4) altering parameters such as temperature, ionic strength, etc.

GENERAL PRINCIPLES OF RECEPTOR RECONSTITUTION

Introduction

We will now examine the process of reconstitution in more detail. In this examination we will focus on two receptor systems for which the most extensive, detailed, and carefully documented reconstitutions exist: the acetylcholine receptor and the receptor for IgE. The nicotinic acetylcholine receptor is quite well studied [Conti-Tronconi and Raftery, 1982]. The binding of the ligand, the neurotransmitter acetylcholine, initiates the depolarization of the membrane via conformational changes in the receptor that lead to cation channel activation. The IgE receptor is found on mast cells and basophils and mediates the antigen-specific degranulation that is central to immediate allergic response [Metzger and Ishizaka, 1982].

The major problem in reconstitution is to first remove a particular receptor from its native membrane and then to insert it into a new membrane with full biochemical integrity and full biological activity. Any loss of this activity is generally referred to as denaturation. The ultimate goal of reconstitution experiments is to study various aspects of a given protein under controlled conditions. Hence if full activity is ascribed to the preparation in its natural biological membrane, and, assuming that the protein remains stable in its natural milieu, any denaturation that occurs is a result of the reconstitution procedure and due to the conditions which are encountered by the receptor during that procedure. In this section we shall concentrate on these conditions and try to examine the effects of some of the variables on the process of reconstitution.

Criteria for Successful Reconstitution

Denaturation is not an all-or-none modification of protein structure. Rather, it represents a summation of structural perturbations at the level of both the peptide skeleton and the amino acid residues. Several parameters can be evaluated in assessing the degree of protein denaturation that has occurred during reconstitution: (1) the degree of association of the purified protein with the lipid bilayer; (2) subunit dissociation (where relevant), which may include altered interactions with cofactors; (3) binding of ligand(s); and (4) preservation of the physiological functions known for the receptor in its natural environment. Denaturation may also be expressed in impaired feedback or in loss of certain biochemical features that are characteristic of receptors: desensitization, down-regulation, shut-off, or activation. The experimental conditions under which denaturation may take place are different for different proteins. Evaluation of the success of any reconstitution procedure should therefore involve careful testing of specific receptor properties, some parameters of which are listed above.

The Basic Features of Reconstitution by Detergent Removal

In order to study a receptor in an isolated system one needs to isolate the protein of interest and then reinsert it into a lipid bilayer structure. Early reconstitution studies used a procedure comprising two separate steps, of which the first one was solubilization of cells or membranes with a detergent and purification with various techniques such as affinity chromatography, ion exchange chromatography, etc. The reconstitution of these proteins was then achieved by mixing the protein (or extract) with solubilized lipids or preformed lipid vesicles, followed by lowering the detergent concentration to accomplish the assembly of the lipids and protein into artificial membrane vesicles. More recently, it has been appreciated that lipids play an important role in conservation of the native properties of many proteins during reconstitution. And only recently was it shown that the conditions used during the solubilization step will, alone, determine the quality of the reconstituted vesicle preparation. In the following section we will dwell on the role of the detergent, the lipids, and the protein in the formation of active reconstituted vesicles. Specifically, we will examine their role in extracting, in preserving the native function of the receptor, and in the formation of the vesicles.

The Role of Lipid and Detergent in the Reconstitution Procedure

Solubilization. The use of detergents in membrane solubilization is a routine, having been practiced for several decades. Several reviews have been published on the subject (see for example Helenius and Simons [1975]), including Hjelmeland and Chrambach, this series. The basic features of this step will be discussed briefly, with emphasis on recent conceptual advances. Upon adding detergent to biological membranes the detergent initially partitions into the membranes until saturation is reached at a specific detergent:lipid ratio. Beyond this concentration the bilayer structure collapses into lipid-detergent or lipid-protein-detergent mixed micelles. This transformation defines solubilization unequivocally. Empirically, this transformation can in principle be assessed by several methods such as determination of the loss of sedimentable material, by observing clarification of turbidity, or by measuring changes in spectral properties of fluorescent or nuclear magnetic resonance probes [Jackson et al, 1982]. Not all methods have been or can be used in every system. Low-speed sedimentation, for instance, cannot be used with sonicated liposomes since, at low detergent concentrations, membrane components will precipitate whereas small liposomes will not. Often turbidity clarification cannot be used as an indication for solubilization, since mixtures of biological membranes and lipid dispersions prepared from total lipid extracts of biological tissues may remain turbid even at detergent concentrations in which some membrane components are fully soluble [Rivnay and Metzger, 1982]. Detergents will be useful in solubilization only if their hydrophilic/lipophilic balance renders them with higher affinity for bilayers than for their own micelles. It has been indicated [Helenius and Simons, 1975] that with such detergents, solubilization takes place at concentrations twice that of the lipid concentration in the solubilization mixture. Yet most of the data are based on approximate values for lipid concentrations and the ratios are expressed as weight ratios. The critical micelle concentration (CMC) of a detergent, ie, the concentration at which detergent monomers will arrange themselves into micelles, is an important physical-chemical characteristic of a given detergent. At the CMC, detergent will start to accumulate in the membranes [Mimms et al, 1981]. Conceivably, however, at increased salt concentrations or in more complex mixtures of several amphiphiles, the effective CMC (the concentration of detergent existing as monomers at the given conditions) may be considerably lower than it would be in a pure solvent [Mimms et al, 1981]. Determination of the effective CMC (CMC_{eff}) may be useful in solubilization under various experimental conditions. Only recently [Jackson et al, 1982; Jackson and Litman 1982; Rivnay and Metzger, 1982], both the lipid concentrations and the CMC of the employed detergent were combined to produce useful parameters: R_{eff} = (detergent −0.022)/(phosphatidylcholine) [Jackson et al, 1982; Jackson and Litman, 1982] or rho (ρ) =(detergent−CMC_{eff})/(phospholipids) [Rivnay and Metzger, 1982]. These two ratios are somewhat different: Whereas in R_{eff} a fixed CMC is used (eg, 22 mM for octylglucoside at 50 mM KCl and at low lipid concentrations), a variable value is employed for the CMC in calculating ρ. This effective CMC may in turn be determined empirically for salt and lipid concentrations identical to those used in the solubilization. We shall use the ρ-parameter to describe the relationship between lipid and detergent throughout this chapter, with the effective detergent CMC determined in saline (pH 7.4) at the phospholipid concentration range of 2–6 mM [Rivnay and Metzger, 1982]. Solubilization of pure lipid vesicles (made of phosphatidylcholine) by octylglucoside occurs at $1.5 < R_{eff} < 3$ [Jackson et al,

1982]. No solubilization occurs at $R_{eff} < 1.5$ and no additional material is solubilized at $R_{eff} \geqslant 3$. The minimal R_{eff} for achieving 90% solubilization is about 2.6. These values are in agreement with data from other detergents like CHAPS and cholate (Rivnay and Metzger, unpublished). According to the latter data, a shift in both turbidity and density of sonicated asolectin liposomes, indicative of solubilization, is accomplished in the range of $\rho = 1$–2.5. In conclusion, ρ can be used as a defined parameter that incorporates the detergent and lipid concentration, the CMC, and the specific experimental conditions. The conditions for solubilization of the receptors for acetylcholine (ACh) and IgE will be discussed in detail in later sections.

The role of lipid and detergent in preserving the function of the native protein.

The detergent. Detergent-induced denaturation occurs in almost every protein during solubilization. It is the result of accumulated conformational changes in the protein, and its detection depends on the assay used. Usually not every possible assay for denaturation is performed, and it is therefore difficult to evaluate the effects of different lipid/detergent conditions on the native structure of receptors, unless such effects are specifically tested or stated. The notion that detergents cannot be regarded as simple amphiphilic substitutes for membrane lipids is not new. In the solubilization of many membrane proteins several detergents were tried initially, and the least denaturing one was normally taken as the detergent of choice. There are many examples of this trial-and-error process, with subsequent choice of the most suitable detergent for a given membrane protein, like octylglucoside for rhodopsin [Stubbs et al, 1976] digitonin for the β-adrenergic receptor [Strauss et al, 1979], CHAPS for an opiate receptor [Simonds et al, 1980], laurylmaltoside for cytochrome oxidase [Simonds et al, 1980], CHAPS for the IgE receptor [Rivnay and Metzger, 1982], cholate for the ACh receptor [Anholt et al, 1981], etc.

The lipids. In later sections we will demonstrate that the activity of reconstituted receptors may be dependent on the ρ-value, which represents the state of the system with regard to the concentration of detergent and lipid. We have to keep in mind, however, that, even for a given detergent, the behavior of proteins at a given ρ-value will also depend on the specific lipid employed. The relative binding of lipid and detergent to sites on the protein is a reflection of the affinities of various chemical functional groups on the amphiphiles for their counterparts on the protein. These affinities will vary depending upon the specific lipid and detergent used.

The role of detergent and lipids in formation of liposomes. Vesicle size (volume) is a very important parameter in many reconstituted receptor systems. It has, for instance, been shown for the reconstituted acetylcholine receptor that small liposomes are not particularly suitable for studying sodium uptake, mediated by the receptor [Anholt et al, 1982]. Total sodium accumulation inside the liposomes remains relatively low, for two reasons: (1) The rate of sodium uptake drops relatively fast because of the relatively small volume-to-surface ratio (ie, a relatively small intraliposomal volume per receptor); and (2) receptor desensitization, which takes place at a certain rate, will limit the total amount of sodium taken up. The effect of this limitation will be more noticeable in small than in large liposomes. Large liposomes are useful not only in studies in which a large volume to surface ratio is desired, but also for experiments requiring a rather low radius of curvature of the reconstituted membrane system.

The various detergents employed for reconstitution differ in structure, charge, and other physical-chemical properties. It is thus anticipated that the mode of their action may be reflected in the characteristics of the resultant vesicles. Electron microscopy of a negatively stained vesicle preparation gives a direct picture of vesicle shape, lamellarity, size and size distribution. The technique has been used in some reconstitution studies to obtain information on vesicle size and shape (Table I). Although in most instances mean diameters are stated, it is generally agreed upon that the size distribution most frequently is a broad one. With methods in which vesicle size is measured indirectly, eg, by density-gradient sedimentation, light scattering, or via determination of the specific entrapped volume (ie, entrapped volume per mole of lipid), spurious results may be obtained if vesicles are leaky or when part of the membranes exist as sheets rather than as vesicles [Mimms et al, 1981]. Moreover, use of the latter methods may result in unreliable numbers for vesicle size, if the vesicle population is highly heterogeneous with respect to size and lamellarity. Cholate reconstitution usually yields vesicles of between 300–1,000 Å diameter, with a small percentage of vesicles having diameters larger than 1,000 Å. This small fraction, however, may contain a considerable part of the total encapsulated volume. Octylglucoside, in general, yields much larger vesicles: 2,250 Å [Mimms et al, 1981], 1,400 Å [Jackson et al, 1982], and 2,500 Å [Gonzalez-Ros et al, 1981].

The structure of the liposomes formed in reconstitution also depends on the types(s) of lipids used. In only a few studies have different lipid species been directly compared in this respect. Popot et al [1981] obtained successful reconstitution of the ACh receptor with soybean lecithin, but not with egg lecithin, which, under the same experimental conditions, formed membrane sheets. Cholesterol has been studied for its effect on vesicle properties. Addition of 25% cholesterol per se does not lead to a change in the average size of the reconstituted vesicles. Following one cycle of freeze-thawing, however, larger vesicles are obtained from the cholesterol-enriched preparation [Anholt et al, 1982]. There are no reports available on direct comparison between mixtures of natural lipids extracted from the receptor source-tissue and commercially available lipids, with respect to size and shape of the vesicles obtained in reconstitution.

Jackson and co-workers [1982] have suggested that, under some conditions, lipid-detergent micelles will precede protein-lipid-detergent micelles in the process of vesicle formation. This may provide an explanation for the observed coexistence of small, pure lipid liposomes and large, protein-containing liposomes [Anholt et al, 1981, 1982; Lindstrom et al, 1980] and for deviations of the lipid: protein ratio in the final liposomes from that in the original mixtures [Jackson et al, 1982]. The influence of the protein:lipid ratio on the size of the vesicles is not clear. Anholt et al [1982] reported that protein (ACh receptor)-containing vesicles had a 50% larger average diameter than pure lipid vesicles formed by cholate dialysis. An opposite trend was found by Herbette et al [1981], with deoxycholate dialysis vesicles containing reconstituted sarcoplasmic reticulum proteins. Vesicles with a low lipid:protein ratio had a smaller average diameter than those with a higher ratio. Mimms et al [1981] also mention a slightly smaller diameter of vesicles formed at a very low ratio of lipid:protein. Several studies [Rivnay and Metzger, 1982; Hazelbauer and Changeux, 1974; Mimms et al, 1981] suggest that in reconstitutions without added, exogenous lipids, in reconstitutions at high protein to lipid ratios, and in reconstitutions from a low ρ-value, the number of vesicles formed is relatively small, the amount of material present in membrane sheets is relatively high, while the size of the liposomes formed is not affected.

For a given detergent and lipid(s) combination, the ratio of detergent to lipid (D:L) does not necessarily affect vesicle size. This has been found for the detergents octylglucoside [Mimms et al, 1981] and cholate and CHAPS (Rivnay and Metzger, unpublished data), under similar experimental conditions. If the system is characterized by a low ρ-value ($\sim$2–5), lipids and protein will form a heterogeneous mixture comprising both unilamellar vesicles and nonvesicular structures. For a given combination of lipid and detergent, the ρ-value may not be the only determinant of vesicle size. Mimms et al [1981] found that

TABLE I. Size Distribution of Reconstituted Vesicles

	Detergent	Lipid[a]	Detergent removal technique	Reconstituted protein[c]	Lipid:Protein (mol:mol)	Diameter (Å)	Reference
1	Na cholate	Nat	Dialysis	AChR		300–1,000	Hazelbauer and Changeux [1974]
2	Na cholate and Triton X-100 contamination	Nat	Dialysis	AChR		1,500–3,000	Michaelson et al [1976]
3	Na deoxycholate	Nat	Dialysis	SR	< 88	1,500–1,700	Herbette et al [1981]
					> 88	2,200–2,800	
4	Na cholate	End		AChR		300–1,000	Sobel et al [1980]
		SBL		AChR		300– 600	
5	Octylglucoside	Nat	Dialysis	AChR		2,500	Gonzalez-Ros et al [1980]
6	Na cholate	SBL	Dial-column[b]	AChR		950	Popot et al [1981]
		SBL	Dialysis	AChR		400	
		EL	Dialysis	AChR		670	
7	Octylglucoside	EL	Dialysis	Glycophorin A		2,250	Mimms et al [1981]
	Octylglucoside	EL	Dialysis	—		2,400	
8	Na cholate	SBL	Dialysis	AChR		300– 500 A few > 1,000	Wu and Raftery [1981a]
9	Na cholate	SBL	Dialysis	AChR		520	Anholt et al [1982]
	Na cholate	SBL + CH	Dialysis	AChR		520	
	Na cholate	SBL (Fr)	Dialysis	AChR		620	
	Na cholate	SBL + CH (Fr)	Dialysis	AChR		760	
10	Octylglucoside	EL	Dialysis	—		1,400	Jackson et al [1982]

[a]Abbreviations in this column: Nat, lipids extracted from the native tissue; End, only the endogenous lipids of the receptor-rich membranes were present; SBL, soybean lecithin; EL, egg lecithin; CH, cholesterol; Fr, liposomes were frozen once.
[b]Dialysis performed until lipid:detergent dropped to 2 followed by Sephadex G-25 filtration.
[c]AChR, receptor for acetylcholine; SR, sarcoplasmic reticulum.

egg lecithin will form 2,400-Å diameter liposomes with octylglucoside at both $\rho = 14$ and $\rho = 3$. Jackson et al [1982], however, using the same constituents, obtained vesicles of 1,400 Å, at a $\rho = 6.3$. It is possible that the size may also be determined by the absolute lipid concentrations; in the two studies cited above the absolute lipid concentrations were 20 and 60 mM, respectively.

The Mode of Detergent Removal

Two methods are widely used for removal of detergents from reconstitution mixtures: dialysis and column chromatography. In dialysis the initial rate of removal depends on the monomer concentration of the detergent (CMC_{eff}); the higher the CMC_{eff}, the higher the initial rate of detergent removal. The porosity of the dialysis membrane (area of pores per unit of surface area) may also affect the initial rate of removal of detergent for a given mixture. Slower removal will occur with increasing lipid concentrations, but this may be overcome by mechanical mixing or by using a dialysis bag with a higher surface-to-volume ratio. In most studies, the preparations were dialyzed against 200–1,000-fold excesses of volume of dialysis buffer, in one or more changes. A rather detailed report on the removal of detergent (cholate and deoxycholate) by dialysis was given by Allen et al [1980]. Though the number of changes of the dialysis buffer was not mentioned, it was demonstrated that cholate was dialyzed away to give a D:L of 1:30 and 1:100 in 50 and 150 h, respectively. Deoxycholate decreased to 1:3 after 50 h but eventually reached (after 150 h) a level similar to that of cholate. Jackson et al [1982] reported, for a mixture of octylglucoside and egg lecithin, a value for D:L of 1:300 with five changes (every 8–12 h) of a 100-fold excess of dialysis buffer. In other studies, detergent removal has been accomplished by combining dialysis with Sephadex column chromatography [Popot et al, 1981], or by the latter method alone [Allen et al, 1980]. Detergent removal by gel chromatography seems to be more efficient than by dialysis, as judged by a considerably lower final D:L ratio in the resulting liposome preparation [Popot et al, 1981; Mimms et al, 1981; Allen et al, 1980]. Sephadex G-200 furthermore proved slightly more efficient than Sephadex G-50 in removing the detergent [Allen et al, 1980]. The liposomes that result from detergent removal on Sephadex generally have a broader size distribution than those obtained through the regular dialysis method [Nozaki et al, 1982; Popot et al, 1981]. At present, it is not unequivocally clear whether a fast rate of detergent removal, as in column chromatography, leads to successful reconstitution.

Sidedness

It has become clear by now that most receptors become reconstituted in vesicles in a topologically correct manner, ie, with the right side out. In a recent careful analysis [Anholt et al, 1981], a correct orientation, as judged by binding of antireceptor monoclonal antibodies and toxins, was found for 65–75% of the reconstituted receptor molecules. Values of 60–90% were found for the IgE receptor by measuring binding of the reconstituted protein to an affinity resin, by trypsin digestion [Rivnay and Metzger, 1982] or by acid-induced dissociation of prebound ligand [Rivnay and Metzger, 1983].

THE RECEPTOR FOR ACETYLCHOLINE

Early studies (before 1978) on the solubilization of the ACh receptor were performed with a variety of detergents, such as polyoxyethylenes like Triton X-100 [Michaelson et al, 1976; Dolly and Barnard, 1977; Michaelson and Raftery, 1974; Huganir et al, 1979] and Lubrol [Chang and Bock, 1979]. Assuming that the CMC for either of these detergents is 0.3 mM and that the membranes used in these studies had a protein:lipid ratio = 2:1 (w/w) [Heidmann et al, 1980], one can estimate that solubilization was usually done at $\rho \geqslant 10$ if the experiments were performed at 2–4 mg protein/ml. Cholate has been used in other studies [Hazelbauer and Changeux, 1974; Briley and Changeux, 1978; Sobel et al, 1980; Nelson et al, 1980; Dalziel et al, 1980; Popot et al, 1981; Wu and Raftery, 1981a,b] and is the most common detergent employed in the last few years. Its use in reconstitution studies of the ACh receptor [Epstein and Racker, 1978; Lindstrom et al, 1980; Anholt et al, 1981, 1982] has been described in relatively great detail. It has been used at concentrations of 2–5% (ie, 23–111 mM). Precise determination of the value of ρ from these studies is difficult because of insufficient data on lipid concentration. Nevertheless, ρ-values in the range of 2–15 can be estimated from much of the published material. Octylglucoside was used in other studies [Gonzalez-Ros et al, 1980, 1981], but determination of ρ from these reports is impossible. Deoxycholate, rarely employed for solubilization of the ACh receptor [Briley and Changeux, 1978], was used at a low ρ (=5).

The only limitation to lowering the detergent concentration is insufficient solubilization of the receptor protein of interest. In a few studies [Epstein and Racker, 1978; Popot et al, 1981; Anholt et al, 1981], detergent concentrations at which only partial extraction occurs, are given. In these studies detergent and lipid concentrations were such that $\rho = 1 \pm 0.5$. It appears that the minimal ρ-value required for solubilization of membranes containing the ACh receptor must be higher than 1.5, consistent with values needed for solubilization of pure phospholipid liposomes. The maximal degree of solubilization obtained in most studies with the ACh receptor amounts to 70% with cholate and to 85–90% with Triton X-100. The maximal possible degree of solubilization was evaluated only in one case [Anholt et al, 1981], where it was noted that Triton X-100 somewhat enhanced the solubilization of the cholate-insoluble fraction, but a clear explanation of the effect is not given. Mescher et al [1981] have indicated that membrane proteins which interact with the cytoskeleton will remain insoluble even at high concentrations of the nonionic detergent Triton X-100. It has not been examined whether this or any kind of protein aggregation may have been responsible for the above-mentioned partial solubilization by cholate.

In most studies on the reconstitution of ACh receptor published until 1980, tissue solubilization is regarded as a separate step in the reconstitution procedure. Efforts to reconstitute the receptor were undertaken under various experimental conditions characterized by a wide range of ρ-values (see Table II). Epstein and Racker [1978] successfully and reproducibly recovered sodium transport activity, using a preparation which was solubilized in the presence of lipids. At the same time Chang and Bock [1979] demonstrated that phospholipids protect the receptor in solution from denaturation as assessed by binding of ligand. The results of these two studies led investigators to carry out solubilization in the presence of exogenous lipid (Table II). In the latter studies, where lipids are included during solubilization, ρ-values range between 10 and 1.5.

Denaturation of the ACh receptor, at several structural and functional levels, has been observed under a variety of conditions. In some studies it is vaguely stated that the receptor is unstable when solubilized in, for instance, 3% cholate (ρ = 5–10) [Sobel et al, 1980]. Wu and Raftery [1981a] mention that the receptor dimer, which is formed by an s-s bond between the δ-chains, spontaneously reduces to the monomer within 2–3 days at p $\sim$ 7. In other studies it is shown that agonist binding parameters are modified upon solubilization with cholate [Briley and Changeux, 1978; Heidmann et al, 1980] with Triton X-100 [Bradley et al, 1976; Dolly et al, 1977], with Lubrol [Chang and Bock, 1979] or with octylglucoside [Gonzalez-Ros, 1981]. In some of these studies solubilized preparations prepared at $\rho \approx 10$ are used, in others purified receptor ($\rho \rightarrow \infty$). Briley and Changeux [1978] found that with receptors which had been exposed to $\rho \approx 80$ (cholate), the carbamyl choline-induced rise in affinity, typical of

TABLE II. Lipid Detergent Conditions in Solubilization of the Acetylcholine Receptor

	Protein conc. (mg/ml)[a]	Inferred[b] endogenous phospholipid conc. (mM)	Added lipids		Detergent		ρ^c	Reference
			Type	Final phospholipid conc.[e]	Type	Conc. (mM)[d]		
1	4.44	2	—	—	Na cholate	119	53	Hazelbauer and Changeux [1974]
2	15	6.8	—	—	Triton X-100	60.6	9	Sobel et al [1977]
					Na cholate	93	12	
3	1.0	0.45	—	—	Na cholate	46	100	Schiebler and Hucho [1978]
4	1.7–3.4	0.77–1.5	—	—	Na cholate	116	83 –166	Briley and Changeux [1978]
	1.8–3.6	0.81–1.6	—	—	Na deoxycholate	8.4	2.8– 5.5	
5	1.67	0.8	Asolectin	21.1	K cholate	45	1.7	Epstein and Racker [1978]
6	2.63	1.2	—	—	Na cholate	23.2	11	Huganir et al [1979]
7	4.6–7.2	2.1–4.2	—	—	Na cholate	69	14 – 28	Sobel et al [1980]
8	3	1.36	Asolectin	< 25	Na cholate	70	>2.3	Popot et al [1981]
9	< 4	1.82	Asolectin	1.69	Na cholate	46	10.5	Anholt et al [1981]
10	2	0.91	Asolectin	3.4	Na cholate	46	8.6	Wu and Raftery [1981a]
11	1.4	0.63	Asolectin	16.9	Na cholate	46	2.1	Wu and Raftery [1981b]
12	< 4	1.82	Asolectin	3.9	Na cholate	46	6.4	Anholt et al [1982]

[a]Concentrations (conc.) corrected for final values in the solubilization mixture.
[b]Assuming protein:lipid (P:L) ratio of 2 (w:w), 70% of lipids are phospholipids and average molecular weight (MW) of 770 for phospholipids.
[c]$\rho = (\text{Det})_{\text{final}} - \text{CMC}_{\text{eff}}/(\text{PL})$.
[d]MW used: Na cholate, 430.6; K cholate, 460.6; Triton X-100, 660; Na deoxycholate, 414.5.
[e]Considering 65% (W:W) of asolectin is phospholipids by phosphorus analysis.

the densitization process, was partially inhibited upon subsequent reconstitution of the receptor into liposomes.

Indications as to how well the native structure and function of the ACh receptor have been maintained throughout the solubilization and reconstitution procedures are given by (1) the degree of association of the protein with the membrane, and (2) the preservation of agonist-induced sodium fluxes or channel activity. Hazelbauer and Changeux [1974] reported a value of 60% incorporation when the receptor was solubilized at $\rho = 53$ and then reconstituted from a $\rho = 10$. In a study in which the entire reconstitution mixture (after dialysis) was analyzed on a sucrose gradient, 80% and 50% of the receptors comigrated with the liposome peak when ρ-values of 8 and 35, respectively, were maintained during the reconstitution process [Schiebler and Hucho, 1978]. Briley and Changeux [1978] found an efficiency of 45% incorporation into liposomes. They carried out the solubilization of the receptor at $\rho \approx 80$, but the reconstitution was performed at a much lower ρ. Michaelson et al [1976] found 80% incorporation of the receptor into liposomes made up of native lipids. They used receptor that had been solubilized in Triton X-100 ($\rho \rightarrow \infty$); unfortunately, the preparation did not possess reproducible sodium channel activity. Receptors, solubilized in octylglucoside, were found associated with liposomes to a degree of 56% [Gonzalez-Ros, 1980]. Popot et al [1981] have shown that the degree of incorporation of receptor into asolectin liposomes depends on the salt (NaCl) concentration. They achieved 77% incorporation at $\rho = 2.4$ in 0.1 M NaCl. However, the incorporation values were only 51% and 36% at NaCl concentrations of 250 and 500 mM, respectively. An explanation for this salt effect is not readily given; it may be that the ultimate degree of association of the protein with the liposomes depends on interaction beween charged groups on the protein and phospholipid head groups. Such an electrostatic interaction would be abolished by salt at high concentration. It is also possible that high salt adversely affects the conformation of the protein. Alternatively, the salt may affect the function of the detergent in the solubilization of the receptor. It is known, for instance, that high salt reduces the CMC_{eff} of cholate. In general it appears that the basic capacity of the receptor to interact with the bilayer was lost only in cases where the protein was purified or subject to conditions characterized by a high ρ-value. In most studies in which lipids were present at low (< 7) ρ-values during solubilization, 85–100% of the protein was found incorporated into liposomes [Wu and Raftery 1981a,b; Anholt et al, 1981, 1982].

As to sodium channel activity, only recently [Anholt et al, 1981; Wu and Raftery, 1981b] have data become available on the effect of lipid-detergent conditions on agonist-induced Na^+ influx. It has become clear that in earlier studies the reconstituted protein failed to exhibit such influx due to exposure of the receptor to high ρ-conditions or to absence of lipids during solubilization. Epstein and Racker [1978] found that, upon solubilization in 4% cholate with 25 mg/ml asolectin ($\rho \approx 3.5$), sodium transport activity was impaired. Later, the same group [Huganir et al, 1979] observed less than 10% loss in channel activity with cholate at $\rho = 2$, 25% loss at $\rho = 11$ with Triton X-100, and complete abolishment of the channel function at $\rho \geqslant 40$ with cholate. No loss of bungarotoxin binding was observed under these conditions. The finding that channel activity, more than other functions of the ACh receptor, was dependent on ρ, was confirmed by Anholt et al [1981] who showed that solubilization with $\rho \triangleq 7$ followed by reconstitution from $\rho \triangleq 1.7$ gave maximal channel activity, whereas if both steps were performed at $\rho \triangleq 7$, the receptor became irreversibly denatured. The activity could not be subsequently recovered upon renewed reconstitution, at $\rho = 1.7$.

In conclusion, it seems that the channel activity of cholate-solubilized ACh receptor, including the desensitization property, is very sensitive to detergent-lipid conditions and can be preserved only when the protein is solubilized at $\rho < 10$ and reconstituted at $\rho \cong 2$. The binding functions (for both toxins and agonists) generally deteriorate less easily, although they were found to be affected in receptor exposed to detergent at high ρ-values or depleted of endogenous lipids [Schiebler and Hucho, 1978; Chang and Bock, 1979]. The basic capacity to interact with the membrane bilayer (as judged by the degree of association of the receptor with the membranes) also seems to be less dependent on ρ than channel activity. Receptor denaturation is usually a slow process; it takes at least several hours [Heidmann et al, 1980; Anholt et al, 1981; Lindstrom et al, 1980; and Huganir et al, 1979].

In ACh receptor-rich membranes, lipids compose about 30% of the total mass of the membranes [Michaelson et al, 1976; Schiebler and Hucho, 1978]. About 70% (by weight) of the lipids consists of phospholipids and about 20% of cholesterol. The specific interrelationship between the ACh receptor functions and lipids has been observed with a variety of techniques such as receptor-monolayer interaction [Popot et al, 1978], digestion of intact membranes with phospholipase A_2 [Andreasen et al, 1979], or the controlled supplementation of defined lipids to delipidated receptor [Chang and Bock, 1979]. It is expected that this receptor, like other membrane proteins, will perform best in an environment similar in composition to its natural environment. Yet, when comparing receptor performance in several different lipid environments, some caution should be taken since reconstitution with various lipid mixtures may result in different types of vesicular structures. Lipid composition may determine size, lamellarity, leakiness, or heterogeneity of vesicles. In one study [Kilian et al, 1980] it has been demonstrated that removal of some neutral lipids from crude, commercial asolectin largely obliterated the agonist-induced $^{22}Na^{+}$ influx via its receptor. Full response was recovered upon addition of 10–20% (w:w) tocopherol or phylloquinone (neutral lipids). If cholesterol (~25%) was also incorporated into the membranes, five-fold enhanced sodium fluxes were found [Dalziel et al, 1980]. Binding of toxin, however, was unchanged. It should be noted that a more defined phospholipid mixture of phosphatidylethanolamine:phosphatidylserine (3:1) was used in these studies instead of crude asolectin or native lipids; no effect of cholesterol was observed when phosphatidylcholine was present in the mixture. The use of native lipids with the purified ACh receptor in octylglucoside [Gonzalez-Ros et al, 1980] may have been responsible for the partial protection and recovery of receptor performance under conditions in which it showed complete denaturation with soybean lipids [Anholt et al, 1981]. A similar finding was reported for receptor solubilized and purified in Triton X-100 [Michaelson et al, 1976]. Although lipids are currently added in all experiments undertaken to reconstitute the ACh receptor, it is important to note that the endogenous lipids in the receptor-rich membranes may be sufficient for preservation of the receptor's characteristic channel activity. Incubation of membranes for 24 h in detergent (cholate) without additional lipids, or in the presence of 1 mg/ml asolectin leads to loss of channel activity, except at relatively high protein concentrations (and concomitantly high lipid concentrations).

THE RECEPTOR FOR IMMUNOGLOBULIN E

This receptor is abundant on mast cells and basophils and binds immunoglobulin E (IgE) with high affinity [Conrad et al, 1976; Kulczycki and Metzger, 1974]. In situ aggregation of this receptor by allergens with which the IgE was elicited, or by other means, leads to cell degranulation and to initiation of an allergic response. For a detailed description of this receptor see reviews by Froese [1980] and by Metzger [1983]. This receptor is currently viewed as a three-polypeptide complex consisting of α, β, and γ subunits, whose structure and topology are only partially characterized. No inherent function of the individuals subunits has been determined other than the binding of IgE by the α-chain. Early efforts to isolate this receptor were done with the detergents NP-40 or Triton X-100. From the different solubilization conditions stated in the literature we calculate ρ-values of 18–36 [Conrad et al, 1976] or values lower than 7.3 [Rossi et al, 1977]. The lowest ρ-value at which complete solubilization was obtained was explicitly reported to be 2.1. Recent experiments [Rivnay and Metzger, 1982] show that 90% solubilization of the IgE-receptor complexes is accomplished at a still lower ρ-value and not only with Triton X-100 but also with cholate, CHAPS, and octylglucoside. The same study suggests that, generally, plasma membrane proteins which are not linked to the cytoskeleton are solubilized in a way similar to the IgE-receptor system on the same cells. Ninety percent of the maximal receptor solubilization can be achieved at a detergent:phospholipid molar ratio of 1.5–2. As will be pointed out below, use of a low detergent concentration minimizes the risk of denaturation of the receptor.

Recent observations suggest that the α-chain of the IgE receptor, which binds the ligand (IgE), may be attached to the membrane only indirectly, by tightly interacting with one or both of the other chains of the complex. Gradual (over 24–48 h) delipidation with lipid-free detergent solution (either CHAPS, octylglucoside, cholate, or Triton X-100) results in elimination of more than 90% of the non-α-components of the receptor [Rivnay and Metzger, 1982]. The remaining lgE α-complexes cannot be reinserted into liposomes. This explains why, in early studies, in which solubilization was peformed at ρ = 10–15 with octylglucoside, and the crude extract was used for reconstitution of the receptor into soybean lecithin, less than 40% of the receptor complexes were incorporated. The dissociation of the non-α-chains from the IgE-α-complex is inhibited in solution at $2 < \rho < 10$, in the presence of tumor-derived lipids (see also the discussion on the role of lipids). The non-α-chains, β and γ, will reinsert even at $\rho = 12$.

The binding activity of the IgE receptor is decreased at concentrations of Triton X-100 above 0.25% [Rossi et al, 1977]. Comparison of the binding of IgE to its receptor in Triton X-100 extracts with that in intact cells reveals similar dissociation rates but a substantial difference in association rates (30-fold faster in the extract) [Wank et al, 1983; Metzger, 1982]. Wank et al [1983] have shown that the rate of dissociation of IgE from its receptor was not dependent on the degree of receptor occupancy on the cell surface. Therefore, the binding between IgE and its receptor is not a diffusion-controlled reaction, so that the binding of ligand to membrane-associated receptors and to solubilized receptors can be directly compared. One explanation for the above-mentioned difference in ligand binding may therefore lie in denaturation caused by the detergent. It should be mentioned that for the IgE receptor no effort has been made to correlate the binding data with the ρ-value characterizing the system.

The IgE receptor from RBL cells is very sensitive to the type of lipids employed for reconstitution. The commercial lipids employed [Rivnay and Metzger, 1982] were soybean and egg lecithin. Similar degrees of receptor incorporation were observed with these lipids and with the RBL-derived lipids, if the exposure to detergent was minimal. However, sensitivity to denaturation by CHAPS, as judged by vesicle incorporation, is greater with soybean lecithin than with tumor derived lipids, either as a function of the ρ-value or the time of incubation. The naturally occurring lipids seem to provide a more efficient protection of receptor integrity than does soybean lecithin. A particularly deleterious combination, in this respect, is soybean lecithin and octylglucoside.

OTHER SPECIFIC EXAMPLES OF RECEPTOR RECONSTITUTION

The following sections of this chapter are

intended to examine other specific examples of receptor reconstitution studies.

Asialoglycoprotein Receptor

This receptor is found on mammalian hepatocytes and is responsible for the receptor-mediated endocytosis of desialylated glycoproteins. The receptor recognizes an exposed terminal galactose of N-acetyl galactosamine component of the carbohydrate chain [Ashwell and Morell, 1974]. It has been purified to homogeneity and isolated free of detergent and lipid [Hudgin et al, 1974]. This receptor has been reconstituted into phospholipid vesicles by two different techniques. First, Baumann and co-workers [1980] reported the use of a detergent-dialysis method. They isolated phospholipid from rat liver membranes according to the method of Bligh and Dyer [1959]. Lipids were stored under nitrogen in methanol. Before reconstitution, phospholipid was dried under nitrogen and dissolved in 1% sodium deoxycholate (DOC), 50 mM Tris HCl, pH 8.0. The lipid was dispersed into buffer by bath sonication (1 min), at room temperature. Purified receptor was added to this mixture and beween 2 to 5 ml were dialyzed for 72 h against 3.5 liters of 10 mM sodium phosphate, 140 mM NaCl, pH 8.0. The reconstitution of the receptor was followed by isolation of the lipid-receptor complexes by isopycnic centrifugation using sucrose. The authors found that the efficiency of reconstitution was dependent upon the lipid-to-protein (L:P) ratio used. At L:P ratios of 3:1 (w:w) about 10% of the radioactive receptor was incorporated into the vesicles. By increasing the ratio to 6:1 the degree of incorporation increased to about 35%–40%. Receptor (500 μg) was added to 3 mg of phospholipid in DOC in a volume of 5 ml. The authors report that at protein concentrations above 0.1 mg/ml the receptor fails to incorporate into the vesicles. The reconstituted vesicles band at a density of 1.08 g/ml in sucrose and are unilamellar structures with diameters of 400–700 Å. This procedure yielded a successful reconstitution as determined by ligand-binding activity. The authors attempted to determine the sidedness of the reconstitution, by comparing the binding activity of the vesicles before and after disruption of the vesicles with 0.1% Triton X-100. They state that 70% of the receptor molecules are correctly oriented.

The authors of this study reported a very intriguing reconstitution of the hepatic asialoglycoprotein receptor into the plasma membranes of mouse L cells, a fibroblast cell culture line lacking the receptor. They used polyethylene glycol to fuse reconstituted phospholipid vesicles with L cells in suspension, and concluded that successful incorporation of the receptor into the L cell plasma membrane had been accomplished.

In another approach to the reconstitution of the asialoglycoprotein receptor, the direct incorporation of the receptor into preformed lipid vesicles was demonstrated [Klausner et al, 1980]. We found that the purified receptor will bind spontaneously to phospholipid vesicles. The preparation of purified receptor exhibited very little binding activity in the absence of detergent. Thus the spontaneous insertion into preformed lipid vesicles could provide a system to assess functional as well as physical reconstitution. Small unilamellar liposomes were made from either dioleoyl, dimyristoyl, or dipalmitoyl phosphatidylcholine. Briefly, 25 mg of lipid were dried under argon and suspended in 5 ml of phosphate-buffered saline with vigorous vortexing at either 30°C (or 45°C for dipalmitoyl phosphatidylcholine). The dispersion was sonicated with a probe for 1 h at the appropriate temperature. The resultant small unilamellar vesicles were separated from larger vesicles on a Sepharose 4B column. Purified receptor protein in phosphate-buffered saline was mixed with the vesicles, and after 30 to 60 min, the vesicle-receptor complexes were separated from unincorporated receptor on a column or isolated by isopycnic centrifugation on KBr density gradients. Regardless of the type of phospholipid used, recombinant formation was observed. The receptor-bilayer association was stoichiometric to L:P (w:w) ratios of 4–5:1. At lower ratios some of the protein failed to combine with the vesicles. All of these spontaneous reconstitutions were done at protein concentrations of 200 μg/ml or less. Once the receptor had interacted with the vesicles, the association was not sensitive to ionic strength (0.001 to 2 M NaCl) or pH (2–10), and the protein could only be removed from the lipid by detergent extraction. The reconstitution occurred equally well with gel-phase as with fluid-phase vesicles. Using these reconstituted vesicles we found that the functional capacity of the receptor had been completely restored. The reconstituted receptor displayed the same ligand specificity, pH sensitivity, and calcium dependence as the native receptor in intact cell membranes.

Insulin Receptor

Several examples in the literature demonstrate the successful reconstitution of this ubiquitous receptor into phospholipid vesicles. One in particular illustrates the use of detergent dilution [Gould et al, 1982]. The insulin receptor was solubilized from a preparation of turkey erythrocyte membranes by treatment of 2.9 mg/ml of membrane protein with 1% β-octylglucoside in 30 mM NaCl, 10 mM sucrose, 1 mM ethylenediaminetetraacetic acid (EDTA), 85 mM Tris-HCl, pH 7.8. The mixture was stirred for 15 min at room temperature. Insoluble membrane components were removed by 1-h centrifugation at 104,000g at 4°C. The supernatant containing the solubilized receptor was mixed with a 20-fold excess (presumably w:w) of phospholipid in 2% octylglucoside. In this study two types of phospholipid mixtures were compared: soybean phosphatidylcholine:bovine brain phosphatidylserine (4:1, w/w) and dimyristoyl phosphatidylcholine:bovine brain phosphatidylserine (4:1, w/w). Each mixture of lipid, detergent, and solubilized protein was passed over a Sephadex G-50 column equilibrated with buffer without detergent at a temperature stated to be above the phase transition temperature of the lipid used. Material eluting in the void volume of the column was dialyzed for 48 h (no volume or temperature given). The vesicles were then pelleted by centrifugation at 104,000g for 1 h at 4°C. This pellet was resuspended in buffer and centrifuged on a 2–30% continuous sucrose gradient until equilibrium was reached. Vesicles of either lipid composition banded at 1.071 g/ml. Both types of vesicle preparations were found to consist almost entirely of unilamellar structures of about 1,000 Å in diameter. These vesicles contained the insulin receptor, as demonstrated by the specific binding of iodoinsulin. In either type of vesicle, 80–90% of the receptor-binding activity was oriented toward the outside of the vesicle. This was demonstrated by examining the total binding activity after solubilization of the vesicles either before or after treatment of the intact vesicles with trypsin. Interestingly, although successful receptor reconstitution was attained with both types of lipid vesicles, the affinity of the reconstituted receptor for insulin was markedly different in the dimyristoyl lecithin and soybean lecithin environments. This is a good example of how reconstitution allows an examination of the effect of lipid structure and composition on receptor function. This study also demonstrates the usefulness of receptor reconstitution even when the receptor is not purified.

β-Adrenergic Receptor

The β-adrenergic receptor system is perhaps the best characterized receptor in terms of the series of events that follow ligand binding [Ross and Gilman, 1980]. After hormone binding to the receptor the hormone-receptor complex interacts with a GTP-binding protein in the membrane. Thus activated, the binding protein interacts with the catalytic component of the system (adenylate cyclase), thereby stimulating the enzyme to synthesize cyclic AMP. Recently, a variety of reconstitution procedures have been used to carefully study this complex system [Pedersen and Ross, 1982]. For most of these studies the receptor is taken from turkey or rat erythrocyte membranes.

Fleming and Ross [1980] report a well-documented reconstitution process using the β-adrenergic receptor from rat erythrocyte membranes. Isolated membranes were stored frozen in 20 mM Hepes (pH 8.0)/2 mM EDTA/0.1 mM phenylmethylsulfonyl fluoride (PMSF)/0.5 mM dithiothreitol, at a protein concentration of 3–7 mg/ml. The membranes were washed and suspended at 10 mg/ml (protein) in 20 mM Hepes (pH 8.0)/2 mM EDTA, and a digitonin solution (70 mg/ml) was added to give a detergent: protein ratio of 3.5:1 (w:w). The mixture was stirred at 0°C for 1 h before being diluted to 5 mg/ml (protein) with saturated ammonium sulfate. After 30 min at 0°C, the material was spun at 50,000 rpm (50 Ti rotor) for 30 min. The supernatant containing the solubilized receptor was kept and stored at −80°C. This procedure resulted in 50% solubilization with a two-fold purification of the receptor. The extract was 5–8 mM in digitonin and contained 1–2 mM endogenous phospholipid. These authors then used a column chromatography method to reconstitute the receptor into vesicles. Dimyristoyl phosphatidylcholine vesicles were sonicated at a lipid concentration of 150 mM to form small unilamellar vesicles. The latter vesicles were added to the membrane extract (all in 55 mM Hepes (pH 8.0)/5.5 mM EDTA/10.6 mM $MgCl_2$/98 mM NaCl/0.89 M $(NH_4)_2SO_4$ to give a final lipid concentration of 5 mM. The mixtures were then chromatographed over a Sephadex G50 column in 50 mM Tris-HCl (pH 7.5)/5 mM $MgCl_2$/100 mM NaCl with no detergent, and the early void volume fractions were collected. These fractions were turbid and contained 50–65% of the receptors, total protein, and phospholipid, but only 10–20% of the digitonin. The detergent concentration was 0.5–1 mM, still higher than the CMC for digitonin (0.08–0.3 mM). This fraction was further treated by rate zonal centrifugation on a linear sucrose density gradient. Reconstituted vesicles were found in a turbid band at 27% sucrose. This band contained 50% of the receptor activity, 6 μmol of lipid/mg protein, and no detectable digitonin (the limit of detection was 40 μM). Electron microscopic studies revealed unilamellar vesicles, 500–900 Å in diameter. The reconstituted vesicles displayed ligand binding properties similar to those found for the intact receptor in native membranes. Interestingly, the antagonist iodohydroxybenzylpindolol, which binds to the native receptor, does not bind to the digitonin-solubilized receptor. However, full binding activity is regained upon reconstitution and detergent removal.

One of the most exciting aspects of reconstitution studies with this receptor comes from several laboratories and involves the reconstitution of the interacting components of the adrenergic receptor-cyclase system. Citri and Schramm [1980] have described such a process. These authors reported the separate solubilization of the receptor and the nucleotide binding subunit (G) and their subsequent reconstitution into phospholipid vesicles. Turkey erythrocyte membranes were washed and treated for 20 min at 4°C in 50 mM phosphate at pH 11.9. This resulted in the irreversible loss of G activity but spared the receptor. The membranes were pelleted at 27,000g for 15 min and resuspended in 20 mM MOPS/1 mM mercaptoethanol at a protein concentration of 1 mg/ml. To this mixture was added a suspension of phospholipid vesicles prepared by the following method: 10 mg/ml of soybean lecithin was sonicated two times for 5 min each, under nitrogen, in 10 mM Tris HCl (pH 7.5)/1 mM EDTA; then 0.2 ml of the vesicle suspension were added per ml of receptor extract. The mixture was incubated at 4°C in 10 mM $MgCl_2$ before addition of 20 μM of the adrenergic ligand isoprenaline. Next the dispersion was incubated for 15 min at 30°C and then centrifuged at 18,000g for 10 min. The pellet was resuspended in 20 mM MOPS (pH 7.0)/0.5M sucrose/0.2 mM EDTA/0.1 mM mercaptoethanol/10 μM PMSF and solubilized by the addition of an equal volume of 12 mg/ml sodium deoxycholate. Insoluble material was removed by a 30-min centrifugation at 200,000g. The solubilized receptor remained in the supernatant at an overall protein concentration of 4 mg/ml.

G protein without receptor was prepared from the same membranes simply by solubilization with cholate. Cholate extracts the former protein but not the receptor. First, membranes at 10 mg of protein/ml were washed in 20 mM Hepes (pH 8.0)/2 mM mercaptoethanol/1 mM $MgCl_2$/0.2 mM EDTA/0.02 mM PMSF. Subsequently, an equal volume of cholate solution (20 mg/ml) was added. Solubilization took place over 30 min at 4°C. Insoluble material was pelleted at 200,000g for 30 min. The G protein was recovered in the supernatant at 0.9 mg protein per ml.

To reconstitute the proteins, the two soluble preparations were mixed in desired proportions, and the majority of detergent was first removed by treatment with SM-2 resin. After 1.2 g of wet resin were added per ml of the mixture, the preparation was then shaken for 1 h at 14 strokes/min. The beads were removed by centrifugation and the supernatant mixed with preformed phospholipid vesicles, made of 10 mg lecithin and 1 mg of stearylamine in 1 ml of 10 mM Tris-HCl (pH 7.5)/1 mM EDTA by sonication under N_2 for 5 min. The author states that the solution is opalescent, indicating the presence of large, multilamellar structures but provides no documentation of the nature of the vesicles [Schramm, 1979]. The vesicle suspension was combined with the resin supernatant to give 0.5 mg phospholipid/ml of supernatant. After a brief incubation at 4°C the suspension was centrifuged at 20,000g for 15 min and the pellet containing the reconstituted proteins resuspended in buffer.

Reconstitution was successful as judged by binding of ligand (receptor activity) and the finding that the hormone induced activation of the G protein, measured by stimulation of adenylate cyclase activity. This latter assay required another kind of reconstitution; the functional integration of proteoliposomes into intact biological membranes [Schramm, 1979; Eimerl et al, 1980]. This type of reconstitution opens a new and exciting area of membrane reconstitution. In this procedure the reconstituted vesicles were pelleted by centrifugation at 18,000g for 10 min. Cells to be fused with the vesicles were suspended over the vesicle pellet and sedimented onto it at room temperature. The two layers were mixed and warmed to 30°C. Subsequently, a salt solution containing 520 mg of polyethylene glycol 6000 in 0.48 ml buffer was added. After 100 s increasing amounts of the buffer without polyethylene glycol were added. This process led to the apparent insertion of the receptor and G protein such that the intrinsic adenylate cyclase of the membranes could be activated. These experiments allowed the evaluation of the effect of the receptor:G stoichiometry on hormone-induced cyclase activation.

Reconstitution of solubilized receptor directly into biological membranes has recently been reported [Jefferey et al, 1980]. In this study washed turkey erythrocyte ghosts were treated with 0.5% digitonin in 20 mM Hepes (pH 8.0)/0.25 M $MgCl_2$/0.125 mM EDTA for 30 min at 4°C. Insoluble material was removed by centrifugation at 105,000g for 1 h. Between 25% and 35% of the β-adrenergic receptor was found in the supernatant. One ml of this supernatant (10 mg of protein/ml) was mixed with 1.5 ml of bovine serum albumin (22.5 mg) and 3.5 ml of buffer (20 mM Na phosphate/2 mM $MgCl_2$ pH 7.0). This was combined with 1 ml of ghosts (10 mg protein) and dialyzed at 4°C for 12–15 h against 200 volumes of 10 mM Na phosphate (pH 7.0)/2 mM $MgSO_4$. The preparation now contained about 7.5% of the receptor present in the initial solubilization supernatant. This system displayed normal binding isotherms for ligand and normal stereospecificity. The reconstituted receptor can only be removed from the acceptor membranes by detergent solubilization, indicating a probably true reconstitution of the receptor as an integral membrane protein. These authors furthermore provided evidence that the implanted receptor biochemically couples to the G protein present in the acceptor membranes. Thus several studies suggest that it is possible to reconstitute receptors into biological membranes. This provides great potential applications for studying the interactions between components of specific receptor systems as well as to monitor the effects of mixing components of different receptor systems. Furthermore, these types of studies provide a fertile experimental approach to the study of receptor mutants.

Thyrotropin Receptor

The receptor for thyroid stimulating hormone (TSH or thyrotropin) is present on the plasma membrane of thyroid follicular cells and the cells of retroorbital tissue. [Tate et al, 1975]. Binding of this hormone to its receptor on thyroid cells initiates the biosynthesis and release of thyroid hormone. The reconstitution of this receptor into liposomes has allowed the evaluation of the lipid specificity of hormone-membrane interaction [Lee et al, 1978]. Thyroid membranes were solubilized with 0.1 M lithium diiodosalicylate. The extract was used as the receptor preparation. It was dialyzed against 0.02 M Tris acetate, pH 7.0. Dipalmitoyl or dioleoyl phosphatidyl choline and cholesterol (PC:Chol = 2:1 molar ratio) were dried under N_2 onto the wall of a glass flask from a 2:1 mixture of chloroform/methanol. Subsequently, lithium diiodosalicylate extract was added, to give about 110 μg protein per 9.5 μmol of phospholipid. The flask was shaken at 4°C overnight. Next, glass beads (3 mm diameter) were added. The flask was placed on a vortex mixer for 5 to 10 min, during which the lipid came off the glass and formed large multilamellar liposomes. The lipid vesicles were pelleted by centrifugation and quantitated in a Coulter counter which was set to register particles of between 10 and 100 μm in diameter [Aloj et al, 1977]. These vesicles proved to be useful in the evaluation of the effect of lipids on TSH binding.

RECONSTITUTION IN PLANAR BILAYERS

Introduction

Planar bilayers serve as exquisitely sensitive tools for the study of transport of ions mediated by channels. The effect of those transport mediators is expressed in terms of extremely small changes in conductance of the planar bilayer in the range of 1–10 picoSiemens (pS, 1 Siemens = 1/ohm). The planar bilayer has also advantageously been used for the study of ion pumps, particularly those activated by light. The particular advantage of the planar bilayer is that the voltage can be manipulated, and voltage-dependent gating of ion transport systems can be studied. Moreover, both sides of the planar bilayer are easily accessible to reagents, and effects on the trans side can be tested when reagent is added to the cis side of the membrane. Topology and orientation of the membrane protein can be monitored and controlled.

There are two types of planar bilayers: (1) the classical black lipid membranes (BLMs) [Mueller et al, 1964] formed by spreading a solution of lipid in organic solvent across a small circular hole in a plastic or teflon septum, and (2) the planar bilayers formed by apposition of monolayers formed at the air-water interface [Montal and Mueller, 1972]. Planar bilayers can in principle also be formed by apposing a giant vesicle (50-μm diameter) to a small hole in a septum, and subsequently puncturing it by applying a high-strength electric field [Darszon et al, 1980]. This last method, although potentially powerful for reconstitution studies, has not yet been shown to be very successful. In the following, we will discuss some salient features about reconstitution in planar bilayers. More background about formation and characteristics of planar bilayers, effects of ionophores and of channels, and about reconstitution can be found in several reviews.

Although membrane proteins can more readily be reconstituted into liposomes (see below), successful reconstitution of membrane proteins (specifically those associated with ion channels) in planar bilayers can yield a greater amount of information about their mode of action. Lipid vesicles or liposomes have been more popular in reconstitution studies because the techniques needed to achieve reconstitution are more accessible. Moreover, there is, in principle, no limitation as to the amount of material used in the reconstituted systems. This makes the lipid vesicle reconstituted protein amenable to physical chemical studies (binding, spectroscopy, etc).

A planar bilayer, on the other hand, with a diameter of 0.1 cm has about 10^{14} molecules of lipid (assuming that the area per lipid molecule is 50 [Å]^2) or about 10^{-10} mol. Obviously there is a limit to the amount of protein and lipid that can be present in the reconstituted planar bilayer. Although spectroscopic studies on the planar bilayer require sophisticated instrumentation, they yield important information about the orientation of molecules in the bilayer. However, most of the important information gleaned from planar bilayer studies is derived from the ability to observe openings and closings of single channels as discrete conductance steps [Ehrenstein and Lecar, 1977]. They probably represent the only reconstitution system where single molecular events can be observed. The studies of discrete conductance steps in bilayers gave an impetus to the development of the patch clamp [Neher et al, 1978] and consequently to the observation of discrete conductance events in biological membranes. Reconstitution in planar bilayers is also very useful for the study of ion pumps [Montal et al, 1981]. Their activation, for instance by light or by ATP, gives rise to the formation of an electrochemical gradient for ions (for instance H^+, Na^+). Short-circuiting that system allows measurement of currents generated by ion pump activation. The planar bilayer is not very useful for reconstitution of transport systems which are electrically silent (eg, the glucose carrier or the anion transport system (band 3) of red blood cell membranes). The lipid vesicle with its larger surface-to-volume ratio is a better system to chemically assay membrane transport.

Reconstitution in Preformed Black Lipid Membranes

Membrane proteins are usually isolated and purified in the presence of detergents. As we discussed earlier, reconstitution most often involves removal of the detergent and replacement with a phospholipid. During this process, lipid vesicles are spontaneously formed, and the reconstitution is completed. Reconstitution of membrane proteins into planar lipid bilayers is quite different. With planar bilayers the membrane protein has to be inserted into a preformed bilayer. There is no possibility for stepwise exchange of lipid for detergent, and obviously the detergent itself perturbs the bilayer. The only successful reconstitutions of membrane proteins into preformed bilayers involve those proteins which do not require detergent in aqueous solution. Those proteins include toxins [Schein et al, 1978], an outer cell wall protein of mitochondria called the voltage-dependent anion channel (VDAC) [Schein et al, 1976], an outer cell wall protein from *E. coli* called porin [Benz et al, 1978] and the asialoglycoprotein receptor protein [Blumenthal et al, 1980].

Insertion of the above proteins into preformed bilayers results in channel formation. On the other hand, the detergent-free soluble asialoglycoprotein receptor seems to interact with planar membranes in a different way [Blumenthal et al, 1980]. As mentioned above, this protein is spontaneously reconstituted into lipid vesicles. We could monitor its spontaneous reconstitution in BLMs by observing a conductance change upon addition of the receptor protein into the solution neighboring the BLM. This conductance increase was voltage-dependent; it was observed when a trans-positive potential was imposed across the BLM. Its nature pointed strongly to a lipid perturbation (rather than to channel-formation) because (1) no discrete step of uniform size was observed, (2) the conductance was not proportional to the first or any higher power of the protein concentration, and (3) the conductance displayed little ion selectivity. Consistent with a lipid-perturbing effect of the asialoglycoprotein receptor was the observation that it lowered and broadened the phase transition of DPPC. For a signal to be transduced by a receptor after ligand binding, a conformational change in the receptor is usually assumed to take place. Given the possibility that such a change could alter lipid-protein interactions, we examined the effect of ligand on the conductance in a receptor-doped BLM. Upon addition of specific ligand to the receptor-doped BLM, we observed a dramatic increase in the conductance. Moreover, the conductance became symmetric in that both a trans-positive and a trans-negative potential induced a conductance increase. We showed that this symmetry represented translocation of part of the receptor protein across the membrane. The BLM provided a system from which with a variety of experimental manipulations information about the conformation and topological orientation of the receptor could be obtained. These experiments also suggest that the transmembrane potential may influence the orientation of at least some membrane proteins. Other membrane-bound proteins such as acetylcholinesterase [Kaufman and Silman, 1980] or dopamine-β-hydroxylase [Kafka et al, 1978] have been "reconstituted" in this fashion and show a ligand-specific effect on the conductance.

The experimental approaches discussed in the preceding section have shed some light on receptor-lipid interaction, receptor-ligand interaction, and changes in receptor topology (with respect to the bilayer). Reconstitution of receptors containing ion channels, however, requires more involved techniques. Those ion channels (porin, VDAC, colicin E) which are initially water soluble and then spontaneously insert into membranes are of relatively low molecular weight. On the other hand, physiological ion channels, whose opening and closing is usually regulated or "gated," are found to be of high molecular weight and can only be purified in a detergent-solubilized form.

It is useful to set up some criteria for successful reconstitution of transport function in BLMs [Shamoo, 1975, 1980]. Those include the pharmacological (ie, effects of ligands and drugs) and transport specificity (ie, ion selectivity) of the native system, the size of the ion channel, the voltage dependence, and the kinetics of opening and closing of the channels. In many cases, the criterion for pharmacological specificity has been satisfied. However, the reconstitution could more readily be interpreted in terms of a ligand-induced lipid perturbation (as we found with the asialoglycoprotein receptor) rather than in terms of complete reconstitution of a receptor-associated ion channel [Blumenthal and Klausner, 1982]. All of the criteria need to be met to justify the conclusion that a successful reconstitution of regulated ion transport has been established. With the currently available techniques of patch-clamping, a precise characterization of physiological channels with respect to channel size, specificity, and kinetics can be achieved in the native membrane. These studies provide the reference characteristics on the basis of which the successfulness of the reconstitution can be judged.

In the following section we will discuss the reconstitutions of receptors that are associated with functional ion channels, into planar bilayers. Such reconstitutions have been relatively successful. They have been brought about either by monolayer apposition or by fusion of vesicles containing the transport system with preformed planar bilayers.

Reconstitution Via the Organic Solvent Route

The original approach to the formation of planar bilayers from apposition of monolayers involved spreading of the lipid dissolved in an organic solvent (hexane) onto a monolayer at the air-water interface. After evaporation of the solvent, a teflon septum is lowered into the aqueous phase, thereby carrying a monolayer of lipid on each face of the septum. The presence of an aperture in the septum allows apposition of the two monolayers to form a bilayer now separating two aqueous compartments. For reassembly of receptors, protein-lipid complexes need first to be transferred into organic solvent [see Boyan and Clement-Cormier, this series]. Gitler and Montal [1972] developed a strategy to induce the partitioning of membrane proteins into an apolar environment by charge neutralization and formation of ion pairs. The success of this approach for reconstitution hinges on the question of whether the membrane protein "survives" in its functional state during this organic solvent treatment. This seemed to be the case for cytochrome c oxidase, rhodopsin, bacteriorhodopsin, the mitochondrial anion channel VDAC, and reaction centers from photosynthetic bacteria.

One could, in principle, make BLMs from protein extracted into an organic solvent according to the procedure of Gitler and Montal [1972]. However, success with such a procedure is not guaranteed. Often the membranes do not thin under those circumstances or they start off with spuriously high conductances. The approach did seem to be successful with bacteriorhodopsin [Bamberg et al, 1982]. Although the above proteins survived the organic solvent treatment according to spectroscopic criteria (eg, as evidenced by the preserved photochemical reactivity of rhodopsin, including the ability of bleached opsin to combine with 11-cis [or 9-cis] retinal to regenerate rhodopsin), it is in most cases not clear whether the transport function remained intact in the planar bilayer.

Successful reconstitution of channels via the organic-solvent route has been achieved with porin and VDAC, maybe because they are single chain proteins and do not require assembly of different subunits for channel formation as is the case for the acetylcholine receptor. Bovin rhodopsin represents an interesting case in this regard. Montal and coworkers [1977] found that rhodopsin, reconstituted into planar bilayers via the organic solvent route, forms voltage-dependent channels with a diameter of about 10 Å. These could, in principle, be involved in light-dependent Ca^{2+} efflux, which is postulated to be the modulator of the Na^{+} dark current in retinal rods [Hagins, 1972]. Since it is still an open question whether such channels physiologically exist, we do not know whether this is a functional reconstitution. According to Ramakrishnan and co-workers [1980] the rhodopsin channels did seem to require aggregation of the protein molecules.

Reconstitution of ion pumps into planar bilayers via the organic solvent route has been carried out with reaction centers from photosynthetic bacteria, and with cytochrome c oxidase and bacteriorhodopsin [Montal et al, 1981]. The (photo)chemistry of those systems could be followed during the extraction and reconstitution steps. With reaction centers absorbance spectra were followed. The dark spectrum exhibits three peaks in the infrared region at 760, 802, and 865 nm. The most prominent light-induced absorbance change is the bleaching at 865 nm that accompanies the primary photochemical reaction. In the case of cytochrome c oxidase the enzymatic oxidation of cytochrome c can be followed. With bacteriorhodopsin the absorbance spectrum or the photochemical cycle can be monitored during all steps of extraction and purification. For reaction centers incorporated into planar bilayers, photoelectric signals that arise from the transfer of electrons from a secondary donor to a secondary acceptor located on opposite sides of the membrane have been observed. This finding constitutes direct evidence for the transduction of light energy into a transmembrane electric current. With planar lipid bilayers containing cytochrome c oxidase, a membrane potential was generated when cytochrome c was added to one compartment. Those results suggest that the cytochrome c oxidase in the planar bilayer is capable of pumping protons across the membrane.

Bacteriorhodopsin seems to be the most resistant pump with respect to denaturation during solvent extraction and reconstitution. Bamberg et al [1979] showed that the incorporation of that protein into planar lipid bilayers could be achieved either by forming BLMs from bacteriorhodopsin extracted into n-alkanes in the presence of lipids, or by forming planar bilayers from those extracts spread onto an air-water interface, or by spreading vesicles, containing purple membrane or monomeric bacteriorhodopsin on an air-water interface and subsequently forming bilayers. On illumination, steady-state photocurrents were observed in all three cases, indicating that those methods had led to a functional, transmembraneous integration of the protein in the BLM. It appeared also possible [Bamberg et al, 1982] to make photoelectrically active planar lipid bilayers by adding vesicles containing monomeric bacteriorhodopsin to the bathing solution. In that case only transient photocurrents were observed, indicating that the vesicles had attached to, but not fused with, the BLM.

Reconstitution Via Apposition of Monolayers Self-Assembled From Protein-Lipid Vesicle Suspensions

Monolayers appear to form spontaneously at the surface of vesicle suspensions [Schindler, 1979]. The vesicles can be of any composition ranging from pure lipid to native biomembrane vesicles. The monolayers can then be adjoined to form a planar bilayer according to Montal and Mueller [1972]. The success of the technique appears to depend on the use of unilamellar vesicles with radius larger than 50 nm. This approach was first successfully applied to the outer membrane of *E. coli* containing the channel protein porin [Schindler and Rosenbusch, 1978]. Subsequently both Schindler's [Schindler and Quast, 1980] and Montal's [Nelson et al, 1980] laboratory reported the successful application of this method in the reconstitution of the acetylcholine receptor channel in planar bilayers. In the study by Schindler and Quast [1980], planar bilayers were formed in mixtures consisting of native membrane vesicles rich in ACh receptor from *Torpedo marmorata* and pure lipid (soybean lecithin: cholesterol, 6:1 molar ratio) vesicles in various ratios. In the reconstitution reported by Nelson and co-workers [1980], planar bilayers were formed from vesicles containing ACh receptor which had been previously solubilized in a lipid-cholate mixture and purified by affinity chromatography using toxin-agarose and concanavalin A (con A)-agarose affinity columns. In both cases the pharmacological specificity, conductance step size, selectivity, and kinetics matched those of the physiological system.

Reconstitution Into Planar Bilayers by Fusion of Vesicles

The interaction of lipid vesicles with each other, with cells, and with planar bilayers is currently being studied in a number of laboratories. One of the objectives of such studies is to gain insight into the mechanism of fusion of biological membranes. In vesicle-cell contact, a number of different ways of interaction, such as fusion, stable adsorption, exchange and endocytosis have been identified [Pagano and Weinstein, 1978]. In the case of planar bilayers all such modes of interaction, except endocytosis, could be involved. Reconstitution of membrane proteins into a planar bilayer can be achieved by both fusion and exchange. For the study of channel proteins in planar bilayers, incorporation of the protein into the bilayer membrane is absolutely required. For ion pumps, however, stable adsorption to the planar bilayer seems to be sufficient to allow one to observe bioelectric currents, driven, for instance, by a light pulse. A number of laboratories [Drachev et al, 1974; Herrmann and Rayfield, 1978; Bamberg et al, 1979] have successfully exploited this method to study the kinetics of light sensitive ion pumps. Apparently the electrochemical gradient set up by the adsorbed vesicles can directly be transferred across the planar bilayer and thus

TABLE III. Transfer of Membrane Channels From Vesicles to Planar Bilayers*

Planar bilayer	Vesicle source	In planar bilayer	Reference
BLM asolectin/PE 1:1	Sarcoplasmic reticulum (rabbit muscle)	Voltage-gated K^+ channel	Coronado and Miller [1979]
BLM PE/PG 2.3:1	Electric organ (*Torpedo california*)	Voltage-gated anion channel	White and Miller [1979]
BLM asolectin/ diphytanoyl PC 9:1 or PS/PC 4:1	VDAC-liposomes (rat mitochondria)	VDAC channel	Cohen et al [1980]
Monolayer apposition asolectin	Porin-liposomes (*E. coli*)	Porin channel	Cohen et al [1982]
BLM PE/PS 2.3:1	T-tubule membranes (rabbit muscle)	Ca^{2+}-dependent K^+ channel	Latorre et al [1982]
BLM PE/PS 1:1	Cardiac sarcolemma (calf ventricle)	95 pS K^+ channel 28 pS K^+ channel 15 pS K^+ channel 55 pS K^+ channel	Coronado and Latorre [1982]
Monolayer apposition (1, 3—SMPC)	Electric organ (*Torpedo marmorata*)	AChR channel	Boheim et al [1981]
Monolayer apposition PE/DPG 4:1	Rat brain	Ca^{2+}-dependent K^+ channel	Krueger et al [1982]

*BLM, black lipid membrane; PE, phosphatidylethanolamine; PG, phosphatidylglycerol; PS, phosphatidylserine; PC, phosphatidylcholine; SMPC, stearoylmyristoylphosphatidylcholine; DPG, diphosphatidylglycerol; AChR, acetylcholine receptor; VDAC, voltage-dependent anion channel.

measured. The process can be enhanced by adding ionophores to the planar bilayer.

Since the fusion of membranes, as occurring in living systems, is a highly complicated and tightly regulated process, it seems, a priori, unlikely that pure lipid membranes or even isolated native membranes will fuse spontaneously without the aid of the biologically developed fusion 'machinery'. The forces between membranes are such that there is a balance of attractive van der Waals forces and electrostatic repulsive forces at an intermembrane distance of about 20 Å [Parsegian et al, 1979]. At smaller distances there is an enormous repulsive energy barrier, for which the hydration of the membrane surface is responsible. There are model systems where those barriers are overcome; they consist of charged phospholipids like phosphatidylserine or phosphatidic acid [Papahadjopoulos et al, 1978]. Ca^{2+} ions can bridge the membranes by complexing the charged groups. The next step involved in fusion has been postulated to consist of either a phase separation of lipids leading to the formation of a defect, or formation of an inverted micelle. This stage is designated an intermediary fusion complex. It has been postulated that an osmotic gradient is necessary to facilitate the breaking up of the intermediary fusion complex [Pollard et al, 1981; Cohen et al, 1982]. Fulfillment of the three conditions outlined above—negatively charged lipids, presence of Ca^{2+}, and an osmotic gradient—seems to be necessary for efficient vesicle-BLM fusion. Another way to facilitate fusion is to incubate the vesicles with the planar bilayer at the phase transition temperature of its lipids [Hanke et al, 1981].

A list of channel proteins which have been transferred from vesicles to planar bilayers is given in Table III. For all cases it has been concluded that transfer was brought about by fusion. Rigorous proof for fusion as the underlying transfer mechanism has been obtained in only one case—the transfer of VDAC, reported by Zimmerberg and coworkers [1980]. In order to speak of fusion, one has to demonstrate that transfer of vesicle contents to the trans side has occurred. In addition, one has to show functional transfer of membrane markers to the planar membrane. Judging from Table III, vesicle-planar bilayer transfer seems to be a promising avenue for incorporation of transport molecules into membranes. Bamberg et al [1982] reported that bacteriorhodopsin could not be incorporated into planar bilayers via vesicle transfer, where it could by other methods. In a number of cases the channel incorporated into the planar bilayer was different from the one expected considering the source of the vesicle material. For instance Coronado and Miller [1979] found a voltage-gated K^+ channel upon incubation of a BLM with sarcoplasmic reticulum vesicles. One would have expected to observe the incorporation of a Ca^{2+} transport system. Similarly, incubation of vesicles from the electric organ of *Torpedo californica* yielded voltage-gated anion channels rather than the expected ACh receptor channels [White and Miller, 1979]. In conclusion, if such pitfalls can be avoided (and if a lucky choice of the type of vesicle and planar bilayer is made), vesicle transfer may prove to be the method of choice for reconstitution of membrane proteins into planar bilayers.

REFERENCES

Allen TM, Romans AY, Kercret H, Segrest JP (1980): Detergent removal during membrane reconstitution. Biochimica et Biophysica Acta 601:328–342.

Aloj SM, Kohn LD, Lee G, Meldolesi MF (1977): The binding of thyrotropin to liposomes containing gangliosides. Biochem Biophys Res Comm 74:1053

Andreasen TJ, Doerge DR, McNamee MG (1979): Effects of phospholipase A_2 on the binding and ion permeability control properties of the acetylcholine receptor. Arch Biochem Biophys 194:468–480.

Anholt R, Lindstrom J, Montal M (1980): Functional equivalence of monomeric and dimeric forms of purified acetylcholine receptors from *Torpedo californica* in reconstituted lipid vesicles. Eur J Biochem 109:481–487.

Anholt R, Lindstrom J, Montal M (1981): Stabilization of acetylcholine receptor channels by lipids in cholate solution and during reconstitution in vesicles. J Biol Chem 256:4377–4387.

Anholt R, Fredkin DR, Deerinck T, Ellisman M, Montal M, Lindstrom J (1982): Incorporation of acetylcholine receptors into liposomes. J Biol Chem 257:7122–7134.

Ashwell G, Morell AG (1974): The role of surface carbohydrates in the hepatic recognition and transport of circulating glycoproteins. Adv Enzymol 41:99–128.

Bamberg E, Apell HJ, Dencher N, Sperling W, Stieve H, Lauger P (1979): Photocurrents generated by bacteriorhodopsin in planar bilayer membranes. Biophys Struct Mech 5:277–292.

Bamberg E, Dencher NA, Fahr A, Heyn MP (1982): Transmembranous incorporation of photoelectrically active bacteriorhodopsin in planar lipid bilayers. Proc Natl Acad Sci USA 78:7502–7506.

Baumann H, How E, Doyle D (1980): Insertion of biologically active membrane proteins from rat liver into the plasma membrane of mouse fibroblasts. J Biol Chem 258:10001–10012.

Benz R, Janko D, Boos W, Lauger P (1978): Formation of large, ion-permeable membrane channels by the matrix protein (porin) of *Escherichia coli*. Biochim Biophys Acta 511:305–319.

Bligh EG, Dyer WJ (1959): A rapid method of total lipid extraction and purification. Can J Biochem Physiol 37:911–917.

Blumenthal R, Klausner RD (1982): The interaction of proteins with black lipid membranes in membrane reconstitution. Cell Surface Rev 8:42–82.

Blumenthal R, Klausner RD, Weinstein JN (1980): Voltage-dependent translocation of the asialoglycoprotein receptor across lipid membranes. Nature 288:333–338.

Boheim G, Hanke W, Barrantes FJ, Eibl H, Sakmann B, Fels G, Maelicke A (1981): Agonist activated ionic channels in acetylcholine receptor reconstituted into planar lipid bilayers. Proc Natl Acad Sci USA 78:3586–3590.

Bradley RJ, Howell JH, Romine WO, Carl GF, Kemp GE (1976): Characterization of a nicotinic acetylcholine receptor from rabbit skeletal muscle and reconstitution in planar phospholipid bilayers. Biochem Biophys Res Commun 68:577.

Briley MS, Changeux J-P (1978): Recovery of some functional properties of the detergent-extracted cholinergic receptor protein from *Torpedo marmorata* after reintegration into a membrane environment. Eur J Biochem 84:429–439.

Chang HW, Bock E (1979): Structural stabilization of isolated acetylcholine receptor: Specific interaction with phospholipid. Biochemistry 18:172–179.

Citri Y, Schramm M (1980): Resolution, reconstitution and kinetics of the primary action of a hormone receptor. Nature 287:297.

Cohen FS, Zimmerberg J, Finkelstein A (1980): Fusion of phospholipid vesicles with planar phospholipid bilayer membranes, II. Incorporation of a vesicular membrane marker into the planar membrane. J Gen Physiol 75:252–269.

Cohen FS, Akabas MM, Finkelstein A (1982): Osmotic swelling of phospholipid vesicles causes them to fuse with a planar phospholipid bilayer membrane. Science 217:458–460.

Conrad DH, Berczi I, Froese A (1976): Characterization of the target cell receptor for IgE. Immunochemistry 13:329–332.

Conti-Tronconi BM, Raftery MA (1982): The nicotinic cholinergic receptor. Annu Rev Biochem 51:491–530.

Coronado R, Miller C (1979): Voltage-dependent caesium blockade of a cation channel from fragmented sarcoplasmic reticulum. Nature 280:807–810.

Coronado R, Latorre R (1982): Detection of K^+ and Ca^+ channels from calf cardiac sarcolemma in planar lipid bilayer membranes. Nature 796:849–851.

Dalziel AW, Rollins ES, McNamel MG (1980): The effect of cholesterol on agonist-induced flux in reconstituted acetylcholine receptor vesicles. FEBS Letters 122:193–196.

Darszon A, Vandenberg CA, Schonfeld M, Ellisman MH, Spitzer NC, Montal M (1980): Reassembly of protein-lipid complexes into large bilayer vesicles: Perspectives for membrane reconstitution. Proc Natl Acad Sci USA 77:239–243.

DeKruijff B, Verkley AJ, Van Echteld CJA, Gerritsen WJ, Mombers C, Noordam PC, DeGier J (1979): The occurrence of lipidic particles in lipid bilayers as seen by pnmr and freeze-fracture electron-microscopy. Biochim Biophys Acta 555:200–209.

Dolly JO, Barnard EA (1977): Purification and characterization of an acetylcholine receptor from mammalian skeletal muscle. Biochemistry 16:5053–5060.

Drachev LA, Jasaitis AA, Koulen AD, Kondrashin HA, Liberman EA, Hayrecek IB, Ostroumov SA, Semenov AY, Skulachev VP (1974): Direct measurement of electric current generation by cytochrome oxidase, H^+-ATPase and bacteriorhodopsin. Nature 249:321–323.

Ehrenstein G, Lecar H (1977): Electrically gated ionic channels in lipid bilayers. Q Rev Biophys 10:1–84.

Eimerl S, Neufeld G, Korner M, Schramm M (1980): Functional implantation of a solubilized β-adrenergic receptor in the membrane of a cell. Proc Natl Acad Sci USA 77:760–764.

Epstein M, Racker E (1978): Reconstitution of carbamylcholine-dependent sodium ion flux and desensitization of the acetylcholine receptor from *Torpedo californica*. J Biol Chem 253:6660–6662.

Eytan GD (1982): Use of liposomes for reconstitution of biological functions. Biochim Biophys Acta 694:185–202.

Fleming JW, Ross EM (1980): Reconstitution of beta-adrenergic receptors into phospholipid vesicles: Restoration of ^{125}I-iodohydroxybenzylpindolol binding to digitonin-solubilized receptors. Cyclic Nucleotide Res 516:407–419.

Froese A (1980): Structure and function of the receptor for IgE. Crit Rev Immunol 1:79–132.

Froman G, Acevedo F, Lundahl P, Hjerten S (1980): The glucose transport activity of human erythrocyte membranes. Reconstitution in phospholipid liposomes and fractionation by molecular sieve and ion exchange chromatography. Biochim Biophys Acta 600:489–501.

Gitler C, Montal M (1972): Formation of decane-soluble proteolipids: Influence of monovalent and divalent cations. FEBS Lett 28:329–332.

Goldin SM (1977): Active transport of sodium and potassium ions by the sodium and potassium ion-activated adenosine triphosphatase from renal medulla. J Biol Chem 252:5630–5642.

Goldin SM, Tong SW (1974): Reconstitution of active transport catalyzed by the purified sodium and potassium ion-stimulated adenosine triphosphatase from canine renal medulla. J Biol Chem 249:5907–5915.

Gonzalez-Ros JM, Paraschos A, Martinez-Carrion M (1980): Reconstitution of functional membrane-bound acetylcholine receptor from isolated *Torpedo californica* receptor protein and electroplax lipids. Proc Natl Acad Sci USA 77:1796–1800.

Gonzalez-Ros JM, Paraschos A, Farach MC, Martinez-Carrion M (1981): Characterization of acetylcholine receptor isolated from *Torpedo californica* electroplax through the use of an easily removable detergent, β-O-octylglucopyranoside. Biochim Biophys Acta 643:407–420.

Gould RJ, Ginsberg BH, Spector AA (1982): Lipid effects on the binding properties of a reconstituted insulin receptor. J Biol Chem 257:477–484.

Hagins WA (1972): The visual process. Annu Rev Biophys Bioeng 1:131.

Hanke W, Eibl H, Boheim G (1981): A new method for membrane reconstitution: Fusion of protein-containing vesicles with planar bilayer membranes below lipid phase transition temperature. Biophys Struct Mech 7:131–137.

Hazelbauer HL, Changeux JP (1974): Reconstitution of a chemically excitable membrane. Proc Natl Acad Sci USA 71:1479–1483.

Heidmann T, Sobel A, Popot J-L, Changeux J-P (1980): Reconstitution of a functional acetylcholine receptor. Eur J Biochem 110:35–55.

Helenius A, Simons K (1975): Solubilization of membranes by detergents. Biochim Biophys Acta 415:29–79.

Herbette L, Scarpa A, Blasie JK, Bauer DR, Wang CT, Fleischer S (1981): Functional characteristics of reconstituted sarcoplasmic reticulum membranes as a function of the lipid-to-protein ratio. Biophys J 36:27–46.

Herrmann TR, Rayfield GW (1978): The electrical response to light of bacteriorhodopsin in planar membranes. Biophys J 21:111–125.

Holloway PW (1973): A simple procedure for the removal of Triton X-100 from protein samples. Anal Biochem 53:304–308.

Hudgin RL, Pricer WE, Ashwell G, Stockert RJ, Morell AG (1974): The isolation and properties of a rabbit liver binding protein specific for asialoglycoproteins. J Biol Chem 249:5536–5543.

Huganir RL, Schell MA, Racker E (1979): Reconstitution of the purified acetylcholine receptor from *Torpedo californica.* FEBS Lett 108:155.

Jackson ML, Litman BJ (1982): Rhodopsin-phospholipid reconstitution by dialysis removal of octylglucoside. Biochemistry 21:5601–5608.

Jackson ML, Schmidt CF, Lichtenberg D, Litman BJ, Albert AD (1982): Solubilization of phosphatidylcholine bilayers by octylglucoside. Biochemistry 21:4576–4582.

Jeffery DR, Charlton RR, Venter JC (1980): Reconstitution of turkey erythrocyte β-adrenergic receptors into human erythrocyte acceptor membranes. J Biol Chem 255:5015–5018.

Kafka MS, Blumenthal R, Walker GA, Pollard MB (1978): The effect of dopamine-β-hydroxylase on the electrical conductance of bimolecular lipid membrane. Membrane Biochem 1:279.

Kagawa Y, Kandrach A, Racker E (1973): Partial resolution of the enzymes catalyzing oxidative phosphorylation. J Biol Chem 248:676–684.

Kagawa Y, Racker E (1971): Partial resolution of the enzymes catalyzing oxidative phosphorylation. J Biol Chem 246:5477–5487.

Kanner B (1970): Solubilization and reconstitution of the γ-aminobutyric transporter from rat brain. FEBS Lett 89:47–50.

Kasahara M, Hinkle PC (1976): Reconstitution of D-glucose transport catalyzed by a protein fraction from human erythrocytes in sonicated liposomes. Proc Natl Acad Sci USA 73:396–400.

Kaufman K, Silman I (1980): The induction of ion channels through excitable membranes by acetylcholinesterase. Naturwissenschaften 67:608–610.

Kilian PL, Dunlap CR, Meuller P (1980): Reconstitution of acetylcholine receptor from *Torpedo californica* with highly purified phospholipids: Effect of α-tocopherol, phylloquinone, and other terpenoid quinones. Biochem Biophys Res Commun 93:409–414.

King TE (1963): Reconstitution of respiratory chain enzyme systems. J Biol Chem 238:4037–4051.

Klausner RD, Bridges K, Tsunoo H, Blumenthal R, Weinstein JN, Ashwell G (1980): Reconstitution of the hepatic asialoglycoprotein receptor with phospholipid vesicles. Proc Natl Acad Sci USA 77:5087–5091.

Knowles AF, Racker E (1974): Properties of a reconstituted calcium pump. J Biol Chem 250:3538–3544.

Krueger BK, French RJ, Blaustein MB, Worley JF (1982): Incorporation of Ca^{2+} activated K^+ channels from rat brain into planar bilayers. Biophys J 37:170a.

Kulczycki A, Metzger H (1974): The interaction of the IgE with rat basophilic leukemia cells. J Exp Med 140:1676–1695.

Latorre R, Vergara C, Hidalgo C (1982): Reconstitution in planar lipid bilayers of a calcium dependent potassium channel from transverse tubule membranes isolated from rabbit skeletal muscle. Proc Natl Acad Sci USA 79:805–809.

Lee G, Consiglio E, Habig W, Dyer S, Hardegree C, Kohn LD (1978): Structure:function studies of receptors for thyrotropin and tetanus toxin: Lipid modulation of effector binding to the glycoprotein receptor component. Biochem Biophys Res Commun 83:313–320.

Lindstrom J, Anholt R, Einarson B, Engel A, Osame M, Montal M (1980): Purification of acetylcholine receptors, reconstitution into lipid vesicles, and study of agonist-induced cation channel regulation. J Biol Chem 255:8340–8350.

MacLennan DH, Khanna VK, Stewart PS (1976): Restoration of calcium transport in the trypsin-treated (Ca^{2+} Mg^{2+})-dependent adenosine triphosphatase of sarcoplasmic reticulum exposed to sodium dodecylsulfate. J Biol Chem 251:7271–7279.

Mescher MF, Jose MJL, Balle SP (1981): Actin-containing matrix associated with the plasma membrane of murine tumour and lymphoid cells. Nature 289:139–144.

Metzger H, Ishizaka T (1982): Transmembrane signaling by receptor aggregation: The mast cell receptor for IgE as a case study. Fed Proc 41:7–34.

Michaelson DM, Raftery MA (1974): Purified acetylcholine receptor: Its reconstitution to a chemically excitable membrane. Proc Natl Acad Sci USA 71:4768–4772.

Michaelson DM, Duguid JR, Miller DL, Raftery MA (1976): Reconstitution of a purified acetylcholine receptor. J Supramol Struct 419:379–385.

Mimms LT, Zampighi G, Nozaki Y, Tanford C, Reynolds JA (1981): Phospholipid vesicle formation and transmembrane protein incorporation using octyl glucoside. Biochemistry 20:833–840.

Montal M, Mueller P (1972): Formation of bimolecular membranes from lipid monolayers, and a study of their electrical properties. Proc Natl Acad Sci USA 69:3561–3565.

Montal M, Darszon A, Trissl HW (1977): Transmembrane channel formation in rhodopsin-containing bilayer membranes. Nature 267:221–225.

Montal M, Darszon A, Schindler H (1981): Functional reassembly of membrane proteins in planar lipid bilayers. Q Rev Biophys 14:1–79.

Mueller P, Rudin DO, Tien HT, Wescott WC (1964): Formation and properties of bimolecular lipid membranes. Recent Prog Surf Sci 1:379.

Neher E, Sakmann B, Steinbach JH (1978): The extracellular patch clamp: A method for resolving currents through individual open channels in biological membranes. Pflugers Arch Ges Physiol 375:219.

Nelson N, Anholt R, Lindstrom J, Montal M (1980): Reconstitution of purified acetylcholine receptors with functional ion channels in planar lipid bilayers. Proc Natl Acad Sci USA 77:3057–3062.

Newman MJ, Wilson TH (1980): Solubilization and reconstitution of the lactose transport system from *Escherichia coli.* J Biol Chem 255:10583–10586.

Nozaki Y, Lasic DD, Tanford C, Reynolds JA (1982): Size analysis of phospholipid vesicle preparations. Science 217:366.

Padan E, Schuldiner S, Kaback HR (1979): Reconstitution of *LAC* carrier function in cholate-extracted membranes from *Escherichia coli*. Biochem Biophys Res Commun 91:854–861.

Pagano RE, Weinstein JN (1978): Interaction of liposomes with mammalian cells. Annu Rev Biophys Bioeng 7:435–468.

Papahadjopoulos D, Portis A, Pangborn W (1978): Calcium-induced lipid phase transitions and membrane fusion. Ann NY Acad Sci 308:50–66.

Parsegian VA, Fuller N, Rand RP (1979): Measured work of deformation and repulsion of lecithin bilayers. Proc Natl Acad Sci USA 76:2750–2754.

Pedersen SE, Ross EM (1982): Functional reconstitution of β-adrenergic receptors and the stimulatory GTP-binding protein of adenylate cyclase. Proc Natl Acad Sci USA 79:7228–7232.

Pollard HB, Creutz CE, Pazoles CJ (1981): Mechanisms of calcium action and hormone release during exocytosis. Recent Prog Horm Res 37:299–332.

Popot J-L, Cartaud J, Changeux J-P (1981): Reconstitution of a functional acetylcholine receptor. Eur J Biochem 118:203–214.

Popot J-L, Demel RA, Sobel A, Van Deenen LM, Changeux J-P (1978): Interaction of the acetylcholine (nicotinic) receptor protein from *Torpedo marmorata* electric organ with monolayers of pure lipids. Eur J Biochem 85:27–42.

Racker E (1972): Reconstitution of cytochrome oxidase vesicles and conferral of sensitivity to energy transfer inhibitors. J Membrane Biol 10:221–235.

Racker E (1973): A new procedure for the reconstitution of biologically active phospholipid vesicles. Biochem Biophys Res Commun 55:224–230.

Racker E (1979): Reconstitution of membrane processes. Methods Enz. 55:699–711.

Ramakrishnan V, Darszon A, Philipp M, Montal M (1980): Rhodopsin in model membranes: The kinetics of channel opening and closing in rhodopsin-containing planar lipid bilayers. Ann NY Acad Sci 358:36–42.

Rivnay B, Metzger H (1982): Reconstitution of the receptor for immunoglobulin E into liposomes. J Biol Chem 257:12800–12808.

Rivnay B, Metzger H (1983): Use of the airfuge for analysis and preparation of receptors incorporated into liposomes: Studies with the receptor for immunoglobulin E. Anal Biochem 130:514–520.

Rogner M, Ohno D, Hamamoto T, Sone N, Kagawa Y (1979): Net ATP synthesis in H^+-ATPase macroliposomes by an external electric field. Biochem Biophys Res Commun 91:362–367.

Ross EM, Gillman AG (1980): Biochemical properties of hormone-sensitive adenylate cyclase. Annu Rev biochem 49:533–564.

Rossi G, Newman SA, Metzger H (1977): Assay and partial characterization of the solubilized cell surface receptor for immunoglobulin E. J Biol Chem 252:704–711.

Schein SJ, Colombini M, Finkelstein A (1976): Reconstitution in planar lipid bilayers of a voltage-dependent anion-selective channel obtained from *Paramecium mitochondria.* J Membrane Biol 30:99–120.

Schein SJ, Kagan BL, Finkelstein A (1978): Colicin K acts by forming voltage-dependent channels in phospholipid bilayer membranes. Nature 276:159–163.

Schiebler W, Hucho F (1978): Membranes rich in acetylcholine receptor: Characterization and reconstitution to excitable membranes from exogenous lipids. Eur J Biochem 85:55–63.

Schindler H, Rosenbusch JP (1978): Matrix protein from *Escherichia coli* outer membranes forms voltage-controlled channels in lipid bilayers. Proc Natl Acad Sci USA 75:3751–3755.

Schindler H, Quast U (1980): Functional acetylcholine receptor from *Torpedo marmorata* in planar membranes. Proc Natl Acad Sci USA 77:3052–3056.

Schindler M (1979): Exchange and interactions between lipid layers at the surface of a liposome solution. Biochim Biophys Acta 555:316–336.

Schramm M (1979): Transfer of glucagon receptor from

liver membranes to a foreign adenylate cyclase by a membrane fusion procedure. Proc Natl Acad Sci USA 76:1174–1178.

Shamoo AE (ed) (1975): Carriers and channels in biological systems. Ann NY Acad Sci 264:1–485.

Shamoo AE (ed) (1980): Second international conference on carriers and channels in biological systems-transport proteins. Ann NY Acad Sci 358:1–387.

Simonds WF, Koski G, Streaty RA, Hjelmeland LM, Klee WA (1980): Solubilization of active opiate receptors. Proc Natl Acad Sci USA 77:4623–4627.

Sobel A, Weber M, Changeux J-P (1977): Large-scale purification of the acetylcholine-receptor protein in its membrane-bound and detergent-extracted forms from *Torpedo marmorata* electric organ. Eur J Biochem 80:215–224.

Sobel A, Heidmann T, Cartaud J, Changeux J-P (1980): Reconstitution of a functional acetylcholine receptor. Eur J Biochem 110:13–33.

Strauss WL, Ghai G, Fraser CM, Venter JC (1979): Detergent solubilization of mammalian cardiac and hepatic β-adrenergic receptors. Arch Biochem Biophys 196:556–573.

Stubbs GW, Smith HG Jr, Litman BJ (1976): Alkyl glucosides as effective solubilizing agents for bovine rhodopsin. A comparison with several commonly used detergents. Biochim Biophys Acta 425:46–56.

Tanford C, Reynolds JA (1976): Characterization of membrane proteins in detergent solutions. Biochim Biophys Acta 457:133–170.

Tate RL, Schwartz HI, Holmes JM, Kohn LD (1975): Thyrotropin receptors in thyroid plasma membranes. J Biol Chem 250:6509–6515.

Verkleij AJ, Mombers C, Gerritsen WJ, Leunissen-Bijvelt L, Cullis PR (1979): Fusion of phospholipid vesicles in association with the appearance of lipidic particles as visualized by freeze fracturing. Biochim Biophys Acta 555:358–361.

Wank SA, Delisi C, Metzger H (1983): Analysis of the rate-limiting step in a ligand-cell receptor interaction: The immunoglobulin E system. Biochemistry 22:954–959.

White MM, Miller C (1979): A voltage-gated anion channel from the electric organ of Torpedo californica. J Biol Chem 254:10161–10166.

Wickner W (1976): Asymmetric orientation of phage M13 coat protein in Escherichia coli cytoplasmic membranes and in synthetic lipid vesicles. Proc Natl Acad Sci USA 73:1159–1163.

Winget GD, Kanner N, Racker E (1977): Formation of ATP by the adenosine triphosphatase complex from spinach chloroplasts reconstituted together with bacteriorhodopsin. Biochim Biophys Acta 460:490–499.

Wu WC-S, Moore H-PH, Raftery MA, (1981): Quantitation of cation transport by reconstituted membrane vesicles containing purified acetylcholine receptor. Proc Natl Acad Sci USA 78:775–779.

Wu WC-S, Raftery MA (1981a): Functional properties of acetylcholine receptor monomeric and dimeric forms in reconstituted membranes. Biochem Biophys Res Commun 99:436–444.

Wu WC-S, Raftery MA (1981b): Reconstitution of acetylcholine receptor function using purified receptor protein. Biochemistry 20:694–701.

Zimmerberg J, Cohen FS, Finkelstein A (1980): Fusion of phospholipid vesicles with planar bilayer membranes, I. Discharge of vesicular contents across the planar membrane. J Gen Physiol 75:241–250.

Molecular and Chemical Characterization of Membrane Receptors, pages 241–259

10

Prospects for the Isolation of Genes for Receptors and Other Proteins of Pharmacological and Neurobiological Interest

Daniel L. Kaufman and Allan J. Tobin

Department of Biology, University of California, Los Angeles, California 90024

INTRODUCTION

Recombinant DNA technology is providing increasingly powerful tools for the isolation of specific genes. The purpose of this paper is to review a number of recent methods that should permit the isolation of genes of neurobiological and pharmacological interest, including neurotransmitter receptors and putative peptide neurotransmitters.

Once a gene of interest is isolated, a wide variety of previously inaccessible questions may be addressed by molecular techniques. The most straightforward set of such questions concerns the structure and organization of the gene itself at the nucleotide level. This information can then be used to evaluate hypotheses concerning the etiology of genetic diseases. For example, mutations within structural genes for receptors or in the adjacent sequences that govern their level of expression have been suggested as possible causes of diseases ranging from Huntington's disease to schizophrenia [Spokes, 1981; Seeman et al, 1980]. The definitive identification of a molecular lesion within a particular gene (or its regulatory sequences) would have important diagnostic and therapeutic value. It would immediately provide the clinician with important clues for conventional diagnosis and therapy. Additionally, it would be possible to diagnose such disorders prenatally by analysis of fetal DNA obtained by amniocentesis in a manner analogous to the diagnosis of sickle cell anemia and thalassemia [Fuchs, 1980]. Finally, completely new approaches of therapy could be developed. Already the understanding of the genetic defects responsible for β-thalassemia has led to the design of a number of molecular manipulations that may soon be clinically useful for the treatment of this disease. These approaches have included (1) stimulating the reexpression of fetal hemoglobin genes by treatment with 5-azacytidine, a drug known to alter patterns of both DNA methylation and gene expression [Ley et al, 1982]; (2) introducing normal hemoglobin genes into hematopoietic stem cells [Salser et al, 1981]; and (3) designing a new tRNA gene that can suppress a nonsense mutation in one form of thalassemia [Temple et al, 1982].

Isolated gene sequences may also be used as molecular probes (1) to study the expression of specific genes during development and the changes in gene expression that occur due to drug treatment, contact with other cells, disease and aging; (2) to identify the cell types within a complex tissue that express specific genes by in situ hybridization [Brahic and Haase, 1978; Hudson et al, 1981; Langer et al, 1981; Roberts et al, 1982; Szabo and Ward, 1982]; and (3) to find the chromosomal location of specific genes [Gerhard et al, 1981].

The availability of purified genes makes possible genetic engineering experiments in which the genotype of a cell line or an entire organism can be altered [Seeman et al, 1980; Pellicer et al, 1980; Anderson and Diacumakos, 1981; Williamson, 1982; Brinster and Palmiter, 1982; Rubin and Spradling, 1982; Palmiter et al, 1982]. Genes have now been introduced into mice, frogs, and *Drosophila*, as well as into a wide variety of tissue culture cells in order to study their regulation during development. A recent dramatic example of genetic engineering has been the introduction of a gene encoding growth hormone into the zygotes of mice, resulting in the development of adult mice weighing 80% more than their unengineered litter mates [Palmiter et al, 1982]. One can anticipate that the introduction of genes into organisms and tissue culture cells will be useful in creating experimental models for neurological disorders caused by alterations in the structure or the level of expression of specific proteins. These genetically engineered animal models could then be used to evaluate alternative methods of treatment for human diseases.

Finally, the ability of bacteria and eukaryotic cells to synthesize the proteins encoded in the isolated genes permits the purification of large quantities of specific proteins—either normal or engineered—for physical and pharmacological studies [Gilbert and Villa-Komaroff, 1980; Itakura et al, 1977; Ulmer, 1983; Guarente et al, 1980]. Such pure proteins would, for example, aid in the elucidation of ligand-receptor interactions, the structure of membrane channels, and in the design of new drugs that might alleviate neurological, endocrinological, and affective disorders.

The application of recombinant DNA technology has already contributed importantly to the understanding of peptide hormones and putative peptide neurotransmitters [Comb et al, 1982; Noda et al, 1982a; Gubler et al, 1982; Kakidani et al, 1982; Rossier, 1982], most strikingly in the unraveling of the structure and alternate routes of expression of the genes for mammalian pro-opiomelanocortin [Nakanishi et al, 1979; Roberts et al, 1979; Herbert et al, 1981], and *Aplysia* egg-laying hormone [Scheller et al, 1982]. These methods have also led to the identification of a nucleotide sequence that is unique to brain messenger RNA (mRNA) which may function as a marker for brain-specific genes [Sutcliffe et al, 1982].

Recently the gene sequences for a neurotransmitter receptor were isolated for the first time. These sequences encode the four subunits of the nicotinic acetylcholine (ACh) receptor of the ray *Torpedo californica* [Ballivet et al, 1982; Noda et al, 1982b; Noda et al, 1982c; Stevens, 1982]. The isolations were accomplished by taking advantage of the extreme specialization of the electric organs of *Torpedo,* in which the receptor polypeptides and their respective mRNAs are abundant. In contrast, mammalian tissue usually contains only low concentrations of receptors and their mRNAs. If, as expected, mammalian and fish ACh receptors are evolutionarily homologous the receptor sequences isolated from *Torpedo* will be immediately useful in obtaining their mammalian counterparts. The isolation of the mammalian ACh receptor genes may help elucidate the nature of neuromuscular diseases, including the autoimmune disease myasthenia gravis, which is known to involve an immunological attack on this receptor [Vincent, 1980].

The nucleotide and amino acid sequences of the *Torpedo* ACh receptor polypeptides suggest that the four subunits have evolved from a single ancestral gene, which has been duplicated several times during evolution [Noda et al, 1982c; Raftery et al, 1980]. This idea has been expanded into the broader suggestion that (1) several ligand-gated ion channels such as the acetylcholine receptors may have evolved from a common ancestor and may form a multigene family; and (2) several voltage-gated ion channels that regulate the passage of a particular ion may have similarly evolved from a common ancestor and form a multigene family [Stevens, 1982]. A model for the development of neurons may be proposed in which cytodifferentiation proceeds in stages by binary decisions: ie, a primary commitment to express particular gene families, followed by secondary choices within each family. The specific type of ion channels and neurotransmitter sensitivity exhibited by differentiated neurons may thus result from the developmental regulation of multigene families. If some channel proteins are evolutionarily related, sequence homologies may exist whereby the isolated gene sequences of one channel may serve as hybridization probes to isolate the genes of other channels.

The key to evaluating hypotheses such as these lies in the analysis of gene structure and of the molecular mechanisms that govern their expression. The application of recombinant DNA techniques will thus lead to a far deeper molecular understanding of how neurons differentiate and maintain their specialized characteristics, as well as to strategies for diagnosis and therapy when neuronal behavior is aberrant.

CLONING AND ISOLATION OF SPECIFIC GENES

The DNA sequences encoding specific proteins may be obtained from three sources: (1) chromosomal DNA, which contains essentially all the genetic information of an organism, (2) mRNA isolated from tissues that synthesize the protein of interest, and (3) chemical synthesis, if the entire amino acid sequence is known [Ulmer, 1983]. In principle the DNA sequence encoding a protein may be obtained from any of these sources. Generally, recombinant molecules containing fragments of genomic DNA are the most easily constructed. Once obtained they should contain all of the DNA sequences present in the organism. In contrast, recombinant molecules carrying sequences derived from mRNA will represent only those genes transcribed in the tissue from which the mRNA was obtained. By using mRNA as the source of genetic information, however, the sequence of interest may be enriched many fold compared to its presence in genomic DNA. Due to differences in eukaryotic and prokaryotic gene structure (Fig. 1), only recombinant molecules carrying sequences derived from mRNA are, in general, able to synthesize authentic proteins from their foreign DNA in bacteria. The third source of specific genes, complete chemical synthesis, is at present applicable to only small genes, such as those for transfer RNA and somatostatin. Short synthetic oligonucleotides corresponding to partial sequences in the protein of interest, however, may be used to isolate the entire gene from a collection of genomic or mRNA sequences. Below we review techniques for the isolation of specific genes from each of these sources. Further descriptions and experimental protocols may be found in several highly useful volumes [Maniatis et al, 1982; Wu, 1979; Grossman and Moldave, 1980].

Cloning of Genomic DNA

The isolation of specific genes from genomic DNA is made possible by the ability to (1) cut large molecules of DNA, such as those that compose the human chromosomes, into fragments using restriction endonucleases (enzymes that cut DNA at specific sequences), nucleases, or mechanical shearing; and (2) isolate and insert these fragments into the DNA of a bacterial plasmid or bacteriophage. Plasmids or bacteriophage DNA are necessary as carriers ("vectors") of foreign DNA because the independent fragments of eukaryotic DNA are incapable of self-replication in the bacterial host, and therefore would not be passed onto daughter cells. Plasmids and bacteriophages are naturally occurring circular nucleic acid molecules that replicate autono-

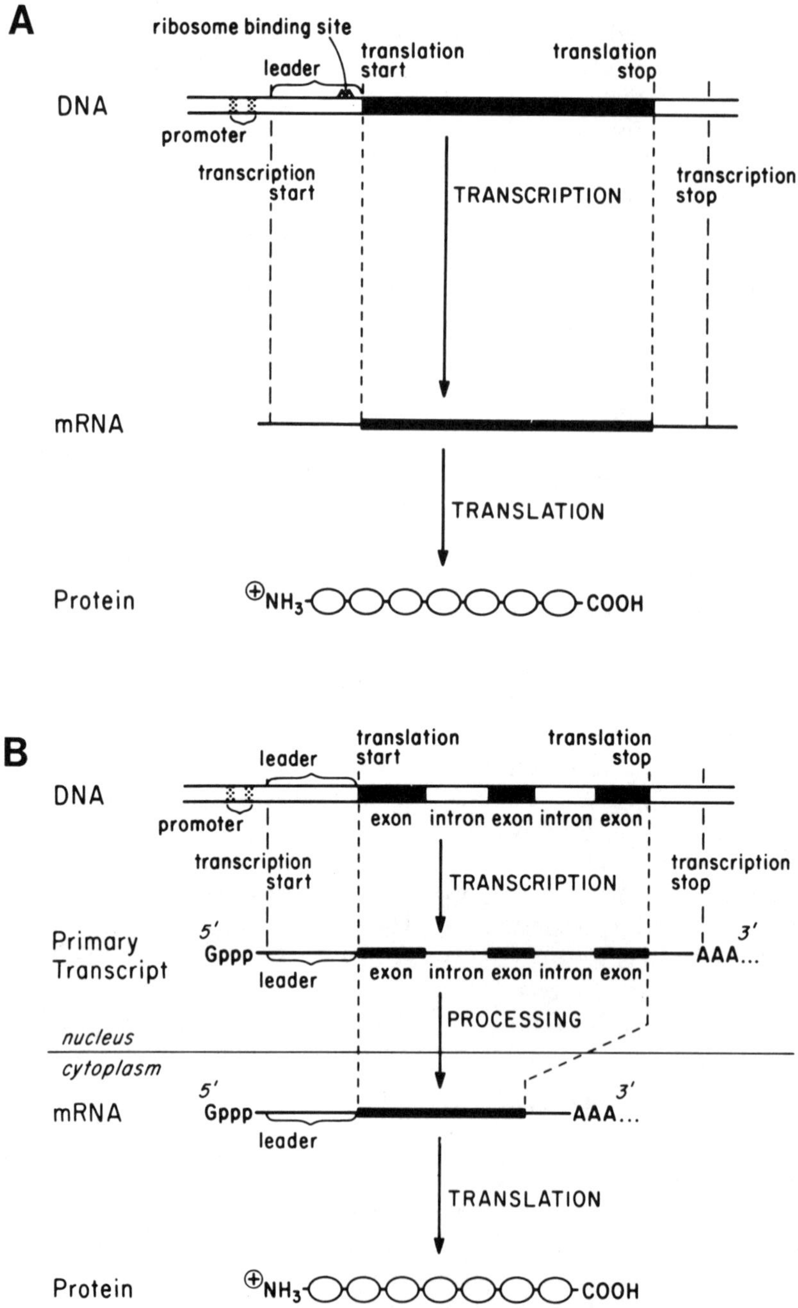

Fig. 1. *A) Information processing in prokaryotes. B) Information processing in eukaryotes.*

mously within their bacterial hosts by virtue of sequences known as origins of replication, which are recognized by the host cell's DNA replication enzymes. Many plasmid and bacteriophage DNAs have now been engineered to facilitate the insertion of foreign DNA and the subsequent isolation of the recombinant molecules of interest. To insert the foreign DNA into a vector, the vector DNA is cleaved open at a unique site by a restriction endonuclease. The vector is then joined to fragments of foreign DNA to form recombinant molecules, each of which contains a different piece of foreign DNA. The recombinant DNA molecules are then used to infect bacteria in such a way that each bacterium receives a single recombinant molecule, becoming "transformed" (in the case of a plasmid) or "transfected" (in the case of a bacteriophage). By growing each bacterium into an individual colony, or "clone," the fragments of foreign DNA are individually propagated and amplified many fold. Plasmid vectors carry genes conferring antibiotic resistance, so that bacteria containing recombinant DNA molecules may be selected by their acquired antibiotic resistance. In the case of bacteriophage vectors, each bacteriophage will produce a "plaque," or clear area, where the bacteriophage have killed the surrounding bacterial cells (Fig. 2).

Gene Libraries

Vectors vary in the size of inserted DNA that they can accommodate. Those with the largest capacity ("cosmids") can accept up to 40,000 base pairs of DNA (40 kilobase pairs = 40 kb). If the human haploid genome (3×10^6 kb of DNA) is broken randomly into fragments of 40 kb (by mechanical shearing or nulease digestion), it would take at least ($3 \times 10^6/40 =$) 7×10^4 independent clones to encompass all its sequences. Such a set of clones is referred to as a "library" or "gene bank." In practice, a larger library is required to ensure that any single sequence has a high probability of being represented. To ensure a 99% probability that a single sequence of interest is represented in a human genomic library composed of 40-kb fragments, it is necessary to construct approximately 4×10^5 clones. Such expanded libraries are not difficult to construct. The major difficulty lies in the identification of a particular clone of interest within any large library.

Screening: Colony and Plaque Hybridization

There are several strategies for the identification of a bacterial colony or phage plaque that contains a desired gene fragment. The most common procedure uses a radioactively labeled DNA or RNA probe that can specifically hybridize to a clone containing complementary sequences [Grunstein and Hogness, 1975; Benton and Davis, 1977; Suggs et al, 1981]. In this method, known as colony hybridization or plaque hybridization, the clones composing a library are transferrred onto nitrocellulose filters, which have a high affinity for DNA. The DNA is released from the bacteria (or bacteriophage) in situ by treatment with alkali and immobilized on the filter, thus forming a "DNA print" of all the colonies or plaques. The DNA on the filter can then be hybridized to a radioactively labeled nucleic acid probe if sufficient lengths of their nucleotide sequences are complementary (Fig. 2).

The probes for screening recombinant clones may be derived from purified mRNAs or be chemically synthesized oligonucleotides whose sequences correspond to known amino acid sequences in the proteins of interest. In either case, the probe will hybridize only to clones containing complementary sequences. The "DNA prints" to which hyridization occurred are then identified by autoradiography. The corresponding colony or plaque can then be picked from the reference plate, tested to verify its identity, and grown to produce large amounts of purified genes. Screening a library by colony or plaque hybridization to a specific probe is the most rapid procedure for the isolation of a single DNA sequence. The

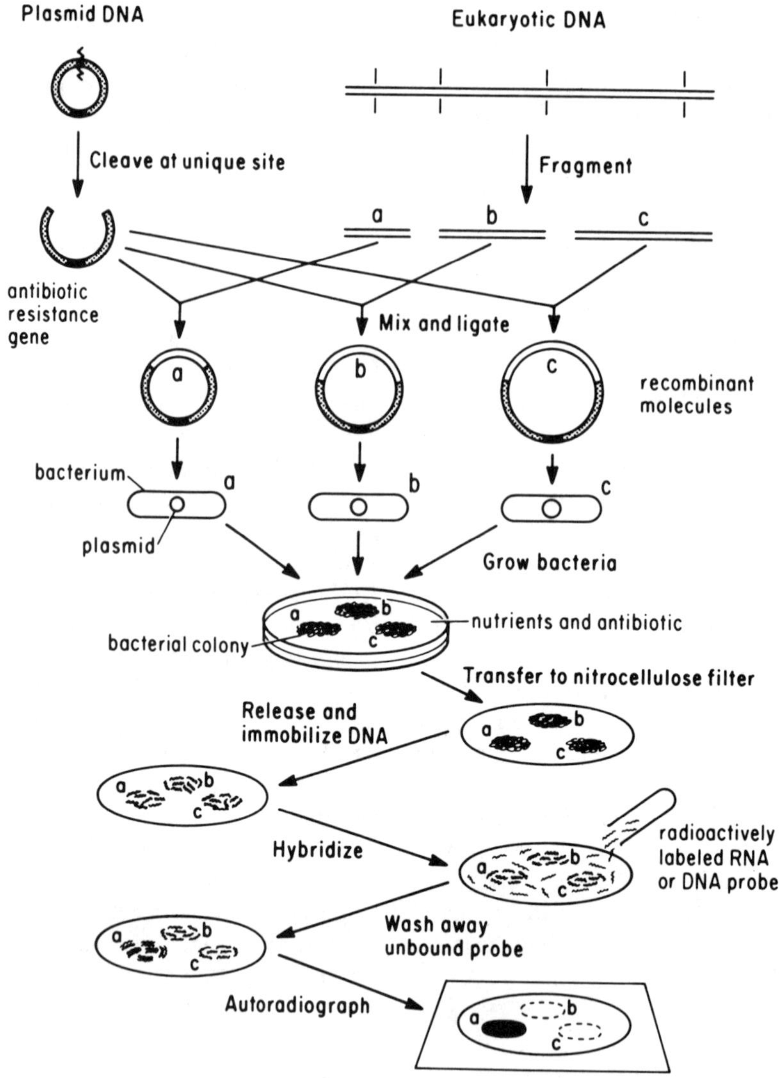

Fig. 2. *Molecular cloning of genomic DNA. Individual fragments of genomic DNA are inserted into a vector (either plasmid or bacteriophage) which is capable of replication in a bacterial host. The recombinant molecules are introduced into bacteria in such a way that each receives a single plasmid containing a unique fragment of foreign DNA. To identify the clone containing a specific foreign gene sequence, out of the many thousands that compose the "library," the method of nucleic acid probe hybridization is especially powerful: The clones are replicated onto nitrocellulose paper, which has a high affinity for DNA, the cells are lysed open, and their DNA is fixed in place. A labeled nucleic acid probe which is complementary to the desired gene sequence is added. The probe may arise from isolated mRNA, another cloned DNA, or be a synthetic oligonucleotide designed from a known amino acid sequence in the protein of interest. Unbound probe is washed away, and the nitrocellose paper is autoradiographed. The exposed spot on the autoradiograph corresponds to the colony containing the desired gene fragment.*

10^5–10^6 colonies or plaques constituting a large library may be grown on a few culture plates and screened in a matter of days.

Screening: Hybrid Selection

If specific nuleic acid probes are unavailable, a cloned DNA sequence may be identified by its ability to bind the mRNA coding for a specific protein in a method called hybrid selection [Ricciardi et al, 1979; Pharnes et al, 1981; Harpold et al, 1978]. In this method, the clones composing a library are divided into groups. These are transferred and lysed on nitrocellulose filters as in colony hybridization. Each filter is then hybridized to total mRNA isolated from a tissue that expresses the protein of interest. Those mRNAs that find complementary DNA sequences will hybridize and remain on the filter while the unbound noncomplementary mRNA is washed away. The hybridized mRNA can then be eluted from the filter and translated into protein in a cell-free system or by microinjection into *Xenopus laevis* oocytes [Gurdon and Melton, 1981]. The protein products may then be analyzed by two-dimensional gel electrophoresis or may be detected immunologically with an appropriate antibody. If a group of clones is found that binds the mRNA for the protein of interest, the group may then be further divided into subsets and the procedure repeated until a single clone containing the desired DNA sequence is isolated. This hybrid selection procedure, however, is arduous, and it is limited to mRNAs that are sufficiently abundant to produce a detectable level of translation product.

Although the first method described, direct clone selection by hybridization to a labeled nuleic acid probe, is rapid and highly sensitive, the needed probes are often difficult to obtain for genes expressed at low levels: either the mRNA of interest must be purified to homogeniety away from the thousands of other mRNAs so that it may serve as a specific hybridization probe, or the protein product must be purified in sufficient quantity to obtain an amino acid sequence from which an oligonucleotide sequence may be designed for use as a hybridization probe. Since many of the most interesting neuronal proteins are likely to be present at low concentrations and arise from rare mRNAs, it will be difficult to prepare nucleic acid probes for screening libraries. There are, however, a vast number of antibodies, both serum and monoclonal, which can specifically recognize rare proteins. These may be used to screen libraries immunologically, if the bacteria composing the libary can be made to synthesize the proteins encoded in the inserted foreign DNA. Conceivably the proteins could also be identified by virtue of their ability to bind specific ligands.

EXPRESSION VECTORS

Both plasmid and bacteriophage "expression" vectors have been developed, which promote the synthesis of polypeptides encoded in the inserted DNA fragments [Itakura et al, 1977; Hans-Urlich and Helinski, 1979; Charnay et al, 1979; Koenen et al, 1982; Ruther et al, 1982; Gray et al, 1982; Helfman et al, 1983; Kaplan et al, 1983; Young and Davis, 1983; Okayama and Berg, 1983; Mulligan and Berg, 1980]. These vectors may be used to create "expression libraries," which may be screened immunologically for specific translation products. The translation products need not be complete or functional proteins: All that is required is the synthesis of a detectable antigenic determinant. Because monoclonal antibodies can be raised against rare proteins [see Fraser and Lindstrom, this volume] this may be the most feasible method for isolating genes whose protein products are too rare to be sequenced. Should protein sequences become available, the expression libraries may be readily screened by synthetic oligonucleotides.

Signals Needed for the Expression of Eukaryotic DNA in Bacteria

Although prokaryotes and eukaryotes share the same genetic code and have essentially the same transcription and translation processes, the genetic signals that direct these processes are distinct. Therefore expression vectors are designed to provide the inserted DNA fragment with the genetic signals necessary for expression in the bacterial host.

The transcription of the inserted DNA requires that it be preceded by a short DNA sequence, a "promoter," which is recognized by RNA polyermase as the point at which to bind and begin transcription. In a typical prokaryotic gene the promoter directs transcription to begin about 40 bases prior to the actual beginning (start codon) of the structural protein. This 40-base "leader" sequence on the mRNA contains the "Shine-Dalgarno sequence," which is essential for attachment of mRNA to ribosomes [Shine and Dalgarno, 1975]. After the leader sequence binds the mRNA to a ribosome, the ribosome moves down the mRNA until a start codon initiates translation. Translation continues until a stop codon is encountered, which triggers the release of the nascent polypeptide.

The promoters of eukaryotic genes do not function in bacteria, nor do eukaryotic mRNAs contain sequences that bind bacterial ribosomes. To enhance expression of the inserted eukaryotic DNA, therefore, expression vectors are constructed to contain strong bacterial promoters and ribosomal binding sites. These sequences have often been taken from the β-galactosidase gene of *Escherichia coli* or from genes of the bacteriophage-λ. Foreign DNA may be inserted directly adjacent to an isolated promoter and ribosomal binding site [Guarente et al, 1980] or may be placed within the coding region of a gene so as to use the promoter and ribosome binding site already present on that gene [Ulmer, 1983; Charnay et al, 1979; Koenen et al, 1982; Ruther et al, 1982; Gray et al, 1982; Helfman et al, 1983; Kaplan et al, 1983; Young and Davis, 1983].

Although insertion within another gene leads to the synthesis of a hybrid or "fusion" protein that is part eukaryotic and part prokaryotic, there are several advantages to the method. First, the host cell contains proteases that often degrade foreign proteins. For example, when the somatostatin gene was cloned into an expression vector that permitted the synthesis of somatostatin from near its own amino terminal, it was rapidly degraded in the cell [Itakura et al, 1977]. When the gene was recloned into the structure of a natural bacterial protein, however, the somatostatin was protected from degradation and the authentic somatostatin was produced by chemical cleavage from the fusion protein. Second, despite its covalent attachment to a bacterial polypeptide, the eukaryotic polypeptide appears to be able to form at least some of its normal antigenic determinants, thus allowing its immunological detection. Furthermore, the prokaryotic polypeptide to which it is fused possesses its own unique properties and antigenic sites, which may be useful in the screening procedure.

The most useful promoters are those in which transcription can be turned off and on at will by the addition of chemicals to the media or by changes in temperature. By initially growing the transformed (or transfected) cells in the uninduced state, one avoids synthesis of a protein product from a cloned sequence that may be deleterious to the host and that may prevent its representation in the libary.

Intervening Sequences

Although expression vectors are designed to provide the signals directing transcription and translation in the bacterial host, certain features of eukaryotic gene organization make the synthesis of an authentic gene product from inserts of genomic DNA unlikely. This is because most eukaryotic proteins are not encoded in continuous stretches of genomic DNA as are prokaryotic proteins, but rather in segments, called "exons," which are separated by segments of noncoding DNA, called "intervening sequences" or "introns" [Chambon, 1981; Brethnach and Chambon, 1981] (see Fig. 1). Several introns may exist within a gene, splitting the coding regions into several exons. In eukaryotic cells, introns are transcribed into RNA along with the exons, but they are not expressed: The intron sequences in the RNA are removed before the transcript leaves the nucleus, and the exon segments are spliced together to form the mature mRNA. Generally, only by copying mRNA into complementary DNA (cDNA) can the uninterrupted protein coding sequence be obtained (see below).

Exons average only 200 nuleotides in length, enough to encode about 65 amino acids, and altogether make up only approximately 5% of the human genome. (The remaining 95% of chromosomal DNA may represent introns, chromosomal structural elements, regulatory sequences, or evolutionary relics such as "pseudogenes.") Thus exons are short stretches of protein-encoding DNA surrounded by a vast amount of noncoding DNA. When fragments of eukaryotic genomic DNA are inserted into an expression vector, the vector directs the transcription and translation of both coding and noncoding DNA. The bacterial host, however, is unable to recognize and remove regions of the transcript which arise from noncoding DNA. As a result, the translation of these transcripts is unlikely to proceed far because noncoding DNA contains frequent translational stop codons (since three of the 64 possible codons are translational stops). Thus, only partial fragments of a protein can be translated from genomic inserts. Although it is unlikely that a complete and authentic protein can be syntheized from genomic DNA inserts, it may be possible to detect the translation product of a single exon immunologically.

Exon Cloning

It may be possible to recognize immunologically a specific protein from the expression of just one of its exon regions, since each exon appears to encode a structural domain of the protein, and may therefore contain an antigenic site [Koenen et al, 1982; Ruther et al, 1982; Sutcliffe et al, 1983; Green et al, 1982; Lerner, 1983]. Once a single cloned exon from a structural gene is immunologically identified, it may serve as a hybridization probe for the isolation of the entire gene from a standard genomic or cDNA library. Several methods have been recently described in which genomic DNA is cut into fragments of about 200 nucleotide pairs and inserted into expression vectors [Koenen et al, 1982; Ruther et al, 1982; Gray et al, 1982]. By cutting the genomic DNA into such small fragments, one produces some inserts that contain only exon sequences alone or exon sequences flanked by small amounts of noncoding DNA. Although proteins are usually produced by the translation of several exon regions, such recombinant clones would synthesize polypeptides encoded by a single exon. The immunological screening of the resulting clones should be done with serum antibodies or with collections of monoclonal antibodies rather than with a single monoclonal antibody, since a single monoclonal antibody will recognize only one antigenic site. Polyclonal antibodies or a collection of monoclonal antibodies to a given protein, will on the other hand, recognize many different antigenic sites and thus increase the chances of detecting antigens formed by the folding of individual exon domains.

If the human genome is divided into 200-nucleotide-pair fragments, approximately 10^8 colonies or plaques would be necessary to ensure a 99% probability that each sequence would be represented at least once. Although each sequence may be represented, several other conditions must be met in order for a translation product to be identifiable: (1) The fragment must be in the correct reading frame, ie, the correct sets of nucleotide triplets must be read. Since there are three possible reading frames and the insert was produced by a random fragmentation, only one third of the inserts will be in the correct reading frame. (2) The fragment must also be in the correct orientation. Half of the time the insert will be inserted in an orientation opposite to that required for translation. Thus only $1/3 \times 1/2 = 1/6$ of the fragments will be in the correct reading frame and orientation for the synthesis of an authentic polypeptide (Fig. 3).

In practice, therefore, one must screen approximately 10^9 clones or plaques to find a

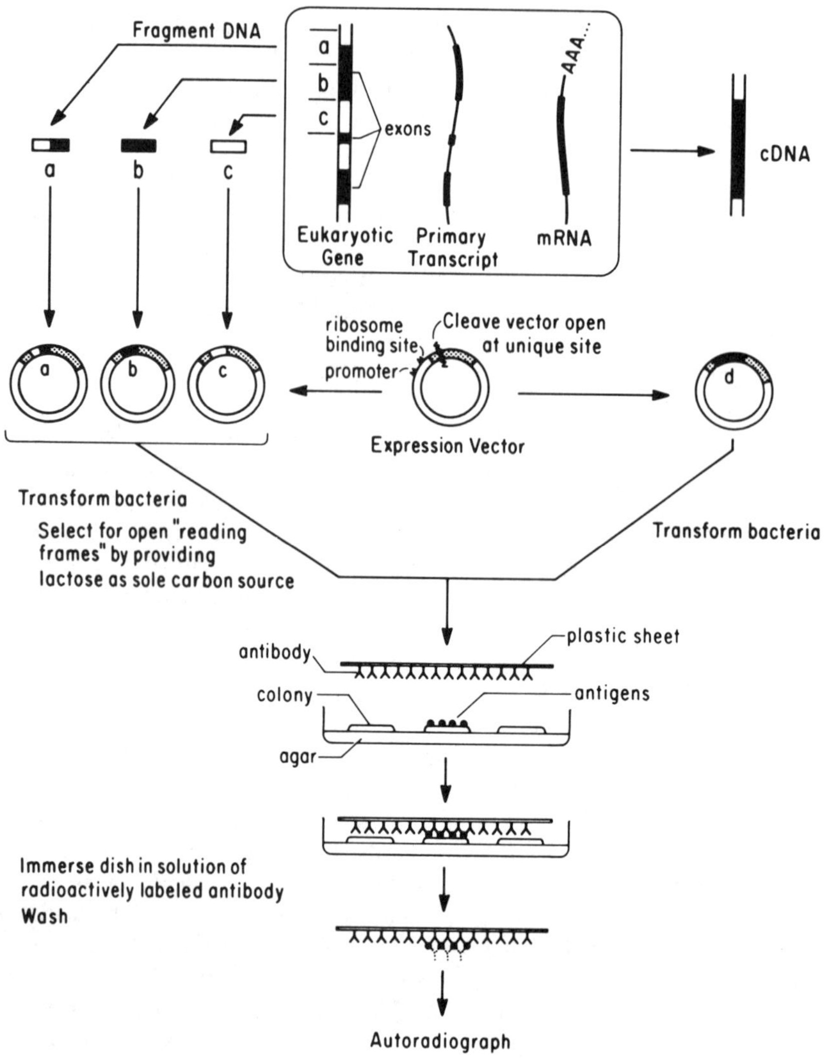

Fig. 3. *Immunological detection of specific gene sequences. DNA sequences derived from mRNA or obtained from the genome are inserted into an expression vector. Expression of authentic polypeptides requires that the insert is in the correct orientation and reading frame. Clones synthesizing a portion of the protein of interest may be detected by replicating the clones, lysing these cells, and placing a plastic sheet which is coated with antibody to the protein of interest over the lysed colonies. Antigen will bind to the antibody and be fixed to the plastic sheet. The sheet is then lifted off and dipped into a solution containing labeled antibody. The antibody will adhere to the antigen and thus mark the position of the antigen-producing clone when autoradiographed.*

single exon sequence. The vast majority of the inserts, however, are likely to contain only noncoding DNA, which makes up about 95% of the genome. Moreover, 5/6 of the remaining coding DNA will be in the wrong reading frame or in the wrong orientation. Thus, few of the clones in such an expression library would be expected to contain "open" reading frames—stretches of protein coding DNA uninterrupted by stop codons. Several screening procedures have been developed to select for clones that contain such open reading frames [Koenen et al, 1982; Ruther et al, 1982; Gray et al, 1982]. In these methods small genomic fragments are inserted into the aminoterminal end of a β-galactosidase gene contained within the expression vector. The activity of this enzyme is generally unaffected by the presence of extra amino acids inserted near its amino terminus. The resulting transcript of this chimeric gene contains a leader sequence, the sequence encoding the first seven amino acid residues of β-galactosidase, followed by the inserted genomic DNA and finally the rest of the β-galactosidase gene. If the inserted DNA fragment contains translational stops, the hybrid transcript will fall off the ribosomes before the bulk of the enzyme is translated, and no active β-galactosidase will be made. The recombinant plasmids or bacteriophage are introduced into *E. coli* that do not themselves possess an active β-galactosidase gene. These bacteria are then grown with lactose as the sole carbon source, so that bacterial growth will depend on the presence of active β-galactosidase to cleave lactose into galactose and glucose. Bacterial growth thus will depend on the presence of vectors carrying a functional β-galactosidase gene, ie, one in which there are no stop codons.

In practice this screening procedure eliminated more than 90% of the clones in a test experiment, presumably those that did not contain open reading frames and therefore did not produce active β-galactosidase [Ruther et al, 1982]. By such an enrichment for exon segments, the number of clones that must be screened immunologically is reduced to 10^7–10^8. Since about 10^4 colonies (arising from plasmid transformation) can be immunologically screened on a single petri dish, approximately 10^3 to 10^4 dishes would have to be screened in order to detect a single exon. Fortunately, recent developments in plating bacteriophage-λ vectors may allow 10^5–10^6 plaques to be screened per plate [Kaplan et al, 1983; Young and Davis, 1983; Kaplan et al, 1981] requiring only 10 to 100 plates for each screen. Once such a library is constructed it can be used to isolate any gene for which an antibody against its protein product is available.

Cloning of Complementary DNA

Despite recent advances that increase the feasibility of screening genomic expression libraries, a more standard approach to the immunological detection of a particular gene would be to make complementary DNA from mRNA. In contrast to genomic DNA fragments, the cDNAs copied from mRNA lack introns, since these were removed in the production of mature mRNA. Furthermore, since most of the genomic DNA is not transcribed into mRNA, far fewer clones would be needed to be screened (by nucleic acid probes or antibodies). A rare mRNA may represent only one part in ten thousand of a total mRNA population, but this is a 100-fold enrichment over its representation in total genomic DNA. (A 3-kb gene is only one part in a million of the total genome of 3×10^6 kb.) Taking account of statistics, reading frame, and orientation, fewer than 10^6 clones would need to be screened immunologically if the mRNA of interest represented 0.01% of the total, as compared to 10^7–10^8 to find a single exon.

Furthermore, before synthesizing complementary DNA, there are many ways by which the concentration of the mRNA of interest may be enriched manyfold: (1) An antibody to the protein of interest may be used to bind polysomes that bear the nascent antigentic polypeptide. The mRNA may then be purified from these polysomes and used to make cDNA [Shapiro and Young, 1981; Korman et al, 1982]. (2) Total mRNA may be fractionated according to size by centrifugation or gel electrophoresis [Ballivet et al, 1982; Maniatis et al, 1982]. Messenger RNA of different sizes can then be translated in vitro and tested for their ability to precipitate antibody or to produce a particular polypeptide spot in a two-dimensional polyacrylamide gel. The mRNA fraction that encodes the protein of interest may then be copied into cDNA. (3) If a partial amino acid sequence of the protein is known, an oligonucleotide may be synthesized that is complementary to the mRNA of interest [Noda et al, 1982a]. When such an oligonucleotide is mixed with total mRNA, it will hybridize only to a complementary mRNA sequence. Because the synthesis of cDNA requires a primer, only the mRNA to which the oligonucleotide probe has hyridized will be transcribed into cDNA. Because the genetic code is degenerate (ie, an amino acid may be represented by several different codons), a mixture of oligonucleotides representing all the possible coding combinations for a small portion of the amino acid sequence is generally used. Although some DNA/RNA hybrids may form on mRNAs other than the one of interest, this method still enriches for the desired cDNA among the population of clones by several hundredfold. If a second synthetic oligonucleotide complementary to another region of the protein is then used as a labeled probe in subsequent colony hybridizaton experiments, the colonies then selected are almost certain to contain the desired sequence. (4) Finally, differential hybridization may be used if two mRNA populations are available that would differ principally in their content of the desired mRNA [Maniatis et al, 1982; Taniguchi et al, 1980; Dworkin and Dawid, 1980; Zimmerman et al, 1980; Rowekamp and Firtel, 1980]. Thus, neuronal mRNA may be hybridized with complementary DNA derived from total mRNA from such tissues as liver. If mRNA that does not hybridize to liver cDNA is cloned, the clones that arise are likely to contain copies of neuron-specific mRNAs. Likewise, the mRNA population of a neural tissue exposed to a drug or hormone may be hybridized to cDNA from the same untreated tissue. The differences in mRNA species are likely to correspond to genes whose transcription was induced (or repressed) by the drug or hormone. Clones derived from unhybridized mRNA are likely to contain copies of genes whose expression was induced by the treatment. Likewise, mRNA populations from tissues at different developmental stages may be hybridized against one another in order to find genes that are expressed in a stage-specific manner.

Construction of cDNA Clones

In order to create a library of sequences encoded in mRNA, the mRNA must be first converted into double-stranded DNA copies, which can then be inserted into a vector. This conversion relies on "reverse transcriptase" (RNA-dependent DNA-polymerase), an enzyme isolated from RNA tumor viruses. This enzyme can synthesize single-stranded cDNA from an mRNA template.

Like other DNA polymerases, reverse transcriptase requires that an oligonucleotide primer be hybridized to the template to provide a growing point for synthesis of the new DNA strand. Since most eukaryotic mRNAs have a long stretch of adenylate residues at their 3′-termini, a primer can be provided by simply hybridizing an oligonucleotide of deoxythymidylate residues (oligo dT) to this poly(A) sequence. Oligo dT will prime the conversion of the entire population of poly(A) containing mRNAs into cDNA. As noted above, if a partial amino acid sequence of the protein of interest is available, an oligonucleotide may be synthesized that will prime only those mRNAs with complementary sequences. The product of this reaction is a hybrid of mRNA and cDNA.

Several techniques are available for the synthesis of the second strand of DNA and the insertion of this molecule into an appropriate vector. All of these methods share similar enzymatic steps, but each method offers particular advantages in terms of yields and ease of subsequent screening. In the most commonly used technique, the mRNA strand is destroyed by alkali, to which the cDNA is resistant. The single stranded cDNA is then converted into a double stranded molecule by a second DNA polymerase reaction. The synthesis of the second strand of cDNA again requires a primer, which is provided by the cDNA itself, since its 3′ end is able to fold back on itself to form a transient loop ("hairpin"). After synthesis of the second strand, this hairpin is removed (because a looped end cannot be joined to a vector) with S_1 nuclease which digests single-stranded DNA, to produce a fully double-stranded cDNA molecule.

Two alternative methods are commonly used to insert double-stranded cDNA into vectors. In the "tailing" method, 10–15 residues of deoxycytidine are added to the 3′ end of each cDNA strand by the enzyme terminal transferase [Rowekamp and Firtel, 1980; Villa-Komaroff et al, 1978]. The vector is cleaved open at a single site by a restriction endonuclease, and 10–15 residues of deoxyguanidine are added to each of its 3′ ends. When the oligo dC-tailed cDNA molecules are mixed with the oligo dG-tailed vector, their complementary tails anneal to form recombinant molecules, which are then used to transform bacteria.

The double-stranded cDNA may also be inserted into a vector by framing it between double-stranded synthetic oligonucleotide "linkers," which contain sites for one or more restriction endonucleases [Helfman et al, 1973; Kurtz and Nicodemus, 1981]. Using DNA ligase, the linkers are attached to both ends of the cDNA. Cleavage of these linkers with a restriction endonuclease produces a cDNA molecule with complementary oligonucleotide sequences extending from its two ends ("sticky ends"). The cDNA is then joined to a vector that has been cut with restriction endonucleases to produce sticky ends complementary to the ends on the cDNA. This method and the Okayama-Berg method (described below) have an advantage over the tailing method in that the orientation of the cDNA insert can be controlled. This is particularly significant in the construction of a cDNA expression library, because when the direction of insertion is random, as in the tailing method, either end of the insert may anneal adjacent to the promoter, and 50% of the clones will be in the wrong orientation for translation. In the linker method this problem is avoided by adding one type of linker before removal of the hairpin loop and a second type of linker after nuclease treatment. The sequential addition of linkers creates a cDNA molecule with defined linkers on each end, permitting its insertion in a specific orientation into a vector with appropriately oriented complementary sticky ends.

Both of the widely used methods described above remove the hairpin primer from the first strand of cDNA by treatment with S_1 nuclease. This treatment, however, invariably removes additional nucleotides from each end of the cDNA, and the final recombinant molecules are therefore less likely to contain full-length representations of the mRNA sequence. To avoid this, two methods have recently been described that eliminate the need for S_1 nuclease digestion, thus improving the yield of full length cDNAs [Land et al, 1981; Okayama and Berg, 1982]. The method developed by Okayama and Berg is particularly efficient in converting mRNA into cDNA. In this procedure an oligo dT tail is added to one end of a plasmid vector. This can hybridize to the poly(A) tail of mRNA, so that the mRNA is directly annealed with the vector in a defined orientation. The vector then serves as a primer for the synthesis of both the first and the second strand of cDNA. This method has been successfully employed for the isolation of cDNA clones encoding all four subunits of the ACh receptor of *Torpedo,* mentioned above [Ballivet et al, 1982; Noda et al, 1982b,c].

Immunological Screening

Complementary DNA libraries may be readily screened by nucleic acid probes using the methods outlined above. Additionally, if an expression vector is used, a gene of interest may be detected by immunological techniques.

In order to synthesize the protein encoded in the cDNA, the insert must be in the correct reading frame and orientation. Furthermore, transcription and translation must proceed through the homopolymers or synthetic linkers that flank the cDNA, as well as through any sequences corresponding to the mRNA leader sequences. Since eukaryotic leader sequences average 60 bases in length, there is some chance ($p \sim 0.6$) that they contain a translational stop codon in the same reading frame as the following coding sequence. Thus when expression is desired, a full-length cDNA is not necessarily the most desirable. The methods employing S_1 nuclease to remove the hairpin automatically generate truncated cDNAs, thus removing some of the leader sequence. Although this may remove some coding sequences as well, all that is required for immunological identification is that the translated sequences be able to fold into recognizable antigens. The success of immunological screening for exons suggests that the polypeptides synthesized from cDNA, which are generally much larger than those encoded by a single exon, are likely to fold into shapes having several antigenic sites.

In methods designed to synthesize full-length cDNA, a population of cDNA molecules is produced in which the majority (80%) of the cDNAs are less than full length (Okayama, personal communication). Thus, some members of these populations may have had their leader sequences partially or completely removed.

A number of immunoassays have been developed to detect clones that synthesize a particular antigen. These techniques are highly sensitive and can detect small fragments of a protein if they possess antigenetic determinants. One of the most widely used methods is that of Broome and Gilbert [1978]. In this method a library of cDNA clones is replicated onto a second culture dish and the replicate clones are then lysed by exposure to chloroform vapor, releasing the cytoplasmic protein. A polyvinyl sheet, to which a specific antibody is bound, is placed in contact with the lysed colonies (or plaques), allowing antigen-antibody complexes to form on the sheet. The sheet is removed and incubated with the same serum antibody. The labeled antibody reacts with the bound antigen at other antigenetic determinants than those bound to the first antibody on the sheet. The spot labeled by antibody is detected by autoradiography and the corresponding clone is picked from the reference plate. Alternatively, radioactively labeled protein A from *Staphylococcus aureus* may be used to detect an unlabeled second antibody [Erlich et al, 1978]. Up to 10,000 colonies per plate may be screened in this fashion (Fig. 3).

Bacteriophage-λ expression vectors have several advantages for immunological screening: (1) Many more plaques may be screened per plate—the microplaque plating methods developed by Kaplan [1981; Kaplan et al, 1983] and Young and Davis [1983] allow 10^5 clones to be screened on a single petri dish. (2) λ-phage cause their host cells to lyse, thus releasing the proteins. By incorporating antibody into the culture dish, the antibody-antigen complex may be detected directly in situ. In the method developed by Kaplan the antibody is covalently linked to the enzyme horseradish peroxidase (HRP) and then incorporated into the agar of the culture dish. When antigen is released from the cells by cell lysis, immunoprecipitation of the antibody with antigen occurs in situ. The culture dish is then immersed in buffer to allow soluble uncomplexed antibody-HRP to diffuse out of the agar, leaving only antibody-HRP-antigen complexes at the site of the plaque which synthesized the antigen. The culture plates are then incubated with chromogenic HRP substrates to produce a colored precipitate at the site of the antibody-HRP-antigen complexes. The colored spots may be picked directly to give viable λ-phage containing the gene fragment of interest.

The use of λ-expression vectors should permit the immunological detection of very rare cDNA inserts among a large population of clones with a minimum of manipulations. λ-Vectors are not well suited to cloning methods that employ homopolymer tails because any nicks in the λ-DNA may also be tailed, causing the vector to become inviable. This problem may be circumvented by using a linker method for insertion of cDNA into the vector.

Prospects for Immunological Identification of Cloned Gene Sequences

How many cDNA clones must be constructed in order to find a particular gene by immunological methods? This number depends upon the relative concentration of the mRNA of interest. Because neurons have a greater molecular complexity than other cell types and because the nervous system is made up of many different types of neurons (as well as of other cell types), the isolation of a particular mRNA sequence from the nervous system is more difficult than from other organs. About 200,000 mRNA species are present in the rat brain, compared to 40,000 and 60,000 in the kidney and liver respectively [Chikaraishi, 1979]. The vast majority of these brain mRNA species are present at less than 0.02 copies per cell. It is thus probable that many mRNA species are entirely absent from most brain cells. On the other hand, they may be present at significant levels within a particular subpopulation.

Because the brain contains so many species of mRNA and because authentic expression requires both the correct reading frame and the proper orientation, the immunological screening of cDNA libraries for genes expressed at low levels (say, less than one part in 10^5) may only be feasible by enriching for the mRNA of interest before construction of the library. As discussed above, antibody can be used either to determine the approximate size of the sought after mRNA, or to isolate polysomes enriched for that mRNA. Size fractionation of mRNA typically enriches for a particular mRNA by a factor of 10. Immunoprecipitation may enrich by a factor of more than 1,000. Korman et al [1982], for example, used immunoprecipitation to purify an mRNA that represented only 0.01% of the total mRNA by over 2,000-fold before constructing a cDNA library.

In view of the complexity of the brain and the presence of the same (or homologous) mRNA species in other tissues, it is often advantageous to use other tissues for the isolation of mRNA. Besides using mRNA from the electric organ of *Torpedo* to isolate the ACh receptor sequences, a pheochromocytoma cell line, derived from the adrenal medulla, was used to isolate the gene for tyrosine hydroxylase [Lamouroux et al, 1982]. Receptors of interest are often found in high concentration in striated, smooth, or cardiac muscle, as well as in neural tissue.

A combination of the judicious choice of tissue from which to isolate mRNAs, and the prepurification of mRNA by size or the antigenicity of nacent polypeptides, should bring most sought-after mRNAs to a concentration at which a manageable number of clones may be constructed and screened immunologically. If the presence of the mRNA can be enriched to better than one part in 10,000, either plasmids or λ-expression vectors would be suitable. For rarer mRNAs, λ-based expression vectors are increasingly advantageous, since they may be screened at higher densities.

CONCLUSIONS

In principle, any gene of interest may be isolated if either a nucleic acid or antibody probe is available. Colony or plaque hybridization with nucleic acid probes is the method of choice, since existing large libraries may readily be screened. Several human genomic libraries, for example, are already available, in which the entire human genome is represented in 3×10^5 cosmids with an average insert size of 40 kb, or 8×10^5 λ-phage, with an average insert size of 17 kb. These libraries are highly likely to contain a copy of any gene of interest.

For genes that are expressed at low levels, however, obtaining a nucleic acid probe is difficult. Since many receptor genes, for example, are likely to be expressed at low levels, it may be difficult to prepare sufficiently pure mRNA from which a nucleic acid probe may be derived. Alternatively, nucleic acid probes may be chemically synthesized to match a known amino acid sequence in the protein. Presently as little as 5 pmol of a pure polypeptide may be used to determine an amino acid sequence from which an oligonucleotide probe may be designed [Hunkapiller and Hood, 1983]. Conceivably, it may be possible to purify receptors by virtue of their ligands [Thoenen, 1982]. Unlike receptors, ligands are often available in large quantities and can be used to raise monoclonal antibodies. By using covalent photoaffinity derivatives of the ligands, the monoclonal antibodies could be used to purify the receptor-ligand complex from a mixed population of proteins. The polypeptide(s) of the receptor could then be sequenced and oligonucleotide probes designed.

Many functionally important proteins, however, have not yet been purified. This means that amino acid sequence data is presently unavailable and that the specificity of available serum antibodies cannot be established. The ability of monoclonal antibodies to recognize a single epitope, however, makes it possible to obtain a set of monoclonal antibodies all of which are specific for a single protein. Thus monoclonal antibodies have been prepared for a number of receptors for which protein sequence data are unavailable. Such monoclonal antibodies may be used to detect genes of interest as long as recognizable antigenic determinants can be produced by an expression library.

In addition to monoclonal antibodies available for specific proteins of known function, many monoclonal antibodies have been recently prepared which recognize surface antigens that are either cell type-specific or reveal a patterning of cells within a particular region of the nervous system. Some of these antibodies may recognize proteins that play an important role in establishing the "hard wiring" of the nervous system during development. For example, these proteins may act as signals for the association, migration, and axonal growth of neurons. Such antibodies could be used to isolate genes encoding these important proteins.

Recent advances in the introduction of genes into eukaryotic tissue culture cells makes it increasingly possible to detect specific genes by their ability to direct synthesis of a functional gene product in recipient cells [Okamaya and Berg, 1983; Smith and Danner, 1981; Klobutcher and Ruddle, 1981; McBride and Peterson, 1980; Paulakis et al, 1981; Jolly et al, 1983]. Tissue culture cells and the oocytes and eggs of *Xenopus laevis* are able to transcribe, translate, and compartmentalize the protein products of introduced genes. For example, mRNAs encoding the four polypeptides of the *Torpedo* ACh receptor have been translated by *Xenopus* oocytes and the polypeptides assembled and inserted into the cell membrane to give a functional receptor capable of binding α-bungarotoxin [Barnard et al, 1982; Sumikawa et al, 1981]. The ability of eukaryotic cells to correctly compartmentalize the protein products of introduced genes offers a novel approach to the isolation of receptor genes. If a heterogenous population of genomic fragments, linked to appropriate regulatory signals and a selectable marker, is introduced into tissue culture cells, cells transformed with DNA fragments encoding receptor genes may be able to synthesize the encoded polypeptide and insert it into the external surface of the cell. Cells expressing the gene of interest could then be selected by fluorescent-activated cell sorting, using approprate antibodies. Selected cells can then be cloned and the gene of interest recovered.

Given these various approaches for the isolation of specific gene sequences, the prospects for the isolation of receptor genes are excellent. The recent isolation of the *Torpedo* ACh receptor genes represents the first step in this direction. The explosive development of new expression vectors and protein-microsequencing techniques will greatly facilitate this effort. The next few years should witness the isolation of many mammalian receptor genes, which will make possible entirely new approaches to understanding neurobiology and endocrinology. These techniques are certain to add exciting new information to an already exciting field.

ACKNOWLEDGMENTS

We thank the following people for their helpful contributions: Lauri Sing, Susanne Huttner, Winston Salser, Craig Venter, Teresa Wood, Carol Wuenschell, James DeKloe, Elisabeth Seftor, Niranjala Tillakaratne, Donald Kaplan, Michael Nunn, and Jennifer Collopy. We are grateful for research support from the Dystonia Medical Research Foundation and the Research Committee of the UCLA Academic Senate.

REFERENCES

Anderson WF, Diacumakos EG (1981): Genetic engineering in mammalian cells. Sci Am 245(1):106–121.

Ballivet M, Patrick J, Lee J, Heinemann S (1982): Molecular cloning of the cDNA coding for the γ subunit of *Torpedo* acetylcholine receptor. Proc Natl Acad Sci USA 79:4466–4470.

Barnard EA, Miledi R, Sumikawa K (1982): Translation of exogenous messenger RNA coding for nicotinic acetylcholine receptor produces functional receptors in *Xenopus* oocytes. Proc R Soc Lond [Biol] 215:241–246.

Benton WD, Davis RW (1977): Screening of λ gt recombinant clones by hybridization to single plaques in situ. Science 196:180.

Brahic M, Haase AT (1978): Detection of viral sequences of low reiteration frequency by *in situ* hybridization. Proc Natl Acad Sci USA 75:6125–6129.

Breathnach R, Chambon P (1981): Organization and expression of eukaryotic split genes coding for proteins. Annu Rev Biochem 50:349–384.

Brinster RL, Palmiter RD (1982): Introduction of foreign genes in animals. Trends Biochem Sci 7:438–440.

Broome S, Gilbert W (1978): Immunological screening method to detect specific translation products. Proc Natl Acad Sci USA 75:2746–2749.

Chambon P (1981): Split genes. Sci Am 245(5):60–71.

Charnay P, Louise A, Fritsch A, Perrin D, Tiollais P (1979): Bacteriophage lambda-*E. coli* K12 vector-host system for gene cloning and expression under lactose promoter control. Mol Gen Genet 170:171–178.

Chikaraishi DM (1979): Complexity of cytoplasmic polyadenylated and nonpolyadenylated rat brain ribonucleic acids. Biochemistry 18:3249–3256.

Comb M, Seeburg PH, Adelman J, Eiden L, Herbert E (1982): Primary structure of human Met- and Leu-enkephalin precursor and its mRNA. Nature 295:663–666.

Dworkin MD, Dawid IB (1980): Use of a cloned library for the study of abundant poly A^+ RNA during *Xenopus laevis* development. Dev Biol 76:449–464.

Erlich HA, Cohen SN, McDevitt HO (1978): A sensitive radioimmunoassay for detecting products translated from cloned DNA fragments. Cell 13:681–689.

Fuchs F (1980): Genetic amniocentesis. Sci Am 242(6):47–53.

Gerhard DS, Kawasaki ES, Bancroft FC, Szabo P (1981): Localization of a unique gene by direct hybridization in situ. Proc Natl Acad Sci USA 78:3755–3759.

Gilbert W, Villa-Komaroff L (1980): Useful proteins from recombinant bacteria. Sci Am 242(4):74–94.

Gray MR, Colot HV, Guarente L, Rosbash M (1982): Open reading frame cloning: identification, cloning and expression of open reading frame DNA. Proc Natl Acad Sci USA 79:6598–6602.

Green N, Alexander H, Olsen A, Alexander S, Schirrick TH, Sutcliffe JG, Lerner RA (1982): Immunogenic structure of the influenza virus hemagglutinin. Cell 28:477–487.

Grossman L, Moldave K (eds) (1980): Methods in Enzymology Vol 65, Nucleic Acids, part 1. New York: Academic Press.

Grunstein M, Hogness D (1975): Colony hybridization: A method for the isolation of cloned DNAs that contain a specific sequence. Proc Natl Acad Sci USA 72:3961–3965.

Guarente L, Roberts TM, Ptashne M (1980): A technique for expressing eukaryotic genes in bacteria. Science 209:1428–1430.

Gurdon JB, Melton DA (1981): Gene transfer in amphibian eggs and oocytes. Annu Rev Genet 15:189–218.

Gubler U, Seeburg P, Hoffman BJ, Gage LP, Udenfriend S (1982): Molecular cloning establishes proenkephalin as precursor of enkephalin-containing peptides. Nature 295:206–208.

Hans-Urlich B, Helinski DR (1979): Use of the lamda phage promoter P_L to promote gene expression in hybrid plasmid cloning vehicles. Methods Enzymol 68:482–492.

Harpold MM, Dobner PR, Evans RM, Bancroft EL (1978): Construction and identification by positive hybridization-translation of a bacterial plasmid containing a rat growth hormone structural gene sequence. Nucleic Acids Res 5:2039–2053.

Helfman DM, Feramisco JR, Fiddes JC, Thomas GP, Hughes SH (1983): Identification of clones that encode chicken tropomyosin by direct immunological screening of a cDNA expression library. Proc Natl Acad Sci USA 80: 31–35.

Herbert E, Birnberg N, Lissitsky J-C, Civelli O, Ohler

M (1981): Proopiomelanocortin: A model for the regulation of expression of neuropeptides in pituitary and brain. Neurosci Comment 1:16–27.

Hudson P, Ponschow J, Shine J, Ryan G, Niall H, Coghlan J (1981): Hybridization histochemistry: Use of recombinant DNA as a "homing probe" for tissue localization of specific mRNA populations. Endocrinology 108:353–356.

Hunkapiller MW, Hood LE (1983): Protein sequence analysis: Automated microsequencing. Science 219: 650–659.

Itakura K, Hirose T, Crea R, Riggs AD, Heyneker HL, Bolivar F, Boyer H (1977): Expression in *E. coli* of a chemically synthesized gene for the hormone somatostatin. Science 198:1056–1063.

Jolly DJ, Okayama H, Berg P, Esty AC, Filpola D, Bohlen P, Johnson GG, Shively JE, Hunkapillar T, Friedmann T (1983): Isolation and characterization of a full-length expressible cDNA for human hypoxanthine phosphoribosyltransferase. Proc Natl Acad Sci USA 80:477–481.

Kakidani H, Furutani Y, Takahashi H, Noda M, Morimoto Y, Hirose T, Asai M, Inayama S, Nakanishi S, Numa S (1982): Cloning and sequence analysis of cDNA porcine β-neo-endorphin/dynorphin precursor. Nature 298:256–249.

Kaplan D, Naonovski L, Collier JR (1981): Chromogenic detections of antigen in bacteriophage plaques: A microplaque method applicable to large scale screening. Gene 13:211–220.

Kaplan DA, Greenfield L, Collier J (1983): Chromogenic method to screen very large populations of bacteriophage plaques for the presence of specific antigen. In Wu R, Grossman L, Moldave K (eds): "Methods in Enzymology 100B." New York: Academic Press, pp 342–368.

Klobutcher LA, Ruddle FH (1981): Chromosome mediated gene transfer. Annu Rev Biochem 50:533–554.

Koenen M, Ruther U, Muller-Hill B (1982): Immunoenzymatic detection of expressed gene fragments cloned in the Lac Z gene of *E. coli.* EMBO J 1:509–512.

Korman AJ, Knudsen PJ, Kaufman JF, Strominger JL (1982): cDNA clones for the heavy chain of HLA-DR antigens obtained after immunopurification of polysomes by monoclonal antibody. Proc Natl Acad Sci USA 79:1844–1848.

Kurtz DT, Nicodemus CF (1981): Cloning of $\alpha 2\mu$ globulin DNA using a high efficiency technique for the cloning of trace messenger mRNA. Gene 13:145–152.

Lamouroux A, Biguet NF, Samolyk D, Privat A, Salomon JC, Pujol JF, Mallet J (1982): Identification of cDNA clones coding for rat tyrosine hydroxylase antigen. Proc Natl Acad Sci USA 79:3881–3885.

Land H, Grez M, Hauser H, Lindenmaier W, Schutz G (1981): 5′ terminal sequences of eukaryotic mRNA can be cloned with high efficiency. Nucleic Acids Res 9:2251–2265.

Langer PR, Waldo AA, Ward DC (1981): Enzymatic synthesis of biotin-labeled polynucleotides: Novel nucleic acid affinity probes. Proc Natl Acad Sci USA 78:6633–6637.

Lerner RA (1983): Synthetic vaccines. Sci Am 248(2):67–74.

Ley TJ, DeSimone J, Anagnov NP, Heller GH, Humphries RK, Turner PH, Young NS, Heller P, Nienhuis AW (1982): 5-Azacytidine selectively increases β-globin synthesis in a patient with β^+ thalassemia. N Engl Med 307:1469–1476.

Maniatis T, Fritsch EF, Sambrook J (1982): "Molecular Cloning." Cold Spring Harbor, NY: Cold Spring Harbor Laboratory.

McBride OW, Peterson JL (1980): Chromosome-mediated gene transfer in mammalian cells. Annu Rev Genet 14:321–345.

Mulligan RC, Berg P (1980): Expression of a bacterial gene in mammalian cell. Science 209:1422–1427.

Nakanishi S, Inove A, Kita T, Nakamura M, Chang A, Cohen S, Numa S (1979): Nucleotide sequence of cloned cDNA for bovine corticotropin β-lipotropin precursor. Nature 278:423–427.

Noda M, Furutani Y, Takahashi H, Toyosato M, Hirose T, Inayama S, Nakanishi S, Numa S (1982a): Cloning and sequence analysis of cDNA for bovine adrenal preproenkephalin. Nature 295:202–206.

Noda M, Takahashi H, Tanabe T, Toyosato M, Furutani Y, Hirose T, Asai M, Inayama S, Miyata T, Numa S (1982b): Primary structure of α-subunit precursor of *Torpedo californica* acetylcholine receptor deduced from cDNA sequence. Nature 299:793–797.

Noda M, Takahashi H, Tanabe T, Toyosato M, Kikyotani S, Hirose T, Asai M, Takashima H, Inayama S, Miyata T, Numa S (1982c): Primary structures of β- and δ-subunit precursors of *Torpedo californica* acetylcholine receptor deduced from cDNA sequences. Nature 301:251–255.

Okayama H, Berg P (1982): High efficiency cloning of full length cDNA. Mol Cell Biol 2:161–170.

Okayama H, Berg P (1983): A cDNA cloning vector that permits expression of cDNA inserts in mammalian cells. Mol Cell Biol 3:280–289.

Palmiter RD, Brinster RL, Hammer RE, Trumbauer ME, Rosenfeld MG, Birnberg NC, Evans RM (1982): Dramatic growth of mice that develop from eggs microinjected with metallothionein-growth hormone fusion genes. Nature 300:611–615.

Paulakis GN, Hizuka N, Gorden P, Seeburg P, Hamer DH (1981): Expression of two human growth hormone genes in monkey cells infected by simian virus 40 recombinants. Proc Natl Acad Sci USA 78:7398–7402.

Pellicer A, Robins D, Wold B, Sweet R, Jackson J,

Lowy I, Roberts JM, Sim GK, Silverstein S, Axel R (1980): Altering genotype and phenotype by DNA-mediated gene transfer. Science 209:1414–1422.

Pharnes JR, Velar B, Felsenfeld A, Ramanathan L, Ferrini U, Appella E, Seidman JG (1981): Mouse α_2 microglobulin cDNA clones corresponding to rare mRNAs. Proc Natl Acad Sci USA 78:2253-2257.

Raftery MA, Hunkapiller MW, Strader CD, Hood LE (1980): Acetylcholine receptor: Complex of homologous subunits. Science 208:1454–1456.

Ricciardi RP, Miller JS, Roberts BE (1979): Purification and mapping of specific mRNAs by hybridization selection and cell free translation. Proc Natl Acad Sci USA 76:4927–4931.

Roberts J, Seeburg PH, Shine J, Herbert E, Baxter JD, Goodman AM (1979): Corticotropin and β-endorphin: Construction and analysis of recombinant DNA complementary to mRNA for the common precursor. Proc Natl Acad Sci USA 76:2153–2157.

Roberts JL, Chen C-LC, Dionne FT, Gee CE (1982): Peptide hormone gene expression in heterogeneous tissues—The pro-opiomelanocortin system. Trends Neurosci 5:314–318.

Rossier J (1982): Pro-enkephalin sequencing and the advent of cDNA technologies. Trends Neurosci 5:179–180.

Rowekamp W, Firtel RA (1980): Isolation of developmentally regulated genes from *Dictostelium.* Dev Biol 79:409–418.

Rubin GM, Spradling AC (1982): Genetic transformation of *Drosophila* with transposable element vectors. Science 218:348–353.

Ruther U, Koenen M, Sippel AE, Muller-Hill B (1982): Exon cloning: Immunoenzymatic identification of exons of the chicken lysozyme gene. Proc Natl Acad Sci USA 79:6852–6855.

Salser W, Tony BD, Stany HD, Brown JK, Mercola K, Bar-Eli M, Uhler MJ (1981): Use of drug resistance, selections in intact animals to insert human globin genes into bone marrow cells of living mice. In Stamatoyannopoulos G, Nienhuis A (eds): "Regulation of Hemoglobin Switching, Vol 2." New York: Alan R. Liss Inc., pp 313–334.

Scheller RH, Jackson JF, McAllister LB, Schwartz JH, Kundel ER, Axel R (1982): A family of genes that codes for ELH, a neuro peptide eliciting a stereotypical pattern of behavior in *Aplysia.* Cell 28:707–719.

Seeman P, Tysone L, Bird ED, Tourtellotte WW (1980): Elevation of brain neuroleptic/dopamine receptors in schizophrenia. In Baxter C, Melnechuk T (eds): "Perspectives in Schizophrenia Research." New York: Raven Press, pp 199–209.

Shapiro SZ, Young TR (1981): An immunochemical method for mRNA purification. J Biol Chem 256:1495–1498.

Shine J, Dalgarno L (1975): Determinant of cistron specificity in bacterial ribosomes. Nature 254:34–38.

Smith HO, Danner DB (1981): Genetic transformation. Annu Rev Biochem 50:41–68.

Spokes EGS (1981): "Huntington's Chorea" In "Metabolic Disorders of the Nervous System." Pitman Books Limited, pp 292–311.

Stevens C (1982): The acetylcholine receptor cloned east and west. Nature 299:776–777.

Suggs SV, Wallace RB, Hirose T, Kawashima E, Itakura K (1981): Use of synthetic oligonucleotides as hybridization probes: Isolation of cloned cDNA sequences for human β_2 microglobulin. Proc Natl Acad Sci USA 78:6613–6617.

Sumikawa K, Houghton M, Emtage JS, Richards BM, Barnard EA (1981): Active multi-subunit ACh receptor assembled by translation of heterologous mRNA in *Xenopus* oocytes. Nature 292:862–864.

Sutcliffe JG, Milner RJ, Bloom FE, Lerner RA (1982): Common 82-nucleotide sequence unique to brain RNA. Proc Natl Acad Sci USA 79:4942–4946.

Sutcliffe JG, Schinnick TM, Green N, Lerner RA (1983): Antibodies that react with predetermined sites on proteins. Science 219:660–665.

Szabo P, Ward C (1982): What's new with hybridization in situ? Trends Biochem Sci 7:425–427.

Taniguchi T, Fuji-Kuriyama Y, Muramatso M (1980): Molecular cloning of a human interferon cDNA. Proc Natl Acad Sci USA 77:4003–4007.

Temple GF, Dozy AM, Roy KL, Kan YN (1982): Construction of a functional human suppressor tRNA gene: an approach to gene therapy for β-thalassemia. Nature 296:537–540.

Thoenen H (1982): Strategies for application of molecular genetics to neurobiology. In Schmitt FO, Bird SJ, Bloom FE (eds): "Molecular Genetic Neuroscience." New York: Raven Press, pp 459–466.

Ulmer KM (1983): Protein Engineering. Science 219:666–671.

Villa-Komoroff L, Efstratiadis A, Broome S, Lomedico P, Tizard R, Naker SP, Chick WL, Gilbert W (1978): A bacterial clone synthesizing pro-insulin. Proc Natl Acad Sci USA 75:3727–3731.

Vincent A (1980): Immunology of acetylcholine receptors in relation to myasthenia gravis. Physiol Rev 60:756–824.

Williamson B (1982): Gene therapy. Nature 298:416–418.

Wu R (ed) (1979): "Methods in Enzymology, Vol 68, Recombinant DNA." New York: Academic Press.

Young A, Davis RW (1983): Efficient isolation of genes by using antibody probes. Proc Natl Acad Sci USA 80:1194–1198.

Zimmerman CR, Orr WC, Leclerc RF, Barnard EC, Timberlake WE (1980): Molecular cloning and selection of genes regulated in *Aspergillus* development. Cell 21:709–715.

Molecular and Chemical Characterization of Membrane Receptors, pages 261–282

11

The Application of Recombinant DNA Techniques to the Study of Cell Membrane Receptors

R.T. Fernley, J.J. Gorman, H.D. Niall, and J.P. Coghlan

Howard Florey Institute of Experimental Physiology and Medicine, University of Melbourne, Parkville, Victoria 3052, Australia

INTRODUCTION

We will discuss possible strategies for the determination of the linear amino acid sequences of proteins composing cell surface receptors for hormones and other ligands. The strategies are based upon recombinant DNA techniques, clearly the most promising means to this end available at present. Success in this task of protein sequencing will have important implications for further studies of receptors at the more sophisticated level of tertiary structure and function and for studies of receptor biosynthesis, gene expression, and evolution. These future possibilities are briefly discussed in a concluding section.

RELEVANT FEATURES OF RECEPTORS

Although receptors for only a few ligands have been studied in a more than cursory way, a general picture has emerged. Cell membrane receptors are protein complexes of high molecular weight, often, though not invariably, consisting of more than one protein subunit, and usually modified posttranslationally by the addition of carbohydrate. They are present in extremely low abundance in target tissues (typically 1–2 pmol/mg of purified membrane protein). It is clear from theoretical considerations and confirmed by some direct studies [Czech, 1982] that receptors consist of several different functional domains which could be encoded by one (necessarily very large) mRNA transcript or by several smaller coordinately expressed transcripts. As Figure 1 illustrates, receptors, to achieve their purpose, must have at least three functionally separate domains directing membrane insertion, ligand binding, and message transduction, respectively. Moreover, if there are multiple subunits, additional domains are needed for subunit/subunit recognition and interaction.

All of this argues very strongly for a high degree of evolutionary conservation. As previously proposed [Niall, 1980], "new" receptors are most likely to evolve from preexisting receptor complexes, since use can be made of functional domains already available. This argument in turn leads us to the conclusion that receptors will turn out to belong to a multigene family that evolved through successive duplications of genes coding for a limited array of functional domains. In this way, diversity of receptors and independence in their regulation can be achieved while retaining a common structural organization.

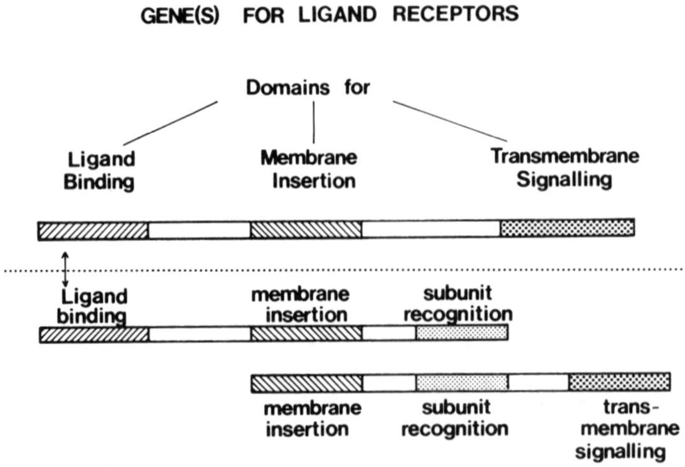

Fig. 1. *Possible organization of receptor functional domains, separately encoded at the genomic level and expressed at the peptide level either as a single molecular species (above dotted line) or as a multiple subunit structure (below dotted line).*

The implications of the above features of receptors for the design of cloning strategies are clear.

1. The low abundance of receptor proteins in target tissues almost certainly means an extremely low abundance of their respective mRNAs. Although no data are available, a reasonable guess would be that receptor mRNAs are in the range of 0.001–0.01% of total mRNA. This eliminates straightaway certain cloning strategies that depend upon much higher abundance of message (see below). Moreover, it seems unlikely that much value will be gained through the physiological manipulation of receptor-specific mRNA levels up or down (eg, by inducing up-or-down regulation of cell surface receptor numbers). Even a two- or threefold rise or fall in specific mRNA abundance will not greatly improve the situation. However, it may still be worth trying to achieve this (see below).

2. The smallest identifiable subunits of many receptors are still quite large (eg, 130,000 and 90,000 daltons for the subunits of the insulin and insulin-like growth factor I (IGF I) receptors). This means they are encoded by large mRNA species which may be difficult to reverse transcribe and clone in their entirety.

3. The low abundance and large size of receptor proteins also mean that approaches based on direct isolation and extensive sequence analysis by traditional protein sequencing methods will not be logistically feasible. However, recently developed microsequencing techniques [Hewick et al, 1981] should be applicable provided receptor proteins can be purified in 50- to 500-pmol quantities.

4. Posttranslational modification of receptor proteins (eg, by disulphide bond formation, carbohydrate addition, combination of subunits, proteolytic processing) may cause technical difficulties. For example, ligands or receptor-specific antibodies may only recognize three-dimensional structural features of mature receptor complexes which are absent in isolated subunits produced by in vitro translation or cloning into host cells.

5. The probability that receptors are encoded by a multigene family at once suggests the existence of certain organizational features already known from other multigene families. These are (A) an extensive degree of sequence homology at the nucleotide and protein level between closely related members of the family (eg, the insulin-related peptide receptors, the opiate receptors); and (B)

the coding of different functional domains on different exons at the level of genomic DNA, with the possibility of a gene-switching mechanism to increase receptor diversity.

All these features of receptors limit and direct our choice of recombinant DNA strategies. In particular, there is a certain resemblance [Czech et al, 1981] between the structure of immunoglobulins and that proposed for the insulin receptor, which leads us to a more detailed examination of the methods successfully used to analyse the immunoglobulin gene family, in the next section.

THE IMMUNOGLOBULIN GENE FAMILY

To analyse the structure and genetic organization of hormone and other ligand receptors, it may be useful to examine how the structure and genetic organization of antibodies was elucidated. The study of antibodies was hindered, not by the minute amounts of protein present, as is the case for receptors, but by their great diversity of structure and the extreme difficulty of obtaining, by conventional biochemical techniques, homogeneous preparations for analysis. This was overcome by the purification of specific unique immunoglobulin molecules from plasma cell tumours (myelomas) in man and in the mouse. The large amounts produced allowed purification and sequence analysis of complete immunoglobulins circulating in plasma and of isolated light chains (Bence-Jones protein) from urine. By the late 1960s the complete amino acid sequence of several immunoglobulins had been completed [Putnam et al, 1967; Edelman et al, 1969]. Additional information was obtained through x-ray crystallographic analysis of Bence-Jones proteins [Schiffer et al, 1973] and Fab′ fragments of immunoglobulins [Poljak et al, 1973; Segal et al, 1974]. The advent of new techniques in molecular biology during the 1970s allowed the gene structure and organization of immunoglobulins to be elucidated [Hozumi and Tonegawa, 1976; Brach and Tonegawa, 1977; Brach et al, 1978; Maxam and Gilbert, 1977; Sanger et al, 1977; Bernard et al, 1978; Seidman et al, 1978a,b; Tiemeier et al, 1978; Max et al, 1979; Sakano et al, 1979; Early et al, 1979, 1980; Korsmeyer et al, 1981; Marx, 1981]. From all these studies a good understanding of immunoglobulin structure and of the mechanisms generating diversity has been achieved.

IMMUNOGLOBULIN STRUCTURE

Each light chain of the immunoglobulin is linked to a heavy chain by one disulphide bond near the carboxyterminus of the light chain, and the two heavy chains are linked together by two disulphide bonds near the centre of the heavy chains in the hinge region (Fig. 2). The light chains consist of two domains, the variable (V) and constant (C) regions of equal size, while the heavy chain consists of four homologous areas of equal size, one V region and three C regions. The heavy chain also contains a small hinge region which joins the VH and CH_1 region to the CH_2 and CH_3 region, and this is not homologous with any other region of the molecule. The x-ray crystallographic studies [Schiffer et al, 1973; Poljak et al, 1973; Segal et al, 1974] showed that each of the light chain regions and the V and first C region of the heavy chain formed separate physical domains in the three-dimensional structure, connected by extended regions of amino acid sequence. The antigen-binding site consists of a cavity bounded by adjacent hypervariable regions of the V domains of both light and heavy chains, near the aminoterminus of each.

IMMUNOGLOBULIN DIVERSITY

Many theories to explain the diversity of antibody structure had been proposed, but until the development of new techniques in molecular biology it was very difficult to test these theories experimentally. The main problem was to explain the seemingly unlimited array of possible antibodies in terms of a

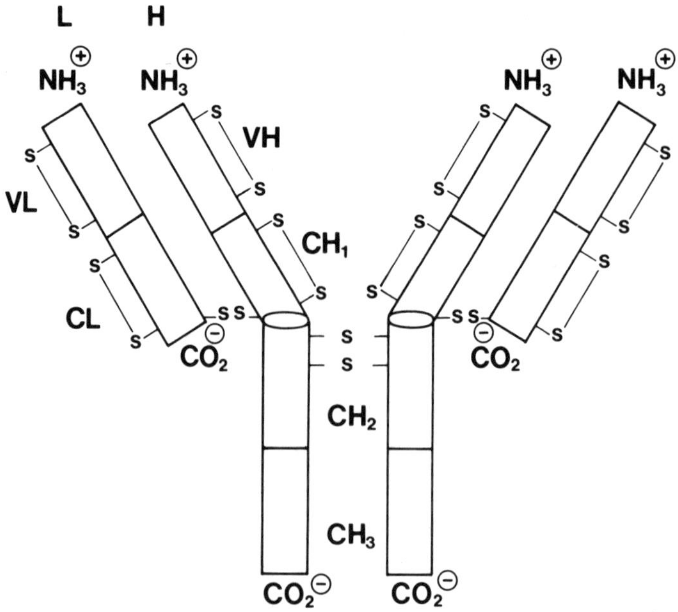

Fig. 2. *Structure of the immunoglobulins. Each light (L) chain is linked to a heavy (H) chain by a disulphide bond, and the two heavy chains are joined together by disulphide bonds.*

finite number of germ line genes. It does not seem relevant to describe the present state of knowledge of immunoglobulin genetics or the methods by which it has been elucidated in any great detail. However, the main feature is the encoding of both light and heavy chains by multiple gene segments separated in genomic DNA. Diversity is achieved by recombining these segments in different ways. For example, there are multiple germ line genes coding for the variable, or V, region of light chains. Each of these may be combined with a particular C region gene through undergoing a recombination event in the cells ultimately committed to producing a certain immunoglobulin [Hozumi and Tonegawa, 1976], so that the particular V gene involved is now adjacent to a C gene in the differentiated cell. Similarly, heavy chains consist of segments encoded by four different sets of genes. In order from the amino terminus, these are VH genes (coding for amino acids 1–101); D, or diversity, genes (amino acids 102–106); JH genes (amino acids 107–123); and a CH gene. In the mouse genome there are five different CH regions, $C\mu$, $C\delta$, $C\gamma$, $C\alpha$ and $C\Sigma$. These are responsible for the production of the five classes of immunoglobin: M, D, G, A, and E respectively. Once an antibody-producing cell assembles a variable regioin by recombining VH, D, and JH regions, this remains constant for that cell but it can be linked to different CH regions during the life of the cell either by RNA splicing or by gene rearrangement to produce different classes of antibody [Korsmeyer et al, 1981; Marx, 1981]. Thus an enormous range of antibodies can be produced by combinations of V, D, J, and C regions. The V region provides diversity of antigen specificity (binding) and the C region diversity in antibody function. An additional dimension of diversity may be provided by somatic mutation involving a hyperviable region of the V genes; however, a full discussion is beyond the scope or needs of this chapter.

Two important aspects have emerged from these studies on immunoglobulins: (1) that different functional regions of the immunoglobulin gene are encoded in separate exons on the genome, and (2) recombination events can combine various exons to produce a functional antibody molecule. These properties may be shared at least in part by the receptor genes. It is possible that the domain of the receptor which binds the hormone is encoded by one exon and the effector domain, translocating extracellular information across the cell membrane is encoded in a second domain. During evolution various hormone-binding domains could have undergone recombination events with effector domains leading to the evolution of new receptor molecules.

CLONING STRATEGIES

There are several possible approaches to the cloning of receptor genes. Since nearly all of them have in common the identification of the relevant cDNA clones within a cDNA bank generated from a receptor-producing tissue, the construction of this bank will first be described.

CHOICE OF TISSUE

Usually in molecular biology the path of least resistance is to be preferred. In other words, a tissue source likely to be enriched for receptor mRNA should be used. The use of mutant cell lines such as the A-431 cells which produce the epidermal growth factor (EGF) receptor [Haigler et al, 1978] would be an advantage. Physiological manipulations which might be expected to increase receptor gene transcription could be worthwhile. As noted above it is unlikely that any feasible increase would be enough to influence the choice of overall cloning strategy, ie, the mRNA will still be in very low abundance. However, the principle is that any enrichment of receptor mRNA will increase to some extent the statistical chances of obtaining receptor-specific clones in the cDNA bank when constructed.

Enrichment of mRNA by sucrose density gradient centrifugation is a possibility, although this presupposes a method of monitoring the distribution in the gradient of the desired mRNA species, usually by a translation-based assay. Specific antibody to receptor protein could also be used to immunoprecipitate polysomes bearing receptor-specific mRNA. The latter method has been successfully used to purify low-abundance mRNA species from the rat liver [Kraus and Rosenberg, 1982].

CONSTRUCTION OF cDNA BANK

It will have to be assumed that despite attempts of the kind just described, the mRNA abundance will be still extremely low (0.01% or less). This situation demands a highly efficient procedure for constructing a cDNA bank. The method must also if possible produce full-length cDNA clones and not select against larger mRNA species of the size range likely to encode receptor proteins. The most widely used procedure to date involves cloning into the bacterial plasmid pBR 322. First, the tissue is extracted with a guanidinium hydrochloride or guanidinium thiocyanate solution to denature RNAse and the homogenate is centrifuged over a cushion of cesium chloride. Total RNA sediments through the cushion leaving most of the protein and DNA in the supernatant. The RNA is further purified by precipitation with ethanol [Chirgwin et al, 1979]. Polyadenylated mRNA is then obtained by adsorption to an oligo(dT)cellulose column under conditions of high ionic strength and elution with low ionic strength buffer. It is desirable though not essential to repeat the oligo(dT)cellulose purification step. If for some reason the tissue being used is in short supply, it is possible to minimize losses of mRNA by cloning total RNA directly.

The next step is to copy the mRNA into cDNA using reverse transcriptase. Usually oligo(dT) is used as a primer to select for polyadenylated mRNA species. This is a critical step since the reverse transcription reaction may not proceed to completion, leading to cDNA species deficient in sequences corresponding to the 5′ end of the mRNA. After removal of the mRNA by digestion with alkali, the second strand of cDNA is synthesized with the aid of the enzyme DNA polymerase I. This reaction requires the end of the first strand to loop around and form, by random hybridization, a short region of double-stranded DNA to act as an initiation site for the DNA polymerase I reaction. The end result is a hairpin loop of single stranded DNA joining the completed double stranded region. This loop has to be digested away with the enzyme S_1 nuclease. The problem is that a significant part of the structure may be lost in this way, again leading to clones deficient at the 5′ end of the coding region. The next step is usually to add homopolymeric sequences of (dC) residues to either end of the double-stranded cDNA (using the enzyme terminal transferase) and to anneal this "tailed" DNA into the plasmid pBR 322 cut open at a single site (often the Pst 1 site) and similarly tailed with (dG) residues. The resulting recombinant plasmids are used to transform a suitable strain of *Escherichia coli* (eg, HB 101 cells) which can be cultured on a large scale. Colonies which are ampicillin sensitive (because the insert interrupts the plasmid penicillinase gene) and tetracycline resistant (because the plasmid pBR 322 bears an intact gene for tetracycline resistance) contain recombinant plasmids and are readily selected for further study, ie, for screening, in this case, for receptor-specific cDNA sequences.

This cloning technique is described in modest detail as it is the most widely used method at present. However, it has several deficiencies. The most telling defect in the context of cloning low abundance mRNAs is that it is not particularly efficient in terms of numbers of clones obtained per μg mRNA. Moreover, "good" batches of G-tailed plasmid are not easy to prepare reproducibly and are not reliably available at present from commercial sources. This situation has provoked considerable efforts to produce better cloning vectors, two of which are now available. One method [Okayama and Berg, 1982] uses a plasmid vector incorporating an oligo(dT) priming site for reverse transcriptase, allowing direct cloning of polyadenylated mRNA. This provides a very efficient incorporation of cDNAs into the vector. Another general procedure [Young and Davis, 1983] is based upon the use of versions of the bacteriophage-λ as cloning vectors. Both of these procedures are capable of providing 1–2 $\times$ 10^5 recombinants per μg mRNA — about 20 times the usual efficiency of the pBR 322 cloning system. Most clones obtained are full-length or nearly so with the Okayama-Berg procedure. While the Young-Davis procedure, or modifications of it, are less efficient in providing full-length clones, screening of the phage library is rapid and simple. On balance, either of these procedures would be more suitable for receptor gene cloning than the pBR 322 system, since it may be necessary to screen of the order of 10^6 recombinants to have a reasonable chance of finding the desired low abundance clones. Details of the two procedures are to be found in the original papers, although the field is advancing so rapidly that improved protocols will probably be available soon.

SCREENING OF cDNA BANK

Let us suppose that $\approx$ 10^6 recombinants from a receptor-containing tissue have been obtained. The next task is obviously to detect the few representative clones containing receptor-specific cDNA inserts. There are three kinds of possible approach: (1) use of synthetic oligodeoxynucleotide probes based upon partial amino acid sequence analysis of receptor proteins; (2) use of mRNA in vitro translation systems to generate receptor proteins detectable by a specific known property (eg, ligand binding or immunoprecipitation by an antireceptor antibody) as described below; (3) screening of groups of cDNA clones for expression of receptor-specific genes again by ligand and/or antibody binding. Of these approaches the first is the most straightforward and is to be preferred.

OLIGODEOXYNUCLEOTIDE PROBE/ MICROSEQUENCING APPROACH

Receptor Purification

The first step is receptor protein purification, not an easy task. The same sort of considerations apply to choice of tissue for extraction of receptors as were discussed above in regard to mRNA enrichment. The only receptor proteins sequenced to date (the acetylcholine receptor subunits) come from the electric organ of the marine species *Torpedo californica,* which has orders of magnitude higher receptor concentrations than mammalian skeletal muscle, although the latter is of more general physiological interest [Conti-Tronconi et al, 1982; Raftery et al, 1980]. As noted above, A 431 cells are particularly rich in epidermal growth factor receptors [Haigler et al, 1978]. Human placenta, being a human tissue readily available fresh and in quantity, is a popular source for purification of a range of receptors, notably the insulin receptor [Siegel et al, 1981; Harrison and Itin, 1980].

In order to settle on a particular source, specific binding of hormone must be demonstrated. Specific binding is defined by several criteria (for review see Kahn [1976]). (1) Binding is rapid and usually reversible, showing a temperature-dependent time course and reversed by altering conditions in such a way as to perturb the binding equilibrium (eg, dilution and pH change effects). (2) Affinity constants for binding should be high, 10^8–10^{11} M^{-1} for most peptide hormones. (3) Binding should be hormone specific in that it is not competitively inhibited by unrelated hormones, and binding of derivatives should closely correlate with their biological activity. (4) Binding is saturable with a finite number of binding sites per cell or unit weight of tissue. These criteria are usually examined by competitive binding assays employing isotopically labelled hormone derivatives of high specific activity. ^{125}I or ^{131}I are usually preferred because of their inherent specific activity; derivatives of approximately 2,200 Ci/mmol may be prepared with one atom of ^{125}I per mole. Peptide hormones tagged with fluorescent labels could also be used in conjunction with fluorescence-activated cell sorting to isolate receptor-bearing cells from mixed cell populations thereby increasing the receptor to cell mass ratio. However, there are limitations in the cell numbers able to be processed which detract from this approach.

Where multiple sources are available for receptor purification, their specific binding capacities may be used as a guide to choosing the tissue likely to result in the best final yield of receptor. The variation in tissue binding capacities of some peptide hormones is given in Table I.

Once a tissue has been selected, a choice of purification strategy must be made. While each kind of receptor may pose particular problems of its own, there are some general principles which derive from the physicochemical nature and low abundance of receptor protein(s) in the starting material.

1. To achieve an adequate yield of pure receptor protein, it is essential to use the least possible number of successive fractionation steps and that each step be efficient in terms of percent recovery of identifiable receptor protein.

2. Again because of the low abundance of receptor protein, schemes based purely on a series of "physicochemical" fractionations (such as gel filtration, ion exchange chromatography, high pressure liquid chromatography [HPLC], gel electrophoresis) are not likely to be adequate. At least one and probably two highly selective steps based on very specific properties of the receptor will be required (eg, affinity chromatography using immobilized ligand, antibody or lectin).

3. Receptor proteins are large, poorly soluble in aqueous media, and initially partially buried in the lipid environment of the cell membrane. Hence the first step (extraction

TABLE I. Representative Binding Capacities of Tissue and Cell Preparations for Peptide Hormones

Hormone	Tissue or cell type	Binding capacity	References
Insulin	Liver plasma membrane preparations	1.5 pmol/mg	Freychet et al [1971]
			Kahn et al [1974]
	Adipose tissue	11,000–50,000 sites/cell	Cuatrecasas [1971]
			Hammond et al [1972]
			Gammeltoft and Gliemann [1973]
	Placental membrane preparations	1 pmol/mg	Posner [1974]
	Cultured human placental cells	50,000 sites/cell	Podskalny et al [1975]
	Human lymphocytes	12,000 sites/cell	Gavin et al [1972]
			Gavin et al [1973]
Epidermal growth factor—Urogastrone	Human fibroblasts	40,000–100,000 sites/cell	Hollenberg and Cuatrecasas [1975]
			Carpenter et al [1975]
			Cohen et al [1975]
	Human glial cells	20,000–35,000 sites/cell	Westermark [1977]
	Cultured rabbit lens epithelium	40,000 sites/cell	Hollenberg [1975]
	Bovine granulosa cells	23,000 sites/cell	Vlodavsky et al [1978]
	Bovine leuteal cells	105,000 sites/cell	Vlodavsky et al [1978]
	Human epitheliod carcinoma cells	2.6×10^6 sites/cell	Haigler et al [1978]
	Cultured rat osteogenic sarcoma cells	25,000–33,000 sites/cell	Ng et al [1982]
	Rat calvarial cells	14,000	Ng et al [1982]
	Cultured pancreatic carcinoma cell line	175,000 sites/cell	Martin (personal communication)
	Cultured kidney carcinoma cell line	1.4×10^6 sites/cell	Martin (personal communication)
Angiotensin	Bovine adrenal fractions	$1.4–6.6 \times 10^{-16}$ mol/mg	Lin and Goodfriend [1970]
	Rat uterine fraction	3.4×10^{-16} mol/mg	Lin and Goodfriend [1970]
	Bovine adrenal tissue fraction	600–788 fmol/mg (high affinity) 5.5–9.7 nmol/mg (low affinity)	Glossmann et al [1974]

Calcitonin	Kidney tissue fractions	2.1–6.7 × 10^{-14} mol/mg	Marx et al [1973]
	Calvarial membranes	6.3 × 10^{-14} mol/mg	Marx et al [1973]
	Isolated trout gill cells	10,000 sites/cell (high affinity)	Fouchereau-Peron et al [1981a]
		42,000 sites/cell (low affinity)	
	Porcine lung membrane fraction	40 × 10^8 sites/mg	Fouchereau-Peron et al [1981b]
	Cultured human breast cancer cells		
	MCF.7	5,000 sites/cell	Martin et al [1980]
	T47D	25,000 sites/cell	Findlay et al [1980a,1981]
	Cultured human lung cancer cells	27,000 sites/cell	Findlay et al [1980b]
	Cultured human lymphocytes	280 sites/cell	Marx et al [1974]

and solubilization), and possibly subsequent steps perforce require the use of reagents (eg, detergents) which may well disrupt the receptor complex, thereby making it unidentifiable during subsequent steps. Thus a strategy must be devised which avoids this problem.

There are two different approaches. In the first, the specific ligand for the receptor is tagged by attachment of a radioactive or fluorescent label and covalently linked to the receptor either while it is still in the cell membrane or at an early stage of purification. The receptor complex (or the subunit thereof which attaches to the ligand) can then be followed through later purification steps by monitoring radioactivity or fluorescence as the case may be. This approach has the advantage that the covalent attachment of label greatly alleviates the problem of receptor disruption during purification. The obvious disadvantage is that there is now no way of testing the receptor function of the final isolated ligand/cell surface protein complex, since the binding site by definition is not available. Thus the absolute specificity of the first binding step at which the ligand is attached is all-important.

In the second approach a reversible affinity binding step is used and conditions devised which will not destroy the binding property of the receptor. The aim here is to obtain a unique protein (or protein complex) which will be demonstrably a part of the receptor in question because it binds the ligand in a believable stoichiometric ratio, eg, 1 or perhaps 2 mol of ligand bound per mole of isolated homogeneous receptor protein. This approach concentrates on the ligand-binding domain of the receptor and may not identify other receptor domains if these are on separate protein molecules.

Covalent affinity labelling for detection of ligand receptors. Insulin [Saviolakis et al, 1981] and epidermal growth factor [Baker et al, 1979; Linsley et al, 1979] form spontaneous covalent attachments to their receptors. Where natural attachment does not occur, or is not apparent, affinity labelling has been exploited to achieve covalent stabilization. Spontaneously reactive bifunctional cross-linking reagents have been used for this purpose. Radiolabelled peptide hormone and receptor in equilibrium are cross-linked in situ by addition of a cross-linker. Many aspects of the subunit structure of insulin receptor have been defined by this approach (Table II). Alternatively, hormone derivatives bearing photoactive cross-linking reagents may be photolytically attached whilst in non-covalent association with their receptors. Affinity labelling may also serve to enhance natural covalent attachment. The advantage of spontaneously reactive reagents is their ease of use. However, the favoured reagents (eg, disuccinimidyl suberate) require close juxtaposition of two reactive nucleophiles such as amino groups. Photoactive reagents are not limited by stereospecific factors as the reactive intermediates generated upon photolysis are sufficiently reactive to insert into adjacent C-H bonds (for reviews see Bayley and Knowles [1977]; Chowdhry and Westheimer [1979]; Ji [1979]). Furthermore, the binding abilities and biological potencies of photoactive hormone derivatives may be assessed prior to photolysis to ensure that they interact specifically with their receptors. The most common photoactivatable group used for photoaffinity labelling is the aryl azido chromophore [Fleet et al, 1972], in some cases with a nitro group on the aromatic ring to lower the wavelength required for photoactivation thereby lessening the chance of photolytic damage of the receptor protein. Aryl azido-bearing peptide derivatives may be produced by working under dimmed light conditions so that the azido is not converted to the reactive nitrene intermediate. The variety of other functional groups which may be attached to the aromatic ring allows for incorporation of an aryl azido group into most of the reactive amino acid side chains and the α-amino and α-carboxyl groups. With this spectrum of reagents it should be possible to produce biologically active photoaffinity probe derivatives of most peptide ligands. To date photoaffinity labelling has been used successfully to identify hormone receptors from many tissues (Table III).

Affinity chromatography. Certain pathological conditions result from autoimmune production of antibodies to hormone receptors [Kahn et al, 1977]. These antibodies may be used for immunoaffinity chromatography. Gamma globulin fractions from pathological sera are immobililzed on solid matrices and used as immunoadsorbants specific for hormone receptors (Table IV). Substantial quantities of insulin receptor have been isolated from human placental membranes using immobilized antibodies obtained from patients with severe insulin resistance [Harrison and Itin, 1980]. Since most cell surface receptors are glycopeptide in nature, lectin-based affinity chromatography is also a useful tool for receptor isolation (Table IV). Monoclonal antibodies to receptor proteins represent another promising approach.

Microsequence Analysis of Receptor Protein

There have been significant recent improvements (Hewick et al, 1981) in the sensitivity of microsequence analysis of proteins by the Edman chemical degradation procedure. It is now possible to obtain some sequence information on as little as 20–50 pmol of protein, although 100 pmol is a more realistic amount for reliable analysis. This is an increase of sensitivity of 10^4-fold compared to the original automated sequencing method as described by Edman and Begg [1967]. In the gas-phase sequenator a miniaturized cartridgelike reaction cell replaces the spinning cup of conventional sequenators and some of the liquid-phase reagents are replaced by gas-phase reagents. The amounts of reagents, solvents, and particularly sample amount are reduced, and highly purified solvents are used to reduce artefacts which interfere in the Edman reaction and identification of the phenylthiohydantoin (PTH) derivatives of amino acids produced in this analytical method. This identification is accomplished by HPLC analysis, which has a sensitivity of about 5–10 pmol of PTH amino acid with the detection

TABLE II. Physicochemical Characteristics of the Insulin Receptor as Determined With the Spontaneously Reactive Bifunctional Cross-Linker Disuccinimidyl Suberate

Tissue	Molecular weight of intact receptor[a]	Subunits identified	Postulated subunit structure	References
Adipocyte membranes	ND[b]	α :140,000	—	Pilch and Czech [1979]
Adipocytes	300,000	α :125,000	$(\alpha)2$	Pilch and Czech [1980]
Adipocytes	ND	α :125,000	—	Heinrich et al [1980]
Adipocytes	350,000	α :125,000 β : 90,000	$(\alpha)2(\beta)2$	Massague et al [1980] Massague et al [1981a] (applies to the three rows grouped by the brace)
Adipocytes	320,000	α :125,000 β : 90,000 β_1: 49,000	$(\alpha)2(\beta)(\beta_1)$	
Adipocytes	290,000	α :125,000 β_1: 49,000	$(\alpha)2(\beta_1)2$	
Cultured human IM-9 lymphocytes	ND	α :130,000 β : 90,000	—	Kasuga et al [1981]
Human placenta	ND	α :135,000 β : 90,000 β_1: 42,000	—	Siegel et al [1981]
Cultured human IM-9 lymphocytes	310,000	α :120,000 β_1: 50,000	$\alpha\,\beta_1$ or $(\alpha)2(\beta_1)2$	Pollet et al [1982]

[a]Where molecular weight of the labelled receptor was determined without cleavage of interchain (subunit) disulphide bonds.
[b]Not determined.

TABLE III. Physiochemical Characteristics of Peptide Hormone Receptors Determined by Photoaffinity Labelling

Hormone	Tissue	Position of substitution of unactivated photoprobe[a]	Molecular weight of receptor: Intact[b]	Molecular weight of receptor: Identified subunit	Postulated subunit structure	References
Insulin	Adipocytes	Amino groups	ND	α :130,000	—	Yip et al [1978]
	Liver	B-29 ϵ-amino	ND	α :130,000	$\alpha\beta$	Yip et al [1980a]
	Adipocytes	B-29 ϵ-amino	ND	β : 90,000		Yip et al [1980b]
	Liver	B-1 α-amino group	ND	α :130,000	—	Yeung et al [1980]
	Liver	Amino groups	300,000	α :130,000	—	Wisher et al [1980]
	Placenta or liver	Amino groups	310,000	α :135,000	—	Jacobs et al [1979]
	Cultured hepatoma cells	Amino groups	ND	α :130,000 β : 95,000 β_1: 40,000	—	Hoffmann et al [1981]
	Liver (membranes)	B2-Napa-des Phe B1	300,000–600,000	ND	—	Kuehn et al [1980]
Epidermal growth factor—Urogastrone	Cultured Swiss mouse 3T3-cells	N-terminal α-amino or Carboxyl groups	190,000	190,000	190,000	Das and Fox [1978] Das et al [1977]
	Human placental membranes	N-terminal α-amino group	180,000 and 160,000	180,000 and 160,000		Hock et al [1980]
Angiotensin	Canine adrenal Canine uterus	N-terminal α-amino group	126,00 and 64,600	α : 65,000 α : 68,000	$(\alpha)_2$ dimer $(\alpha)_2$ dimer	Capponi and Catt [1980]
Calcitonin	Cultured human breast cancer cells	γ-Carboxamide of Glutamine 14	85,000	85,000	—	Gorman et al [1981] Moseley et al [1982]
Parathyroid hormone 1–34	Renal cortical membranes	ϵ-amino groups of lysine	ND	70,000	—	Coltrera et al [1981]
	Canine renal cortical membranes	Indole of tryptophan 23	ND	60,000	—	Draper et al [1982]

Chorionic gonadotrophin	Granulosa cells	Amino groups	ND	96,000 66,000 63,000		Ji and Ji [1980]
Leutrophin	Granulosa cells	Amino groups	250,000 370,000	120,000 96,000 76,000 73,000	Multimer of identified subunits	Ji et al [1981]
Nerve growth factor	Rabbit superior cervical ganglia membranes	Amino groups	128,000 97,000	143,000 112,000	143,000	Massague et al [1981b]

[a]Refers to position of attachment of photoprobe used to produce the photoaffinity hormone derivative.
[b]Where molecular weight of the labelled receptor was determined without cleavage of interchain (subunit) disculphide bonds.

system usually employed. When such a small amount of protein is required for analysis, gel electrophoresis can be used as a final purification step. Human lymphoblastoid interferon was eluted from a sodium dodecyl sulfate (SDS) gel and sequenced in this manner [Zoon et al, 1980]; the first 20 residues wre identified using from 20 to 500 pmol of protein. Such an approach would be very suitable for obtaining limited sequence data on receptors, the use of SDS gel electrophoresis being especially suited for purifying hydrophobic membrane protein. It is of interest to note that Harrison and Itin [1980] purified the human insulin receptor probably to homogeneity, obtaining about 190 μg of protein, more than sufficient for such sequence determinations. Very recently this methodology has been successfully used to determine the aminoterminal sequences of the four polypeptides of the acetylcholine receptor protein, thus allowing identification of cDNA clones coding for these subunits as described in the next section.

Exploitation of Sequence Information

The preceding two sections describe approaches to isolation and partial sequence information of receptor proteins. It is necessary to determine only enough of the amino acid sequence to allow the chemical synthesis of two or perhaps three different sets of oligodeoxynucleotide primers predicted to hybridize to the receptor mRNA or cDNA. Contiguous sequences of 4–6 amino acids are needed, preferably comprised of those with unique codons (Trp and Met) or with a limited number of codons (Asp, Asn, Glu, Gln, Cys, Tyr, Lys, Phe, His are each specified by only two possible codons). Sequences with Ser, Arg or Leu should be avoided as they each have six possible codons. The phosphite [Matteucci and Caruthers, 1980] or phosphotriester [Markham et al, 1980] methods are suitable for preparing mixtures of the requisite synthetic oligodeoxynucleotides (11–20 units in length).

An example will explain the general approach to be used which was first applied by

TABLE IV. Physiochemical Characteristics of Peptide Hormone Receptors Isolated by Affinity Chromatography or Immunoprecipitation

Hormone	Tissue	Affinity medium	Molecular weight of receptor: Intact[a]	Molecular weight of receptor: Identified subunits	Postulated subunit structure	References
Insulin	Rat liver	Insulin immobilized on agarose[b,c]	310,000	α :135,000 β_1: 45,000	$(\alpha)(\beta_1)$	Cuatrecasas [1972] Jacobs et al [1977] Jacobs et al [1979]
	Rat liver	Insulin immobilized on agarose[b,d]	310,000 180,000	α :135,000 β_1: 45,000	—	Jacobs et al [1980]
	Human placental membranes	Insulin immobilized on agarose[b,c]	350,000	α :135,000 β : 90,000 β_1: 45,000	—	Siegel et al [1981]
	Human placental membranes	Wheat germ lectin immobilized on agarose followed by either immunoprecipitation with receptor autoantibody or chromatography on agarose receptor autoantibody matrix[c]		α :126,000 β : 90,000 β_1: 42,000	—	Harrison and Itin [1980]
	Cultured human IM-9 lymphocytes	Wheat germ lectin immobilized on agarose followed by immunoprecipitation with receptor autoantibody	—	α :126,000 β : 90,000 67,000[i]	—	Harrison et al [1982]
	Adipocyte membranes	Protein A immobilized on agarose[e]	—	α :135,000	—	Heinrich et al [1980]
EGF—Urogastrone	A-431 cells	EGF-immobilized on agarose[b,f]	170,000 150,000	170,000 150,000	Single polypeptide chain 170,000	Cohen et al [1980] Cohen et al [1982]
	Human placental membranes	Cibacron Blue and wheat germ lectin immobilized on agarose[g]	160,000	160,000	Single polypeptide chain 160,000	

	Human placental membranes	Anti-EGF antibody immobilized on agarose[h]	180,000 160,000	180,000 160,000	180,000	Hock et al [1980]
Human chorionic gonadotrophin (HCG)	Rat testis	HCG immobilized on agarose[c]	194,000	α : 90,000	$(\alpha)_2$194,000	Dufau et al [1975]
Thyrotropin	Tryptic digest on bovine thyroid plasma membranes	Thyrotropin immobilized on agarose[c]	—	15,000 30,000	—	Tate et al [1975]

[a]Where the molecular weight of the isolated receptor was determined without cleavage of interchain (subunit) disulphide bonds.
[b]Affinity chromatography was subsequent to diethylaminoethyl (DEAE)-cellulose ion-exchange step.
[c]Isolated receptor remained functionally active for hormone binding.
[d]Lectin affinity chromatography also used.
[e]Antiinsulin antibodies were used to complex with affinity-labeled receptor and the complex absorbed onto immobilized protein A.
[f]It was postulated that the smaller component (Mr = 160,000) was derived by proteolysis of the larger component (Mr = 180,000).
[g]Affinity chromatography steps subsequent to CM-cellulose ion-exchange step.
[h]Photoaffinity-labelled receptor was absorbed onto the immunoadsorbent matrix containing anti-EGF antibody.
[i]Not identified.

Noyes et al [1979] and Agarwal et al [1981]. Haley et al [1982] used the amino acid sequence in porcine relaxin, Trp-Val-Glu-Ile, and synthesized the eight possible oligonucleotides coding for this sequence (deleting the last nucleotide for Ile). These oligomers were mixed with mRNA isolated from ovaries of pregnant pigs at a stage of pregnancy when the hormone was known to be actively biosynthesized. Reverse transcriptase was then added under conditions where the "correct" oligodeoxynucleotide (ie, the one of the eight possibilities which was actually complementary to the relaxin-specific mRNA) would prime the synthesis of a radioactively labelled cDNA complementary to the relaxin mRNA. The reaction mixture(s) were then analysed by gel electrophoresis and the primed cDNA identified by autoradiography and eluted. This cDNA was then used to probe an ovarian clone bank and identify clones containing relaxin cDNA sequences.

This approach has proved to be quite effective in obtaining clones to relatively low abundance (0.1–0.2%) mRNAs. However, for very rare mRNAs this approach will not work since insufficient primed cDNA would be obtained to identify on electrophoresis gels above background. Thus, it is not applicable to receptor cloning unless a 10–100-fold enrichment of receptor mRNA is achieved through approaches outlined above. Usually this will not be possible, and a variation on this primer extension technique must be used in which similar sets of oligodeoxynucleotides complementary to short, known amino acid sequences are used to probe clone banks directly. Clone banks constructed by one of the means described above are probed with the oligodeoxynucleotide mixtures, the latter being end-labelled with ^{32}P using γ-ATP and the enzyme polynucleotide kinase. Since the probe mixtures may contain up to 32 components, only one of which will be specific for the desired receptor clones, it is obvious that many "false-positive" clones will be radioactively labelled by chance cross-hybridization with one or other of the synthetic DNA probes. However, selection of clones which hybridize to two or more probes representing different amino acid sequence regions has been proved effective in isolating rare species from large clone banks.

The primary structures of the four polypeptide chains of the acetylcholine receptor from *Torpedo californica* have been deduced in this manner [Noda et al, 1982,1983]. A library of cDNA clones derived from the mRNA of the electric organ of the ray was constructed according to the method of Okayama and Berg [1982].

For the α-subunit, for example, the clone bank was probed with a mixture of 32 nucleotides, 14 bases long (14-mers). These represented all the possible coding sequences for amino acids 25–29 (His-Thr-His-Phe-Val) of the α-subunit, omitting the final nucleotide for Val. Fifty-seven positive clones were identified of the 2×10^5 transformants screened. These clones were further screened with a mixture of 32 16-mers representing the sequence Phe-Asn-Tyr-Asn-Lys-Val at position 13–18 of the α-subunit. Of these 57 clones, 20 hybridized to the second probe and all these contained sequences for the α-subunit.

A further impressive demonstration of the power of this approach was recently reported by Furutani et al [1983], who cloned and sequenced the cDNA for the ovine corticotropin-releasing factor precursor from hypothalamic tissue. The message for this hormone precursor is extremely rare. In a tour de force of enterprise and industry, a total of almost 3.8 million clones were screened in a series of experiments using several different probes before a clone containing the entire coding sequence was identified. To put this achievement in perspective, some hormones studied previously (pituitary growth hormones, placental lactogen) have messages so abundant that screening a few hundred or at most a thousand recombinants is quite adequate to find the desired clone.

This study of Furutani and co-workers is most encouraging for those interested in cloning receptors as the message abundances are probaby similar. The Japanese group used both the Okayama-Berg method and the pBR 322 procedure described above. A number of clones deficient in 5′ regions of sequence were found. As noted above, this is a common problem in the present cloning procedures. The S_1 nuclease digestion used in the formation of double-stranded cDNA for insertion into a vector often removes substantial portions of the 5′ end of cDNA — the area in the mRNA which codes for the N-terminus of the protein. Thus, probes synthesized to N-terminal regions of the protein may not be able to detect any clones in the clone bank.

Of the cloning procedures available, the Okayama-Berg method [1982] is the most effective in generating full-length or near full-length inserts in large numbers. However, it may be necessary to adopt a second approach in which the protein is cleaved with specific reagents, the peptides separated by HPLC and sequenced as described previously. This provides amino acid sequences in the interior of the molecule — areas more likely to be represented in the cDNA clone bank. In addition, if cyanogen bromide is used to cleave at Met residues [Gross and Witkop, 1956] or o-iodosobenzoic acid to cleave at Trp residues [Mahoney et al, 1981], then the internal sequence obtained will include a Met or Trp residue (implied), thus reducing the number of oligonucleotides to be synthesized.

If these additional experiments are necessary, significantly more purified receptor protein will be required — probably of the order of 0.5 - 1 nmol — to allow for losses during chemical cleavage and peptide isolation.

A combination of specific mRNA enrichment, high-efficiency cloning methods, and screening of the clone bank with at least two sets of oligonucleotide probes offers the most promising route to obtaining clones for receptors. Alternatively a genomic DNA library itself could be used instead of a cDNA library. Genomic DNA fragments cloned into λ-phage or cosmids could be efficiently screened with the oligonucleotide probes. However, this approach would be complicated by the probable distribution of receptor coding regions in several different exons separated by large intervening sequences. However, a genomic region cross-hybridizing to the probes could be used to screen tissue-derived cDNA libraries. Longer probes (preferably greater than 20 nucleotides) are desirable in screening genomic libraries to reduce the level of nonspecific hybridization.

Other Approaches

If it is not possible to obtain amino acid sequence information and use the strategy just outlined, it may still be possible to identify receptor-specific clones by taking advantage of in vitro translation or expression systems, with the major qualification that these systems must generate a receptor protein which is still recognizable, ie, it must be produced in a form which will still bind to its specific ligand or to an antibody known to be receptor-specific. These procedures still depend on obtaining a large cDNA library statistically likely to contain receptor clones. These clones may be screened, in batches, for their ability to block mRNA directed translation of receptor protein, assuming the latter can be detected by immunoprecipitation or by ligand binding. The clones would do this by specific binding of receptor mRNA by complementary cDNA sequences in the group of clones being screened. The mRNA is thus not available for translation. This is the so-called hybridization-arrest-of-translation system [Paterson et al, 1977].

In a related technique successive groups of clones from the bank are immobilized on nitrocellulose filters, and the filters exposed to total mRNA from the same tissue. If a receptor-specific clone is present in the group bound to the filter, the corresponding mRNA will hybridize to it. The mRNA is subsequently eluted from the filter, translated and the translation products screened for receptor protein. This is called the positive-selection system [Ricciardi et al, 1979].

There are some advantages to these methods: eg, the clone in the cDNA bank need not be full-length as long as it is long enough to hybridize strongly to a significant part of the message. However, if the mRNA is very low in abundance it may not be possible to detect the translation product reliably above background, and the approach will fail. Of the translation systems available, microinjection of mRNA into *Xenopus laevis* oocytes produces the most translation product, though it is laborious and not suitable for large-scale screening. The rabbit reticulocyte lysate system [Allen and Schweet, 1962] may be the best compromise. Wheat germ translation systems have a low background but are less efficient in translating larger messenger species.

In general the translation-based approaches are less sensitive and probably less suitable than the microsequencing/oligodeoxynucleotide probe method.

Another possibility is to clone mRNA into an expression vector and screen for production of the receptor by the host cells of the cDNA library. Again, antibody or ligand must be used to detect receptor protein in the cultures. The expression vector bacteriophage-λ gt 11 would be suitable, in transformed *Escherichia coli* cells, but there are other possible systems.

It is hard to generalize about these strategies. They all require a great deal of work, and one needs to be convinced that the translation or expression system under consideration would generate an identifiable product before starting to screen perhaps hundreds of thousands of clones. It would be prudent to make sure that the detection system used does not depend critically upon glycosylation or other posttranslational modification.

FUTURE STUDIES

It seems certain that before long several receptors and their genes will be characterized to the extent of nucleotide and amino acid sequence determination of their respective structures. Clearly this is an important and necessary step in our understanding of receptor function, but equally it is only a beginning. Many new avenues of investigation will open up. It will be possible to generate relatively large amounts of receptor proteins by recombinant DNA techniques. These can be used for studies of tertiary structure by x-ray crystallography or by high resolution nuclear magnetic resonance. The interaction of different subunits with one another and any coincident allosteric changes in subunit architecture can be studied. Availability of cDNA probes for receptor mRNA will allow studies of receptor gene expression and receptor biosynthesis during ontogeny and in different physiological situations. In time, a knowledge of receptor gene organization and of the basis of receptor diversity will emerge, with an insight into the evolutionary origins of receptors. It should be possible to study in detail receptor-effector coupling and thus the basis of hormone and neurotransmitter modulation of target cells. Finally, it may be possible to discover the molecular basis of clinically important defects in receptor function.

REFERENCES

Agarwal K L, Brunstedt J, Noyes BE (1981): A general method for detection and characterization of an mRNA using an oligonucleotide probe. J Biol Chem 256:1023–1028.

Allen EH, Schweet RS (1962): Synthesis of hemoglobin in a cell-free system. J Biol Chem 237:760–767.

Baker JB, Simmer RL, Glenn KC, Cunningham DD (1979): Thrombin and epidermal growth factor become linked to cell surface receptors durning mitogenic stimulation. Nature 278:743–745.

Bayley H, Knowles JR (1977): Photoaffinity labeling. In Jakoby WB, Wilchek M (eds): "Methods in Enzymology," New York: Academic Press, Vol 46, pp 69–114.

Bernard O, Hozumi N, Tonegawa S (1978): Sequences of mouse immunoglobulin light chain genes before and after somatic changes. Cell 15:1133–1144.

Brack C, Tonegawa S (1977): Variable and constant parts of the immunoglobulin light chain gene of a mouse myeloma cell are 1,250 non translated bases apart. Proc Natl Acad Sci USA 74:5652–5656.

Brack C, Hirama M, Lenhard-Schuller R, Tonegawa S (1978): A complete immunoglobulin gene is created by somatic recombination. Cell 15:1–14.

Capponi AM, Catt KJ (1980): Solubilization and characterization of adrenal and uterine angiotensin II receptors after photoaffinity labeling. J Biol Chem 255:12081–12086.

Carpenter G, Lembach KJ, Morrison M, Cohen S (1975): Characterization of the binding of ^{125}I-labeled epidermal growth factor to human fibroblasts. J Biol Chem 250:4297–4304.

Chirgwin JM, Przybyla AE, MacDonald, RJ, Rutter WJ (1979): Isolation of biologically active ribonucleic acid from sources enriched in ribonuclease. Biochemistry 18:5294–5299.

Chowdhry V, Westheimer FH (1979): Photoaffinity labeling of biological systems. Annu Rev Biochem 48:293–325.

Cohen S, Carpenter G, Lembach KJ (1975): Intreraction of epidermal growth factor (EGF) with culture fibroblasts. Adv Metabol Disorders 8:265–284.

Cohen S, Carpenter G, King L Jr (1980): Epidermal growth factor-receptor-protein kinase interactions: Co-purification of receptor and epidermal growth factor-enhanced phosphorylation. J Biol Chem 255:4834–4842.

Cohen S, Ushiro H, Stoscheck C, Chinkers M (1982): A native 170,000 epidermal growth factor receptor-kinase complex from shed plasma membrane vesicles. J Biol Chem 257:1523–1531.

Coltrera MD, Potts JT, Rosenblatt M (1981): Identification of a renal receptor for parathyroid hormone by photoaffinity radiolabeling using a synthetic analogue. J Biol Chem 256:10555–10559.

Conti-Tronconi BM, Hunkapiller MW, Lindstrom JM, Raftery MA (1982): Subunit structure of the acetylcholine receptor from *Electrophorus electricus*. Proc Natl Acad Sci USA 79:6488–6493.

Cuatrecasas P (1971): Insulin-receptor interactions in adipose tissue cells: Direct measurement and properties. Proc Natl Acad Sci USA 68:1264–1268.

Cuatrecasas P (1972): Affinity chromatography and purification of the insulin receptor of liver cell membranes. Proc Natl Acad Sci USA 69:1277–1281.

Czech MP, Massague J, Pilch PF (1981): The insulin receptor: Structural features. Trends Biochem Sci 6:222–225.

Czech MP (1982): Structural and functional homologies in the receptors for insulin and the insulin-like growth factors. Cell 31:8–10.

Das M, Fox CF (1978): Molecular mechanism of mitogen action: Processing of receptor induced by epidermal growth factor. Proc Natl Acad Sci USA 75:2644–2648.

Das M, Miyakawa T, Fox CF, Pruss RM, Aharonov A, Herschman HR (1977): Specific radiolabeling of a cell surface receptor for epidermal growth factor. Proc Natl Acad Sci USA 74:2790–2794.

Draper MW, Nissenson RA, Winer J, Ramachandran J, Arnaud CD (1982): Photoaffinity labeling of the canine renal receptor for parathyroid hormone. J Biol Chem 257:3714–3718.

Dufau ML, Ryan DW, Baukal AJ, Catt KJ (1975): Gonadotropin receptors: solubilization and purification by affinity chromatography. J Biol Chem

250:4822–4824.

Early PW, Davis MM, Kaback DB, Davidson N, Hood L (1979): Immunoglobulin heavy chain gene organization in mice: Analysis of a myeloma genomic clone containing variable and α constant regions. Proc Natl Acad Sci USA 76:857–861.

Early P, Huang H, Davis M, Calame K, Hood L (1980): An immunoglobulin heavy chain variable region gene is generated from three segments of DNA:V_H, D and J_J. Cell 19:981–992.

Edelman GM, Cunningham BA, Gall WE, Gottlieb PD, Rutishauser U, Waxdal MJ (1969): The covalent structure of an entire γG immunoglobulin molecule. Proc Natl Acad Sci USA 63:78–85.

Edman P, Begg G (1967): Protein sequenator. Eur J Biochem 1:80–91.

Findlay DM, deLuise M, Michelangeli VP, Ellison M, Martin TJ (1980a): Properties of a calcitonin receptor and adenylate cyclase in BEN cells, a human cancer cell line. Cancer Res 40:1311–1317.

Findlay DM, Michelangeli VP, Eisman JA, Frampton RJ, Moseley JM, MacIntyre I, Whitehead R, Martin TJ (1980b): Calcitonin and 1,25-dihydroxyvitamin D_3 receptors in human breast cancer cell lines. Cancer Res 40:4764–4767.

Findlay DM, Michelangeli VP, Moseley JM, Martin TJ (1981): Calcitonin binding and degradation by two cultured human breast cancer cell lines (MCF 7 and T 47D). Biochem J 196:513–520.

Fleet GWJ, Knowles JR, Porter RR (1972): The antibody binding site: Labelling of a specific antibody against the photo-precursor of an aryl anitrene. Biochem J 128:499–508.

Fouchereau-Peron M, Moukhtar MS, Benson AA, Milhaud G (1981a): Characterization of specific receptors for calcitonin in porcine lung. Proc Natl Acad Sci USA 78:3973–3975.

Fouchereau-Peron M, Moukhtar MS, Le Gal Y, Milhaud G (1981b): Demonstration of specific receptors for calcitonin in isolated trout gill cells. Comp Biochem Physiol 68A:417–421.

Freychet P, Roth J, Neville DM Jr (1971): Insulin receptors in the liver: Specific binding of [^{125}I] insulin to the plasma membrane and its relation to insulin bioactivity. Proc Natl Acad Sci USA 68:1833–1837.

Furutani Y, Morimoto Y, Shibahara S, Noda M, Takahashi H, Hirose T, Asai M, Inayama S, Hayashida H, Miyata T, Numa S (1983): Cloning and sequence analysis of cDNA for ovine corticotropin-releasing factor precursor. Nature 301:537–540.

Gammeltoft S, Gliemann J (1973): Binding and degradation of ^{125}I-labelled insulin by isolated rat fat cells. Biochim Biophy Acta 320:16–32.

Gavin JR III, Roth J, Jen P, Freychet P (1972): Insulin receptors in human circulating cells and fibroblasts. Proc Natl Acad Sci USA 69:747–751.

Gavin JR III, Gorden P, Roth J, Archer JA, Buell DN (1973): Characteristics of the human lymphocyte insulin receptor. J Biol Chem 248:2202–2207.

Glossmann H, Baukal AJ, Catt KJ (1974): Properties of angiotensin II receptors in the bovine and rat adrenal cortex. J Biol Chem 249:825–834.

Gorman JJ, Folk JE, Moseley JM, Findlay DM, Martin TJ (1981): Enzymatic production of photosensitive peptide hormone derivatives. Proc Seventh Am Peptide Symp:131–134.

Gross E, Witkop B (1962): Nonenzymatic cleavage of peptide bonds: The methionine residues in bovine pancreatic ribonuclease. J Biol Chem 237:1856–1860.

Haigler H, Ash JF, Singer SJ, Cohen S (1978): Visualization by fluorescence of the binding and internalization of epidermal growth factor in human carcinoma cells A-431. Proc Natl Acad Sci USA 75:3317–3321.

Haley J, Hudson P, Scanlon D, John M, Cronk M, Shine J, Tregear G, Niall H (1982): Porcine relaxin: Molecular cloning and cDNA structure. DNA 1:155–162.

Hammond JM, Jarett L, Mariz IK, Daughaday WH (1972): Heterogeneity of insulin receptors on fat cell membranes. Biochem Biophys Res Comun 49:1122–1128.

Harrison LC, Itin A (1980): Purification of the insulin receptor from human placenta by chromatography on immobilized wheat germ lectin and receptor antibody. J Biol Chem 255:12066–12072.

Harrison LC, Itin A, Kasuga M, Van Obberghen E (1982): The insulin receptor of the human lymphocyte: down regulation of 126,000 and 90,000 glycosylated subunits. Diabetologica 22:233–238.

Heinrich J, Pilch PF, Czech MP (1980): Purification of the adipocyte insulin receptor by immunoaffinity chromatography. J Biol Chem 255:1732–1737.

Hewick RM, Hunkapiller MW, Hood LE, Dreyer WJ (1981): A gas-liquid solid phase peptide and protein sequenator. J Biol Chem 256:7990–7997.

Hock RA, Nexo E, Hollenberg MD (1980): Solubilization and isolation of the human placenta receptor for epidermal growth factor-urogastrone. J Biol Chem 255:10737–10743.

Hofmann C, Ji TH, Miller B, Steiner DF (1981): Photoaffinity labeling of the insulin receptor in H4 hepatoma cells: Lack of cellular receptor processing. J Supramol Struct 15:1–13.

Hollenberg MD (1975): Receptors for insulin and epidermal growth factor: relation to synthesis of DNA in cultured rabbit lens epithelium. Arch Biochem Biophys 171:371–377.

Hollenberg MD, Cuatrecasas P (1975): Insulin and epidermal growth factor. Human fibroblast receptors related to deoxyribonucleic acid synthesis and amino acid uptake. J Biol Chem 250:3845–3853.

Hozumi N, Tonegawa S (1976): Evidence for somatic rearrangement of immunoglobulin genes coding for variable and constant regions. Proc Natl Acad Sci

USA 73:3628–3632.

Jacobs S, Schecter Y, Bissell K, Cuatrecasas P (1977): Purification and properties of insulin receptors from rat liver membranes. Biochem Biophys Res Commun 77:981–988.

Jacobs S, Hazum E, Shechter Y, Cuatrecasas P (1979): Insulin receptor: Covalent labeling and identification of subunits. Proc Natl Acad Sci USA 76:4918–4921

Jacobs S, Hazum E, Cuatrecasas P (1980): The subunit structure of rat liver insulin receptor: Antibodies directed against the insulin binding subunit. J Biol Chem 255:6937–6940.

Ji TH (1979): The application of chemical crosslinking for studies on cell membranes and the identification of surface receptors. Biochim Biophys Acta 559:39–69.

Ji I, Ji TH (1980): Macromolecular photoaffinity labeling of the lutropin receptor on granulosa cells. Proc Natl Acad Sci USA 77:7167–7170.

Ji I, Yoo BY, Kaltenbach C, Ji TH (1981): Structure of the lutropin receptor on granulosa cells: Photoaffinity labeling with the α subunit in human choriogonadotropin. J Biol Chem 256:10853–10858.

Kahn CR, Freychet P, Roth J, Neville DM Jr (1974): Qualitative aspects of the insulin receptor interaction in liver plasma membranes. J Biol Chem 249:2249–2257.

Kahn CR (1976): Membrane receptors for hormones and neurotransmitters. J Cell Biol 70:261–286.

Kahn CR, Megyesi K, Bar RS, Eastman RC, Flier JS (1977): Receptors for peptide hormones. Ann Int Med 86:205–219.

Kasuga M, Van Obberghen E, Yamada KM, Harrison LC (1981): Autoantibodies against the insulin receptor recognize the insulin binding subunits of an oligomeric receptor. Diabetes 30:356–357.

Korsmeyer SJ, Hieter PA, Ravetch JV, Poplack DG, Waldmann TA, Leder P (1981): Developmental hierarchy of immunoglobulin gene rearrangements in human leukemic pre-B-cells. Proc Natl Acad Sci USA 78:7096–7100.

Kraus JP, Rosenberg LE (1982): Purification of low-abundance messenger RNAs from rat liver by polysome immunoadsorption. Proc Natl Acad Sci USA 79:4015–4019.

Kuehn L, Meyer, H, Rutschmann M, Thamm P (1980): Identification of photoaffinity labeled insulin receptor proteins by linear polyacrylamide gradient gel electrophoresis under non-denaturing conditions. FEBS Lett 113:189–192.

Lin SY, Goodfriend TL (1970): Angiotensin receptors. Am J Physiol 218:1319–1328.

Linsley PS, Blifeld C, Wrann M, Fox CF (1979): Direct linkage of epidermal growth factor to its receptor. Nature 278:745–748.

Mahoney WC, Smith PK, Hermodson MA (1981): Fragmentation of proteins with o-iodosobenzoic acid: Chemical mechanism and identification of o-iodoxybenzoic acid as a reactive contaminant that modifies tyrosyl residues. Biochemistry 20:443–448.

Markham AF, Edge MD, Atkinson TC, Greene AR, Heathcliffe GR, Newton CR, Scanlon D (1980): Solid phase phosphotriester synthesis of large oligodeoxyribonucleotides on a polyamide support. Nucl Acids Res 8:5193–5205.

Martin TJ, Findlay DM, MacIntyre I, Eisman JA, Michelangeli VP, Moseley JM, Partridge NC (1980): Calcitonin receptors in a cloned human breast cancer cell lines (MCF 7). Biochem Biophys Res Commun 96:150–156.

Marx JL (1981): Antibodies: Getting their genes together. Science 212:1015–1017.

Marx SJ, Woodward C, Aurbach GD, Glassman H, Keutmann HT (1973): Renal receptors for calcitonin: Binding and degradation of hormone. J. Biol Chem 248:4797–4802.

Marx SJ, Aurbach GD, Gavin JR III, Buell DW (1974): Calcitonin receptors on cultured human lymphocytes. J Biol Chem 249:6812–6816.

Massague J, Pilch PF, Czech MP (1980): Electrophoretic resolution of three major insulin receptor structures with unique subunit stoichiometries. Proc. Natl Acad Sci USA 77:7137–7141.

Massague J, Pilch PF, Czech MP (1981a): A unique proteolytic cleavage site on the β-subunit of the insulin receptor. J Biol Chem 256:3182–3190.

Massague J, Guilette BJ, Czech MP, Morgan CJ, Bradshaw RA (1981b): Identification of a nerve growth factor receptor protein in sympathetic ganglia membranes by affinity labeling. J Biol Chem 256:9419–9424.

Matteucci MD, Caruthers MH (1980): The synthesis of oligodeoxypyrimidines on a polymer support. Tet Letters 21:719–722.

Max EE, Seidman JG, Leder P (1979): Sequence of five potential recombination sites encoded close to an immunoglobulin k constant region. Proc Natl Acad Sci USA 76:3450–3454.

Maxam AM, Gilbert W (1977): A new method for sequencing DNA. Proc Natl Acad Sci USA 74:560–564.

Moseley JM, Findlay DM, Martin TJ, Gorman JJ (1982): Covalent cross-linking of a photoactive derivative of calcitonin to human breast cancer cell receptors. J Biol Chem 257:5846–5851.

Ng KW, Partridge NC, Niall M, Martin TJ (1983): Epidermal growth factor receptors in clonal lines of a rat osteogenic sarcoma and in osteoblast rich bone cells. Calcif Tissue Int 35:298–303.

Niall HD (1980): The evolution of peptide hormones. In Cumming IA, Funder JW, Mendelsohn FAO (eds): "Proceedings of the VI International Congress of Endocrinology." Australian Academy of Science: Elsevier/North Holland, pp 13–18.

Noda M, Takashashi H, Tanabe T, Toyosato M, Furutani Y, Hirose T, Asai M, Inayama S, Miyata T, Numa S (1982): Primary structure of α-subunit precursor of Torpedo californica acetylcholine receptor deduced from cDNA sequence. Nature 299:793–979.

Noda M, Takahashi H, Tanabe T, Toyosato M, Kikyotani S, Hirose T, Asai M, Takashima H, Inayama S, Miyata T, Numa S (1983): Primary structures of β and δ-subunit precursors of *Torpedo californica* acetylcholine receptor deduced from cDNA sequences. Nature 301:251–255.

Noyes BE, Mevarech M, Stein R, Agarwal KL (1979): Detection and partial sequence analysis of gastrin mRNA by using an oligodeoxynucleotide probe. Proc Natl acad Sci USA 76:1770–1774.

Okayama H, Berg P (1982): High-efficiency cloning of full-length cDNA. Mol Cell Biol 2:161–170.

Paterson BM, Roberts BE, Kuff EL (1977): Structural gene identification and mapping by DNA. mRNA hybrid-arrested cell-free translation. Proc Natl Acad Sci USA 74:4370–4374.

Pilch PF, Czech MP (1979): Interaction of cross-linking agents with the insulin effector system of isolated fat cells. Covalent linkage of ^{125}I-insulin to a plasma membrane receptor protein of 140,000 daltons. J Biol Chem 254:3375–3381.

Pilch PF, Czech MP (1980): The subunit structure of the high affinity insulin receptor. Evidence for a disulfide-linked receptor complex in fat cell and liver plasma membranes. J Biol Chem 255:1722–1731.

Podskalny JM, Chou JY, Rechler MW (1975): Insulin receptors in a new human placenta cell line: Demonstration of negative cooperativity. Arch Biochem Biophys 170:504–513.

Poljak RJ, Amzel LM, Avey HP, Chen BL, Phizackerley RP, Saul F (1973): Three-dimensional structure of the Fab′ fragment of a human immunoglobulin at 2.8 Å resolution. Proc Natl Acad Sci USA 70:3305–3310.

Pollet RJ, Kempner ES, Standaert ML, Hasse BA (1982): Structure of the insulin receptor of the cultured human lymphoblastoid cell IM-9. Evidence suggesting that two subunits are required for insulin binding. J Biol Chem 257:894–898.

Posner BI, (1974): Insulin receptors in human and animal placental tissue. Diabetes 23:209–217.

Putnam FW, Titani K, Wikler M, Shinoda T (1967): Structure and evolution of kappa and lambda light chains. Cold Spring Harbor Symp Quant Biol 32:9–30.

Raftery MA, Hunkapiller MW, Strader CD, Hood LE (1980): Acetylcholine receptor: complex of homologous subunits. Science 208:1454–1456.

Ricciardi RP, Miller JS, Roberts BE (1979): Purification and mapping of specific mRNA's by hybridization-selection and cell-free translation. Proc Natl Acad Sci USA 76:4927–4931.

Sakano H, Rogers JH, Hüppi K, Brack C, Traunecker A, Maki R, Wall R, Tonegawa S (1979): Domains and the hinge region of an immunoglobulin heavy chain are encoded in separate DNA segments. Nature 277:626–633.

Sanger F, Nicklen S, Coulson AR (1977): DNA sequencing with chain-terminating inhibitors. Proc Natl Acad Sci USA 74:5463–5467.

Saviolakis GA, Harrison LC, Roth J (1981): The binding of ^{125}I-insulin to specific receptors in IM-9 human lymphocytes. J Biol Chem 254:4924–4928.

Schiffer M, Girling RL, Ely KR, Edmundson AB (1973): Structure of a λ-type Bence Jones protein at 3.5 Å resolution. Biochemistry 12:4620–4631.

Segal DM, Padlan EA, Cohen GH, Rudikoff S, Potter M, Davies DR (1974): The three dimensional structure of a phosphorylcholine-binding mouse immunoglobulin Fab and the nature of the antigen binding site. Proc Natl Acad Sci USA 71:4298–4302.

Seidman JG, Leder A, Edgell MH, Polsky F, Tilghman SM, Tiemeier DC, Leder P (1978a): Multiple related immunoglobulin variable-region genes identified by cloning and sequence analysis. Proc Natl Acad Sci USA 75:3881–3885.

Seidman JG, Leder A, Nau M, Norman B, Leder P (1978b): Antibody diversity. Science 202:11–17.

Siegel TW, Ganguly S, Jacobs S, Rosen OM, Rubin CS (1981): Purification and properties of the human placental insulin receptor. J Biol Chem 256:9266–9273.

Tate RL, Holmes JM, Kohn LD, Winand RJ (1975): Characteristics of a solubilized thyrotropin receptor from bovine thyroid plasma membranes. J Biol Chem 250:6527–6533.

Tiemeier DC, Tilghaman SM, Polsky FI, Seidman JG, Leder A, Edgell MH, Leder P (1978): A comparison of two cloned mouse β-globulin genes and their surrounding and intervening sequences. Cell 14:237–245.

Van Obberghen E, Kasuga M, LeCam E, Itin A, Hedo JA, Harrison LC (1981): Biosynthetic labelling of insulin receptor: Studies of subunits in cultured human 1M-9 lymphocytes. Proc Natl Acad Sci USA 78:1052–1056.

Vlodavsky I, Brown KD, Gospodarowicz D (1978): A comparison of the binding of epidermal growth factor to cultured granulosa and luteal cells. J Biol Chem 253:3744–3750.

Westermark B (1977): Local starvation for epidermal growth factor cannot explain density-dependent inhibition of normal human glial cells. Proc Natl Acad Sci USA 74:1619–1621.

Wisher MH, Baron MD, Jones RH, Sönksen PH (1980): Photoactive insulin analogues used to characterise the insulin receptor. Biochem Biophys Res Commun 92:492–498.

Yeung CWT, Moule ML, Yip CC (1980): Photoaffinity labeling of insulin receptor with an insulin analogue

selectively modified at the amino terminal of the B-chain. Biochemistry 19:2196–2203.
Yip CC, Yeung CWT, Moule ML (1978): Photoaffinity labeling of insulin receptor of rat adipocyte plasma membrane. J Biol Chem 253:1743–1745.
Yip CC, Yeung CWT, Moule ML (1980a): Photoaffinity labeling of insulin receptor proteins of liver plasma membranes preparations. Biochemistry 19:70–76.
Yip CC, Moule ML, Yeung CWT (1980b): Characterization of insulin receptor subunits in brain and other tissues by photoaffinity labeling. Biochem Biophys Res Commun 96:1671–1678.
Young RA, Davis RW (1983): Efficient isolation of genes by using antibody probes. Proc Natl Acad Sci USA 80:1194–1198.
Zoon KC, Smith ME, Bridgen PJ, Anfinsen CB, Hunkapiller MW, Hood LE (1980): Amino terminal sequence of the major components of human lymphoblastoid interferon. Science 207:527–528.

Index